The Logica Yearbook
2008

The Logica Yearbook
2008

edited by

Michal Peliš

ISBN 978-1-904987-46-8

College Publications
Scientific Director: Dov Gabbay
Managing Director: Jane Spurr
Department of Computer Science
King's College London, Strand, London WC2R 2LS, UK

http://www.collegepublications.co.uk

Original cover design by Laraine Welch
Printed by Lightning Source, Milton Keynes, UK

Table of Contents

Preface

The international symposium *Logica* organized by the Institute of Philosophy of the Academy of Sciences of the Czech Republic has a relatively long history. It began in 1987 when the first conference of series was held in Liblice Chateau. In the beginning the conferences were of mostly local importance, but over the years they have acquired the status of a prestigious international conference with a multidisciplinary flavour.

The annual symposia cover a broad field of logical topics and aim to promote dialogue between various branches of logic. *Logica* hosted many presentations by top specialists in mathematical and philosophical logic as well as analytical philosophy and linguistics. The professional orientation of the conference can be illustrated by mentioning several names of scholars that the conference welcomed as the invited speakers: Nuel Belnap, Simon Blackburn, Robert Brandom, Melvin Fitting, Yuri Gurevich, Petr Hájek, Rom Harré, Jaakko Hintikka, Wilfrid Hodges, David Lewis, Per Martin-Löf, Barbara Partee, Graham Priest, Greg Restall, Gabriel Sandu, and Stewart Shapiro.

Among the central points of the 'publication policy' behind the proceedings of the conference is the rule that the volume has to appear before beginning of the next year's conference. This 'rush', however, does not affect, we believe, the quality of the preparation of the volumes. Until last year they were published by the Institute's publishing house *Filosofia*; the present volume is the first prepared for *College Publications*. It contains a majority of the papers presented at the symposium *Logica 2008*, which took place from June 16 to 20 in the former Franciscan monastery of Hejnice, the Czech Republic, where the participants spent five days not only in the lecturing hall but also in many informal discussions during the breaks, lunches and social events. As you can see *The Logica Yearbook 2008* brings together various texts from mathematical and philosophical logic, history and philosophy of logic, and natural language analysis. As has become usual for the *Yearbook* series, the articles have not been sorted by subject — they are ordered alphabetically by author and it is up to the reader to pick and choose.

Editor's acknowledgements

Both the *Logica* symposium and this book series are the result of a joint effort of many people, who deserve my deep thanks. Among them are the main organizers Vladimír Svoboda and Timothy Childers from the Department of Logic of the Institute of Philosophy and Pavel Baran, the director of the Institute of Philosophy. The conference was promoted also by the Grant Agency of the Czech Republic, which provided significant support by financing the grant project no. 401/07/0904. We are further indebted to Marie Vučková, Head of the Foreign Relations Department of the Institute, for organizational support. The organization would be impossible without the help of Petra Ivaničová during and before the conference. Our stay in Hejnice Monastery was made pleasant through the efforts of Father Miloš Raban and the staff of the monastery. Special thanks also go to the Bernard Family Brewery of Humpolec, traditional sponsor of the social programme of the symposium. I would also like to thank to Marie Benediktová for the layout of this volume. Many thanks go to College Publications and its managing director Jane Spurr. Last but not least we would like to thank all the conference participants and to authors of the articles for their outstanding cooperation during the editorial process.

Prague, May 2009 Michal Peliš

Fuzzy Logics Interpreted as Logics of Resources

Libor Běhounek*

Girard's linear logic (1987) is often interpreted as the logic of resources, while formal fuzzy logics (see esp. Hájek, 1998) are usually understood as logics of partial truth. I will argue that deductive fuzzy logics can be interpreted in terms of resources as well, and that under most circumstances they actually capture resource-aware reasoning more accurately than linear logic. The resource-based interpretation then provides an alternative motivation for formal fuzzy logics, and gives an explanation of the meaning of their intermediary truth values that can be justified more easily than their traditional motivation based on partial truth.

1 Linear and substructural logics

Recall that linear logic and its variants are representatives of basic substructural logics (see, e.g., Restall, 2000, Paoli, 2002, Ono, 2003), i.e., logics that result from discarding some of the structural rules from the Gentzen-style calculi LK and LJ for classical and intuitionistic logic. In particular, *linear* logic **LL** discards the rules of contraction (C)

$$\frac{\Gamma, A, A, \Delta \Longrightarrow \Sigma}{\Gamma, A, \Delta \Longrightarrow \Sigma} \qquad \frac{\Gamma \Longrightarrow \Sigma, A, A, \Pi}{\Gamma \Longrightarrow \Sigma, A, \Pi}$$

and weakening (W)

$$\frac{\Gamma \Longrightarrow \Sigma}{A, \Gamma \Longrightarrow \Sigma} \qquad \frac{\Gamma \Longrightarrow \Sigma}{\Gamma \Longrightarrow \Sigma, A}$$

from the calculus LK for classical logic. *Intuitionistic* linear logic **ILL** discards the same rules (C, W) from the calculus LJ for intuitionistic logic. *Affine* linear logic **ALL** and *intuitionistic affine* linear logic **IALL** discard only the rule of contraction (C) from the calculi LK and LJ, respectively, but retain the rule of weakening (W).

* The work was supported by Grant No. IAA900090703 "Dynamic formal systems" of the Grant Agency of the Academy of Sciences of the Czech Republic and by Institutional Research Plan AV0Z10300504.

Recall further that substructural logics work in general with two conjunctions: the *lattice conjunction* $\wedge$ (also called weak, additive, or extensional conjunction) and *fusion* & (also called group, strong, multiplicative, or intensional conjunction). Similarly there are in general two disjunctions (lattice $\vee$ and strong) as well as two implications, two negations, etc., but the latter split connectives will not play a significant role in our account, as we shall mainly deal with intuitionistic substructural logics (which lack strong disjunction) and commutative fusion (then both implications coincide). Since in such substructural logics implication internalizes the sequent sign $\Longrightarrow$ and & the comma on the left-hand side of sequents (cf. Ono, 2003), the validity of the sequent $A_1, \ldots, A_n \Longrightarrow B$ is equivalent to the validity of the formula $A_1 \mathbin{\&} \ldots \mathbin{\&} A_n \to B$. Consequently, the rule of contraction corresponds to the validity of $A \to A \mathbin{\&} A$ and the rule of weakening to the validity of $A \mathbin{\&} B \to A$.

The algebraic semantics of substructural logics is that of *residuated lattices* (see, e.g., Jipsen & Tsinakis, 2002; Ono, 2003; Galatos, Jipsen, Kowalski, & Ono, 2007), i.e., lattices endowed with an additional monoidal operation $*$ (representing &) monotone w.r.t. the lattice order $\leq$, and its two *residuals* $/, \backslash$ (representing implications) that satisfy the residuation law

$$x * y \leq z \quad \text{iff} \quad y \leq x\backslash z \quad \text{iff} \quad x \leq z/y.$$

If $*$ is commutative, the two residuals $/, \backslash$ coincide and are usually denoted by $\Rightarrow$. The set of designated elements is $\{x \mid x \geq 1\}$, where 1 is the neutral element of the monoidal operation $*$. If convenient, residuated lattices may be expanded (to Ono's **FL**-algebras) by a constant 0 for falsity, which makes it possible to define negation as $x \Rightarrow 0$.

The term *substructural logics* will in this paper denote logics of classes of residuated lattices, following the stipulative definition by Ono (2003). In particular, (affine) intuitionistic linear logic is the logic of all (bounded integral) commutative residuated lattices,[1] and (affine) linear logic is the logic of those that furthermore satisfy the law of double negation.

2 Linear logic as the logic of resources

The reason why linear logic has been regarded as the logic of resources is illustrated by Girard's (1995) well-known 'Marlboro–Camels' example:

[1] A residuated lattice is called *commutative* if its monoidal operation $*$ is commutative; it is called *bounded integral* if $0 \leq x \leq 1$ for all elements x. We shall usually work with commutative residuated lattices only.

Consider the propositions

$$\begin{aligned} D &= \text{“I pay \$1.”}, \\ M &= \text{“I get a pack of Marlboro.”}, \\ C &= \text{“I get a pack of Camels.”} \end{aligned}$$

Then the sequent

$$D \to M, D \to C \Longrightarrow D \to M \mathbin{\&} C$$

expressing the inference

If I pay \$1, I get a pack of Marlboro
If I pay \$1, I get a pack of Camels
∴ If I pay \$1, I get a pack of Marlboro and I get a pack of Camels

is derivable by the rules of classical as well as intuitionistic logic. The inference is, however, viewed as counter-intuitive, if the conclusion is straightforwardly understood as getting *both* packs. The disputable sequent is not derivable in linear logic, though: linear logic only derives the sequent

$$D \to M, D \to C \Longrightarrow D \mathbin{\&} D \to M \mathbin{\&} C$$

which under a similar interpretation captures the fact that I need to pay *two* dollars to get both packs of cigarettes.

In this sense, linear logic is said to regard formulae as 'resources', which are 'spent' when used as premises of implications (in the Marlboro–Camels example, the premise D is spent by being detached from $D \to M$ to obtain M, and cannot be used again for $D \to C$ to obtain $M \mathbin{\&} C$). More formally, since premises cannot in linear logic be contracted (due to the lack of the rule (C)), they act as tokens for 'resources' needed to support the conclusion: a sequent is valid in linear logic only if it has the needed *amounts* of premises required for arriving at the conclusion.[2] In other words, linear logic 'counts' premises of sequents as if they represented resources needed for 'buying' the conclusion (where different propositional letters would represent different *types* of resources, while their occurrences in the sequent would represent *tokens* or units of that type).

Nevertheless, this feature of linear logic is due solely to the absence of the rule (C) of contraction, and therefore is common to *all* contraction-free substructural logics. It is not clear why exactly linear logic should

[2] *Exactly* the needed amounts in **LL** or **ILL**; *at least* the needed amounts in their affine versions (it being an effect of weakening that we need not spend all premises). In logics with both (C) and (W), e.g., classical or intuitionistic logic, each premise required for arriving at the conclusion only needs to be present at least once.

be more adequate as a logic of resources than any other contraction-free logic. Rather, it is to be expected that different contraction-free logics will correspond to different assumptions on the structure of resources. In the following sections I will argue that linear logics are in fact adequate only for very general structures of resources, while under most common circumstances, stronger logics are appropriate.

3 The structure of resources

As a first task, we need to refine our conception of resources. Since we aim at an *informal* semantic explanation of certain logics, instead of giving a formal definition we shall just list a few examples indicating what kind of resources we have in mind, and specify the mathematical properties they are assumed to satisfy.

Our notion of a resource will be rather broad: it can include any kind of things that can be counted or measured, that can be acquired and expended, or used for any purpose. Among the resources we consider are, e.g.: money (costs, prices, debts, etc.); goods (packs of cigarettes, clothes, cars, etc.); industrial materials (chemicals, natural raw materials, machine components, etc.); cooking ingredients (flour, salt, potatoes, etc.); computer resources (disk space, computation time, etc.); penalties (which can be regarded as a kind of costs incurred); sets, multisets, or sequences (tuples or vectors) of the above; etc.

It can be observed that all of these (as well as many other) kinds of resources exhibit the structure of a residuated lattice. In particular, there is:

- A partial order $\preceq$ comparing the amounts of the resources. For instance, 300 g of flour is more than 200 g of flour; two pens and three pencils are more than one pen and three pencils; etc. For the sake of compatibility with further definitions, we shall understand $x \preceq y$ as "the resource x is *larger* than or equal to y." The order need not be linear, as for instance two pens are not comparable with three pencils (if different items are counted separately). However, it can be assumed that $\preceq$ is a *lattice order*, as this is true for all prototypical cases: by definition, it amounts to supposing that for any two resources x, y (for instance: x = 2 pens and 3 pencils; y = 1 pen and 4 pencils), there is the least resource that is at least as large as both (in this case, 2 pens and 4 pencils) and the largest resource that is at most as large as both (here, 1 pen and 3 pencils). Even though there may exist resources that do not satisfy this assumption, we leave them aside in our considerations.

- A monoidal operation $*$ of *composition* (or fusion) of resources. For example, 300 g of flour and 200 g of flour is 500 g of flour; 2 pens and 3 pencils plus 1 pen and 3 pencils are 3 pens and 6 pencils; etc. Putting the resources together can be assumed to be associative (i.e., we presume that the total sum does not depend on the order of summation). The kinds of resources we consider always have a neutral element e, the *empty resource*, which does not change the amount when added to another resource: e.g., 0 g of flour; 0 pens and 0 pencils; etc. Even though composition of resources need not be commutative (consider, e.g., the order of adding ingredients when cooking), for the sake of simplicity of exposition we shall only consider commutative $*$ here (generalization to non-commutative $*$ is always straightforward).

- Finally, resources of all typical kinds can be 'subtracted' or 'evened up', i.e., their composition has the *residual* operation $\Rightarrow$ expressing the remainder, or the difference of amounts: $x \Rightarrow y$ is the least resource to be added to x in order to get a resource at least as large as y.[3] For example, if x = 200 g of flour and y = 300 g of flour, then $x \Rightarrow y$ is 100 g of flour, as one needs to add 100 g of flour to 200 g of flour to get at least 300 g; while if x = 2 pens and 3 pencils, and y = 1 pen and 3 pencils, then $x \Rightarrow y$ is 0 pens and 0 pencils (i.e. the empty resource e), as we need not add anything to x to get at least y.

All kinds of resources we consider thus have the structure of a (commutative) residuated lattice $\mathrm{L} = (L, \wedge, \vee, *, \Rightarrow, e)$. Particular kinds of resources can have additional properties: for example, most usual kinds of resources satisfy the so-called divisibility condition $x * (x \Rightarrow y) = x \wedge y$.

Since we aim at a simple resource-based interpretation of existing logical calculi rather than development of an expressively rich logic of resources for computer science, we do not consider such phenomena as, e.g., resource dynamics or possible non-totality of $*$ (which are modeled by such systems as the logic of bunched implications, computation logics, or synchronous and asynchronous calculi — see, e.g., Pym & Tofts, 2006 for references), but only reconstruct and refine the assumptions on resources that are adopted by linear logic.

4 Formulae as resources

There are at least two possible representations of resource-based semantics of substructural logics. One of them takes resources (i.e. elements of the

[3] I.e. $x \Rightarrow y = \sup\{z \mid z * x \preceq y\}$, which is an equivalent formulation of the residuation law in complete lattices. For incomplete lattices, a more cautious formulation based on Dedekind–MacNeille cuts is due, namely $\{z \mid z * x \preceq y\} = \{z \mid z \preceq x \Rightarrow y\}$, which is a general equivalent of the residuation law.

residuated lattice L described in Section 3) directly as semantic values assigned to propositional formulae. Recall that a logical calculus can have interpretations other than propositional: cf., e.g., the interpretation of the Lambek calculus as the categorial grammar (where the semantic values of formulae are grammatical categories), or the Curry–Howard combinatorial interpretation of the implicational fragment of intuitionistic logic (where formulae are interpreted as types and proofs as programs). In a similar vein, we can interpret the algebraic semantics of substructural logics under the "formulae-as-resources" paradigm as follows:

- The semantic value of a formula φ is a *resource* $\|\varphi\| \in \mathrm{L}$.
- The Tarski condition $\|1\| = e$ of the algebraic semantics interprets the formula 1 as the empty resource (or 'being for free').
- Similarly, the clause $\|\varphi \mathbin{\&} \psi\| = \|\varphi\| * \|\psi\|$ says that conjunction represents the *fusion of resources.*
- The value of implication, $\|\varphi \to \psi\| = \|\varphi\| \Rightarrow \|\psi\|$, is the resource needed to get at least $\|\psi\|$, given the resource $\|\varphi\|$.
- Finally, the lattice connectives $\wedge, \vee$ represent the meet and join of resources (with respect to the size order $\preceq$ of resources).

The formula φ is regarded as valid under a given evaluation iff $e \preceq \|\varphi\|$, i.e., iff it represents a resource that is for free or even cheaper.

5 Resources as possible worlds

Another way how to interpret substructural logics in terms of resources (cf. Pym & Tofts, 2006) is to regard the structure L of resources as a Kripke frame $(L, \succeq)$ endowed with a monoidal structure $(*, e)$. Unlike in the "formulae-as-resources" paradigm, formulae are here interpreted as propositions, and resources only serve as indices that may (or may not) validate them. The forcing relation $r \Vdash \varphi$, "the resource $r \in \mathrm{L}$ *supports* the formula φ," is required to satisfy the following conditions:

- $e \Vdash 1$,
- $r \Vdash \varphi \mathbin{\&} \psi$ iff $\exists s, t \in \mathrm{L}\colon\ r \preceq s * t$ and $s \Vdash \varphi$ and $t \Vdash \psi$,
- $r \Vdash \varphi \to \psi$ iff $\forall s \in \mathrm{L}\colon$ if $s \Vdash \varphi$, then $r * s \Vdash \psi$,
- $r \Vdash \varphi \wedge \psi$ iff $r \Vdash \varphi$ and $r \Vdash \psi$ ("shared resources" — contrast the clause for &),

- $r \Vdash \varphi \vee \psi$ iff $\exists s, t \in \mathrm{L}$: $r \preceq s \vee t$ and ($s \Vdash \varphi$ or $s \Vdash \psi$) and ($t \Vdash \varphi$ or $t \Vdash \psi$),

and the condition of persistence (if $r \preceq s$ and $s \Vdash \varphi$, then $r \Vdash \varphi$), expressing that "larger resources suffice as well." The formula φ is defined to be valid under $\Vdash$ iff $e \Vdash \varphi$, i.e., iff supported even by the empty resource.[4]

6 The role of tautologies

In the above semantics, tautologies w.r.t. a class $\mathcal{K}$ of (commutative) residuated lattices are defined as the formulae φ that get a value $\|\varphi\| \succeq e$ under all evaluations of propositional letters in any residuated lattice $\mathrm{L} \in \mathcal{K}$ (resp. are supported by e under all $\Vdash$ in every Kripke frame $\mathrm{L} \in \mathcal{K}$). The tautologies of substructural logics thus represent combinations of resources that are always "for free or cheaper".

More importantly, since all residuated lattices validate

$$e \preceq r \Rightarrow s \quad \text{iff} \quad r \preceq s,$$

tautologies of the form $\varphi \to \psi$ internalize sound rules of resource transformations that "preserve expenses" (in the sense of $\preceq$). Inference in substructural logics can thus be understood as inference *salvis expensis*, in a similar manner as inference *salva veritate* in classical logic.[5]

Classes of residuated lattices admitted as possible structures of resources then determine particular logics of resources in the above sense. In particular, by the known completeness theorem, **ILL** is the logic of all commutative residuated lattices, and so it is an adequate logic if just the general structure of a commutative residuated lattice is assumed for admissible kinds of resources. Its variants **IALL**, **ALL**, and **LL** restrict the structure of resources to narrower classes of commutative residuated lattices, and other substructural logics correspond to further specific classes of residuated lattices of resources.[6]

In the following sections I will argue that most typical kinds of resources satisfy the so-called *prelinearity condition*, and so are in fact governed by deductive fuzzy logics rather than linear logics.

---

[4] As this is not the aim of this paper, we omit the details on the correspondence between the Kripke-style and algebraic semantics of substructural logics. For more information see (Ono & Komori, 1985).

[5] Note that the general validity of $\|\varphi\| \preceq \|\psi\|$ defines the *local* consequence relation (expressed, i.a., by sequents in Section 1), while Hilbert-style calculi for substructural logics usually capture the *global* consequence relation "$e \preceq \|\psi\|$ whenever $e \preceq \|\varphi\|$".

[6] For example, classical logic can be interpreted as the logic distinguishing just two sizes of resources: empty $e = \|1\|$ and non-empty $f = \|0\| \prec e$.

7 Deductive fuzzy logics

Deductive fuzzy logics can be delimited as logics of (classes of) *linearly* ordered residuated lattices (Běhounek & Cintula, 2006; Běhounek, 2008). Among the extensions of **ILL** they can be characterized as those that satisfy the axiom of *prelinearity* (Pre): $((A \to B) \wedge 1) \vee ((B \to A) \wedge 1)$, or in the presence of weakening, equivalently $(A \to B) \vee (B \to A)$.

Let us call residuated lattices for which a substructural logic $\boldsymbol{L}$ is sound, *$\boldsymbol{L}$-algebras*. The prelinearity axiom ensures that a deductive fuzzy logic $\boldsymbol{L}$ is sound and complete, not only w.r.t. the class of *all* $\boldsymbol{L}$-algebras (the *general completeness theorem*), but also w.r.t. the class of all *linear* $\boldsymbol{L}$-algebras (the *linear completeness theorem*). The linear completeness theorem characterizes deductive fuzzy logics among substructural logics; the finitary ones are moreover characterized by the *linear subdirect decomposition property*, which says that each $\boldsymbol{L}$-algebra is a subdirect product[7] of linear $\boldsymbol{L}$-algebras. (See Cintula, 2006 for details.)

Besides the general and linear completeness theorems, most important deductive fuzzy logics furthermore enjoy the *standard completeness theorem*, i.e. the completeness w.r.t. a set of (selected) $\boldsymbol{L}$-algebras on the unit interval $[0, 1]$ of reals (with the usual ordering $\leq$), called the *standard* $\boldsymbol{L}$-algebras. Since $\boldsymbol{L}$-algebras on $[0, 1]$ are fully determined by the monoidal operation $*$, standard-complete deductive fuzzy logics can be defined as logics of (sets of) such monoidal operations $*$ on $[0, 1]$. For example,

- *Łukasiewicz logic* **Ł** is the logic of the Łukasiewicz t-norm $x * y = (x + y - 1) \vee 0$,
- *Gödel–Dummett logic* **G** is the logic of the minimum, i.e. of $x*y = x \wedge y$,
- *Product fuzzy logic* **Π** is the logic of the ordinary product of reals, $x * y = x \cdot y$,
- Hájek's *basic fuzzy logic* **BL** is the logic of all continuous t-norms,[8]
- *Monoidal t-norm logic* **MTL** is the logic of all left-continuous t-norms,
- *Uninorm logic* **UL** is the logic of all left-continuous uninorms, etc.

For more information on these logics see (Hájek, 1998; Esteva & Godo, 2001; Metcalfe & Montagna, 2007).

[7] I.e. a subalgebra of the direct product with all projections total.

[8] A commutative associative monotone binary operation on $[0, 1]$ with a neutral element $e \in [0, 1]$ is called a *uninorm*. A *t-norm* is a uninorm with $e = 1$.

The weakest deductive fuzzy logic extending a substructural logic $\boldsymbol{L}$ is often[9] obtained by adding the prelinearity axiom (Pre) to $\boldsymbol{L}$: for instance,

$$\mathbf{ILL} + (\text{Pre}) = \mathbf{UL}$$
$$\mathbf{IALL} + (\text{Pre}) = \mathbf{MTL}$$

are the weakest deductive fuzzy logics extending intuitionistic linear logics, or the logics of linear commutative (bounded integral) residuated lattices. (For **LL** and **ILL**, the double negation law is to be added to **UL** resp. **MTL**.)

8 Fuzzy logics as logics of costs

Since deductive fuzzy logics are logics of (special classes of) residuated lattices, they can be interpreted as logics of resources in the same way as other substructural logics. Specifically, by the linear completeness theorem (see Section 7), deductive fuzzy logics are sound and complete w.r.t. particular classes of *linear* residuated lattices, and so they are adequate for resources that are linearly ordered by $\preceq$. In other words, deductive fuzzy logics are those logics of resources in which we can assume that all resources are *comparable.*

Prototypical linearly ordered resources are *costs*, that is, resources converted to money. Even though resources in general need not be comparable (cf. the examples in Section 3), their costs (if specified) can always be compared, as money (of a single currency) forms a linear scale.[10] Besides money, there are many other kinds of resources that are linearly ordered, e.g., gallons of fuel, computation time, operational memory, etc. Irrespective of their nature, we shall call all linearly ordered resources *costs*, to distinguish them from resources that are not linearly ordered. For convenience, costs with values in the interval $[0, +\infty]$, e.g., monetary prices (where 0 is "gratis" and $+\infty$ may represent the price of unattainable goods), will be called *prices.*

Deductive fuzzy logics can thus be regarded as *logics of costs*, in the same sense as linear logics are regarded as logics of resources. Different ways of adding up costs — given by the fusion operation — yield different deductive fuzzy logics. The most typical examples are given below:

- If prices are summed by ordinary addition, we obtain the *product logic* $\mathbf{\Pi}$, since the residuated lattice $[0, +\infty]$ with the fusion $+$ and the lattice order $\geq$ is isomorphic (via the function $p \mapsto 2^{-p}$) to the standard product algebra $[0, 1]$ with the fusion $\cdot$ and the lattice order $\leq$.

[9] Always if modus ponens is the only derivation rule of $\boldsymbol{L}$ (Cintula, 2006).

[10] This idea is due to Petr Cintula (pers. comm.).

Note that in the standard product algebra, 0 represents the infinite cost and 1 the null cost. If the infinite cost is not considered, the standard product algebra without 0 (called the standard *cancellative hoop*) and its logic **CHL** (*cancellative hoop logic*, see Esteva, Godo, Hájek, & Montagna, 2003) are obtained.[11]

- If prices are bounded by a value $a \in (0, +\infty)$ and summed by bounded addition truncated at a, we obtain the *Łukasiewicz logic* **Ł**, since the residuated lattice $[0, a]$ with bounded addition and $\geq$ is isomorphic via $p \mapsto (a - p)/a$ to the standard [0,1] algebra for Łukasiewicz logic. The bound a (corresponding to 0 in the standard algebra) appears naturally if, e.g., a fixed maximum price is set, if there is a maximal possible cost in the given setting, or if the price a is in the given context unaffordable.

- If prices are combined by the maximum, *Gödel logic* **G** (or its hoop variant) is obtained (by the same isomorphism $p \mapsto 2^{-p}$ as in the case of addition). The maximum may seem a strange operation for summation of prices, but it occurs naturally whenever the costs can be shared by the summands. For example, if temporary results can be erased before the computation proceeds, the memory needed for temporary results is only the maximum (rather than sum) of their sizes.

Logics of other particular t-norms are obtained by using variously distorted 'addition' of prices. For instance, the logic of an ordinal sum of the three basic t-norms corresponds to using different summation rules (of the three described above) in different intervals of prices. The logic **MTL** is obtained if all monotone commutative associative left-continuous operations with the zero price acting as the neutral element are admitted as 'addition' of prices; similarly for **BL** and continuous such operations, etc. The logic **UL** and other uninorm logics only differ by permitting also negative prices, which express gains rather than costs.

9 Fuzzy logics as logics of resources

In spite of the linear completeness theorem, which makes it possible to regard deductive fuzzy logics as logics of linearly ordered costs, algebras for deductive fuzzy logics need not be linear (consider, e.g., their direct

[11] If the costs come in packages (e.g., if one has to buy a whole pack of cigarettes even if one needs only a few), the algebra is in general just a **ΠMTL**-chain instead of a product algebra, and the resulting logic in general only extends the logic **ΠMTL** (Esteva & Godo, 2001) or its hoop variant. A similar effect of packaging, which destroys the divisibility of the algebra (see Section 3), can be observed in other algebras of costs as well. (This observation is based on remarks by Rostislav Horčík and Petr Cintula.)

products). By the general completeness theorem (see Section 7), a deductive fuzzy logic $\boldsymbol{L}$ is also sound and complete w.r.t. the class of all $\boldsymbol{L}$-algebras: thus $\boldsymbol{L}$ can also be interpreted as the the logic of all kinds of resources that form the structure of a (possibly non-linear) $\boldsymbol{L}$-algebra.

Let us restrict our attention to finitary deductive fuzzy logics only, as they include all prototypical cases; for the sake of brevity, let us call them just *fuzzy logics* further on. By the linear subdirect decomposition theorem (see Section 7), any $\boldsymbol{L}$-algebra for a fuzzy logic $\boldsymbol{L}$ can be decomposed into a subdirect product of linear $\boldsymbol{L}$-algebras. Fuzzy logics can thus be characterized as logics of such resources that either are linearly ordered, or can at least be decomposed into linearly ordered components. In other words, a sound and complete resource-based semantics of fuzzy logics need not be just that of costs, but also that of resources representable as *tuples* (possibly infinitary) of costs.

It can be observed that many kinds of non-linear resources can actually be represented as tuples of linearly ordered values. For example, ingredients for making pizza and those for making spaghetti are not subsets of each other, thus cooking ingredients do not form a linearly ordered residuated lattice.[12] Nevertheless, they can be decomposed into (potentially infinitely many) linearly ordered components, as the *amounts* of each individual item on an ingredient list are always comparable; and indeed it can be checked that the prelinearity axiom is valid in this residuated lattice.[13]

In fact, most typical resources (including those mentioned in Section 3) *are* indeed decomposable in this way into linear components. Even many resources for which such a decomposition is not known (e.g., human intelligence) can at least be believed to be linearly decomposable (into some unknown and very fine linear components). It is actually rather hard to find a kind of resources that demonstrably cannot be so decomposed.

Thus we can conclude that all typical kinds of resources are linearly decomposable, and therefore they satisfy the axiom of prelinearity, which is not valid in linear logic nor in its affine or intuitionistic variants; consequently, they are actually governed by *deductive fuzzy logics* rather than linear logics. Linear logics are thus only adequate for a very general structure of resources, which admits even the rare kinds of resources that are not decomposable into linearly ordered components. As regards most usual kind

---

[12] The elements of the residuated lattice of all possible ingredient lists (such as can be found in recipe books) are tuples of quantities of particular ingredient types (e.g., [300 g of flour, 2 tomatoes, 2 lt of oil], zero amounts omitted). The tuples are naturally ordered by inclusion (i.e. pointwise by component sizes), and fusion represents adding up amounts of each ingredient.

[13] Since the fusion of amounts is (unbounded) addition in each component and infinite amounts do not occur, by extending the considerations of Section 8 the residuated lattice can actually be identified as a cancellative hoop, and the logic of cooking ingredients as the cancellative hoop logic **CHL**.

of resources, linear logic is too weak for them, as it does not validate the law of prelinearity they obey. Assuming commutativity of fusion, the weakest logic adequate for typical resources is the uninorm logic **UL** (or **MTL** if weakening is assumed, i.e., if the empty resource is the smallest). Specific structures of typical resources are governed by even stronger fuzzy logics — in particular, product logic **Π** if resources are combined by addition in each linear component, Łukasiewicz logic **Ł** if the addition is bounded, and Gödel logic **G** in the case of shared resources (i.e., if they componentwise combine by the maximum).

Thus it turns out that despite the common opinion, it is actually fuzzy logics, rather than linear logics, that could be categorized as typical logics of resources.[14] The interpretation in terms of resources and costs moreover provides an alternative motivation for deductive fuzzy logics and an explanation of the meaning of their intermediary truth-values that can in some respects be more easily justified than the standard account based on degrees of partial truth.

Libor Běhounek
Institute of Computer Science
Academy of Sciences of the Czech Republic
Pod Vodárenskou věží 2, 182 07 Prague 8, Czech Republic
behounek@cs.cas.cz

References

Běhounek, L. (2008). On the difference between traditional and deductive fuzzy logic. *Fuzzy Sets and Systems*, *159*(10), 1153–1164.

Běhounek, L., & Cintula, P. (2006). Fuzzy logics as the logics of chains. *Fuzzy Sets and Systems*, *157*(5), 604–610.

Cintula, P. (2006). Weakly implicative (fuzzy) logics I: Basic properties. *Archive for Mathematical Logic*, *45*(6), 673–704.

Esteva, F., & Godo, L. (2001). Monoidal t-norm based logic: Towards a logic for left-continuous t-norms. *Fuzzy Sets and Systems*, *124*(3), 271–288.

Esteva, F., Godo, L., Hájek, P., & Montagna, F. (2003). Hoops and fuzzy logic. *Journal of Logic and Computation*, *13*(4), 531–555.

Galatos, N., Jipsen, P., Kowalski, T., & Ono, H. (2007). *Residuated lattices: An algebraic glimpse at substructural logics.* Amsterdam: Elsevier.

Girard, J.-Y. (1987). Linear logic. *Theoretical Computer Science*, *50*(1), 1–102.

[14] The price paid for the more accurate account is a more complex proof theory, as prelinearity destroys the good proof-theoretical properties of linear logics.

Girard, J.-Y. (1995). Linear logic: Its syntax and semantics. In J.-Y. Girard, Y. Lafont, & L. Regnier (Eds.), *Advances in linear logic: Proceedings of the Workshop on Linear Logic, Cornell University, June 1993* (pp. 1–42). Cambridge University Press.

Hájek, P. (1998). *Metamathematics of fuzzy logic.* Dordercht: Kluwer.

Jipsen, P., & Tsinakis, C. (2002). A survey of residuated lattices. In J. Martinez (Ed.), *Ordered algebraic structures* (pp. 19–56). Dordrecht: Kluwer.

Metcalfe, G., & Montagna, F. (2007). Substructural fuzzy logics. *Journal of Symbolic Logic*, *72*(3), 834–864.

Ono, H. (2003). Substructural logics and residuated lattices—an introduction. In V. F. Hendricks & J. Malinowski (Eds.), *50 years of Studia Logica* (pp. 193–228). Dordrecht: Kluwer.

Ono, H., & Komori, Y. (1985). Logics without the contraction rule. *Journal of Symbolic Logic*, *50*(1), 169–201.

Paoli, F. (2002). *Substructural logics: A primer.* Kluwer.

Pym, D., & Tofts, C. (2006). A calculus and logic of resources and processes. *Formal Aspects of Computing*, *18*(4), 495–517.

Restall, G. (2000). *An introduction to substructural logics.* New York: Routledge.

Strong Paraconsistency and Exclusion Negation

Francesco Berto*

1 *True* or *not*?

Strong paraconsistency, also called *dialetheism*, is the view according to which there are dialetheias, that is, sentences A such that both A and $\neg A$ are true,[1] and it is rational to accept and assert them (an eminent case being allegedly provided by the various versions of the Liar paradox). One could therefore picture dialetheism as disputing the Law of Non-Contradiction (LNC). As a matter of fact, though, all the main formulations of the LNC are not disputed by a typical dialetheist, in the sense that she is committed to *accept* them. The dialetheic attitude of the dialetheist is expressed by typically accepting, and asserting, *both* the usual versions of the LNC and sentences inconsistent with them.

Of course, this calls for a drastic revision of our standard notions of *truth* and *negation*. Philosophers often disagree on the content of basic logical and metaphysical concepts (such as *identity*, *existence*, *necessity*, etc.), or on the validity of some very basic principles of inference (such as Contraposition or the Disjunctive Syllogism). It is well known that this kind of discussion often faces impasses, or turns into a hard conflict of intuitions. It is very difficult to establish when some party or other begins to beg the question. One wonders whether a non-standard explanation of a basic logical notion involves a real disagreement with a classical account of *that* notion, or its principles simply describe a different thing using the same name or symbol (the famous "change of subject" Quinean motto).

* The ideas on the NOT-operator developed in this paper have been hinted at in (Berto 2007, Ch. 14), and exposed extensively in (Berto 2008) — I am grateful to the Editors of the Australasian Journal of Philosophy for the permission to reuse some of that material. Thanks to Graham Priest, Francesco Paoli, Ross Brady, Max Carrara, Vero Tarca, Luca Illetterati, and Diego Marconi, for helpful comments, and to the participants to Logica 2008 for the lively discussion of the talk given there.

[1] See (Berto & Priest, 2008).

This seems to be decidedly the case with strong paraconsistency. When someone claims that both A and not-A are true, one wonders what is meant by "true"; and, of course, by "not":

> The fact that a logical system tolerates A and $\sim A$ is only significant if there is reason to think that the tilde means 'not'. Don't we say 'In Australia, the winter is in the summer', 'In Australia, people who stand upright have their heads pointing downwards', 'In Australia, mammals lay eggs', 'In Australia, swans are black'? If 'In Australia' can thus behave like 'not' (...), perhaps the tilde means 'In Australia'?[2]

Is paraconsistent negation just an In-Australia operator? In a thoroughly argued essay, Catarina Dutilh Novaes has recently suggested that, critics notwithstanding, the real philosophical challenge for paraconsistent logics does not consist in providing a plausible account for negation, but for the notion of contradiction.[3] Attacks to paraconsistency delivered by claiming that paraconsistent negation is not negation, according to Dutilh Novaes, "can be neutralized if it is shown that the conflation between contradiction and negation is not legitimate," and that "paraconsistent negation is in principle as real a negation as any other;"[4] for, as the nice survey of the history of logical negation provided in her paper shows, there is no unique real negation around.[5]

On the contrary, it is the notion of contradiction which spells trouble for (strong) paraconsistentists. The concept of contradiction "can be [defined] without using the negation: A and B are contradictory propositions iff $A \vee B$ holds and $A \wedge B$ does not hold, regardless of the form of A and B." Therefore "contradiction is the property of a pair of propositions which cannot both be true and cannot both be false at the same time;" since two propositions that are contradictories according to classical logic can both be true according to the (strong) paraconsistentist, one concludes that the latter simply rejects the classical notion of contradiction. So "paraconsistent logicians must give an account of what contradiction amounts to within a paraconsistent system."[6]

A strong paraconsistentist may object to the characterization of the notion of contradiction just given, for the definition uses a negation in the *definiens*: contradictories "can*not* both be true and can*not* both be false." Now is that "not" a classical or a paraconsistent negation? (Strong) paraconsistent logicians such as Graham Priest prefer to assert that negation

[2] (Smiley, 1993, p. 17).
[3] See (Dutilh Novaes, 2007).
[4] (Dutilh Novaes, 2007, pp. 479 and 482).
[5] On this, see also (Wansing, 2001).
[6] (Dutilh Novaes, 2007, pp. 479 and 483).

is a contradictory-forming operator, but define contradictoriness without adopting a negation in the *definiens*: A and B are contradictories iff, if A is true then B is false; and if A is false then B is true. Many paraconsistent negations, then, turn out to be contradictory-forming operators; for in such logics as LP (Priest's Logic of Paradox)[7] and FDE (Belnap and Dunn's First Degree Entailment), negation actually is an operator that truth-functionally switches truth and falsity: if A is true, then $\neg A$ is false; if A is false, then $\neg A$ is true; if A is both true and false, then $\neg A$ is, too (and if the semantics admits truth-value gaps, we may also have that A is neither true nor false; then, $\neg A$ is neither true nor false, too). Now the debate has been moved back to the notions of truth and falsity: do they overlap? Can some truth-bearer bear both?

If we want to have a non-question-begging debate on dialetheias and the LNC, instead of concentrating on truth and falsity we may go back to negation. Or, at least, this is the way pursued in this paper. Must there be a unique good account of negation? Perhaps, as Dutilh Novaes forcefully argues, not. We may have distinct intuitions on different sentential and predicate negations, which may be characterized by different theories. This does not entail, though, that no non-question-begging debate is feasible. On the contrary, I think it is possible to characterize a negation (I shall label it "NOT") with the following pleasant features:

1. its definition does not refer to the contentious concept *truth*;

2. it has a strong pre-theoretical motivation, because of its indispensable expressive function in language and communication; and

3. it is fully accepted also by dialetheists, because it is based on a deep metaphysical intuition they show to fully share: the intuition of *exclusion*.

If the characterization of NOT proposed in the following is sufficient to confer a determinate meaning to the negation in question, we can conveniently phrase a formulation of the LNC via such a negation. This LNC might be indisputable also from the dialetheist's point of view. "Indisputable" should be understood in the following sense: the dialetheist is forced to accept it, without also accepting something inconsistent with *it*. It might be a version of the LNC on which both the orthodox friend and the dialetheic foe of consistency can agree in this sense.

[7] See (Priest, 1979), (Priest, 1987).

2 The exclusion problem and Priest's pragmatic way out

I will start with a problem facing strong paraconsistency, which has been variously recognised in the literature. I believe it to be the main theoretical trouble for dialetheism, and I have elsewhere proposed to call it the *Exclusion Problem.*[8] It goes as follows.

When you say: "A", and a dialetheist replies: "$\neg A$", she might not have managed to rule out what you have said, precisely because of the features of her paraconsistent negation. In the dialetheic framework, $\neg A$ does not rule out A on logical grounds: it may be the case both that A and that $\neg A$, so the dialetheist may accept them both. Also saying "A is false," and even "A is not true," need not rule out A on the dialetheist's side. In many paraconsistent logics, beginning with LP, given any set of sentences S, it is logically possible that every sentence of S is true. This happens in the so-called *trivial model* of LP: if all atomic sentences are both true and false, then all sentences (truth-functionally) are. In a nutshell: nothing is ruled out *on logical grounds only* in the dialetheic framework. Many authors have inferred that dialetheism faces the risk of ending up inexpressible.[9]

According to Priest, though, these troubles with ruling out things can be solved by turning into the realm of pragmatics. In order to help the dialetheist rule out something, he has provided an interesting treatment of the notion of *rejection.* Let us call *acceptance* and *rejection* two mental states a subject x has towards (the proposition expressed by) a sentence. Acceptance and rejection are polar opposites: to reject something is to positively refuse to believe it. *Assertion* and *denial,* on the other hand, are (typically) linguistic acts or, equivalently, illocutionary forces attached to utterances. Roughly, assertion and denial are the linguistic counterparts of acceptance and rejection. Acceptance and assertion, and, respectively, rejection and denial, are often conflated by philosophers, and anyway for most of our purposes we can run linguistic acts and the corresponding mental states together. Let's have two sentential operators, "$\vdash_x$" and "$\dashv_x$", whose reading is, respectively, "rational agent x accepts/asserts (that)" and "rational agent x rejects/denies (that)." The standard treatment has it that rejection/denial is equivalent to the acceptance/assertion of negation:

$$\dashv_x A \leftrightarrow \vdash_x \neg A. \tag{1}$$

If we understand it in terms of linguistic acts, (1) is the claim, famously held by Frege and Peter Geach, according to which to deny something just is to assert its negation. But Priest says that accepting $\neg A$ is different from rejecting A: a dialetheist can do the former and not the latter —

[8] See (Berto, 2006), (Berto, 2007, Ch. 14).
[9] See, e.g., (Parsons, 1990), (Batens, 1990), (Shapiro, 2004).

exactly when she thinks that A is paradoxical. When A is a dialetheia, the natural assumption (1) breaks down, and negation and denial come apart. A denial/rejection of A becomes a non-derivative mental or linguistic act, in that it is directly aimed at A (or at the content of A, or at the proposition expressed by A, etc).[10]

Given that (1) can fail, the dialetheist can accept both A and $\neg A$, but she does not need to accept and reject A. Actually, according to Priest she *cannot* even do that: Priest considers acceptance and rejection as reciprocally *incompatible*, even though A and $\neg A$ are not:

> Someone who rejects A cannot simultaneously accept it any more than a person can simultaneously catch a bus and miss it, or win a game of chess and lose it. If a person is asked whether or not A, he can of course say 'Yes and no'. However this does not show that he both accepts and rejects A. It means that he accepts both A and its negation. Moreover a person can alternate between accepting and rejecting a claim. He can also be undecided as to which to do. But do both he can not.[11]

And this is how the dialetheist can manage to rule out something, and to express this. Although the she cannot rule out A by simply saying "$\neg A$", she can *reject* A. So the pragmatic incompatibility of acceptance/assertion and rejection/denial plays a pivotal role in Priest's reply to the Exclusion Problem.

3 NOT

This shows that even dialetheists have an intuition of *exclusion*, or *incompatibility* between something and something else. So I propose to search for an operator (arguably, a negation) that allows us to capture and express the intuition. We need to start from this very notion precisely because we want to avoid explicitly employing the concepts of *truth* and *falsity* to characterize such an operator. The dialetheist casts doubts on *their* being exclusive, by pointing out that some truth-bearers, notably, the Liars, fall under both concepts simultaneously. Truth tables or truth conditions for negation can give us no sense of the connection between negation and exclusion unless we *already* share the intuition that truth and falsity rule out each other.

This brings us back to the issue of the notion of contradiction, raised by Dutilh Novaes. Specifically, we should refrain from expressing exclusion via the traditional concept of *contrariness*: defining A and B as contraries iff "$A \wedge B$" is logically false won't help when discussing with the dialetheist. But

[10] See (Priest, 2006, p. 104).

[11] (Priest, 1989, p. 618).

one may try with the intuitive notion of exclusion itself, taken as primitive. Animals and infants perceive incompatibilities in the world long before they have developed or mastered an articulated language to express them. One of the uses of a linguistic item that counts as a negation can therefore be explained, as Huw Price has claimed, as "initially a means of *registering* (publicly or privately) a perceived incompatibility." And if "incompatibility [is] a very basic feature of a speaker's (or proto-speaker's) experience of the world,"[12] then one can explain the negation we are looking for in terms of incompatibility. We only need to assume that ordinary speakers and rational agents have some acquaintance with exclusions — things of the world ruling out each other: they can recognize them in the world, and in their commerce with the world.

I shall talk of *material exclusion* or, equivalently, of *material incompatibility*. One may characterize it in terms of concepts, properties, states of affairs, propositions, or worlds, depending on one's metaphysical preferences.[13] Material exclusion bears this name to stress the fact that it is not a merely logical, in the sense of formal, notion: it is based on the material content of the involved concepts, or properties, etc. Some examples: phenomenological colour incompatibilities, such as *being (solidly) Red* and *being (solidly) Green*; concepts that express our categorization of physical objects in space and time, such as x *being here right now* and x *being way over there right* now, for a suitably small x.[14] Or x *being less than two inches long* and x *being more than three feet long*.[15] But also Priest's above x*'s catching the bus* and x*'s missing the bus* will do.

Ok, this was the intuition. How do we formalize it? A feasible formal account may adapt the idea developed by Michael Dunn that "one can define negation in terms of one primitive relation of incompatibility (...) in a metaphysical framework."[16] So let us talk in terms of *propositions* (that which is expressed by a sentence) and build a small algebra. Think of a structure $\langle U, \subset, V, \bullet \rangle$ where U is a set of propositions; $\subset$ and $\bullet$ are binary relations defined on U; and V is a unary operation on subsets of U. $\subset$ is to be thought of as a pre-order, and "$p \subset q$" can be read as "The proposition p entails the proposition q". Given a set of propositions $P \subseteq U$, VP is the

[12] (Price, 1990, pp. 226–228).

[13] For instance, we may view it as the relation that holds between a couple of properties P_1 and P_2 iff, by having P_1, an object has dismissed any chance of simultaneously having P_2. Or we may also claim that material incompatibility holds between two concepts C_1 and C_2 iff the very instantiating C_1 by a puts a bar on the possibility that a also instantiates C_2. Or we may say that it holds between two states of affairs S_1 and S_2 iff the holding of S_1 precludes the possibility that S_2 also holds (in world w, at time t, etc.).

[14] See (Tennant, 2004, p. 362).

[15] See (Grim, 2004, p. 63).

[16] (Dunn, 1996, p. 9).

(possibly infinitary: more on this soon) disjunction of all the propositions in P. And $\bullet$ is precisely our primitive relation of material exclusion.

A proposition may have one or more incompatible peers: it may rule out a whole assortment of alternatives. Patrick Grim, for instance, talks about the *exclusionary class* of a given property. The exclusionary class of a proposition p, then, is the set

$$E = \{x; x \bullet p\}.$$

Then NOT-p is nothing but $\mathrm{V}E$. If E has finite cardinality, then NOT-p is just an ordinary disjunction: $q_1 \vee \cdots \vee q_n$ where $q_1, \ldots, q_n$ are all the members of E.

Suppose there are infinitely many propositions incompatible with p. This is a heavy metaphysical assumption, but let us grant it. Then, NOT-p turns out to be an infinitary disjunction. If one has (understandable) problems with infinitary disjunctions, we cannot avoid quantifying on propositions:

$$\text{NOT-}p =_{\text{df}} \exists x(x \wedge x \bullet p). \tag{2}$$

Both in the finitary and infinitary case, it is clear in which sense NOT-p is the logically weakest among the n incompatibles: it is entailed by any q_i, $1 \leq i \leq n$, such that $q_i \bullet p$. One may express the point via the following equivalence:

$$x \subset \text{NOT-}p \quad \text{iff} \quad x \bullet p. \tag{3}$$

Putting NOT-p for x, and by detachment, we get:

$$\text{NOT}\,p \bullet p, \tag{4}$$

NOT-p is incompatible with p. The right-to-left direction of (3), then, tells us that NOT-p is the weakest incompatible, i.e., it is entailed by any incompatible proposition.[17]

Which logic should be read off the algebra depends on which algebraic postulates we want to add. Depending on the choices we make, NOT will become palatable for some logicians, even though others will be disappointed. One may assume, reasonably enough, that $\bullet$ is symmetric. But if in the algebraic framework NOT is stipulated as an operation of period two, i.e.

$$\text{NOT-NOT-}p = p, \tag{5}$$

this is likely to be rejected by an intuitionist, though not by many paraconsistent logicians. The intuitionist may also object to the fact that NOT has

[17] Variations on the theme of the characterization of negation via incompatibility, and on negation as the minimal incompatible, can be found in (Brandom, 1994, pp. 381ff.); (Harman, 1986, pp. 118-20); (Peacocke, 1987).

been defined using other operators, which goes against the independence of logical constants in a constructivist framework (a remark I owe to Francesco Paoli). Or, if we make the *prima facie* natural assumption that:

$$\text{If} \quad p \subset q \text{ and } x \bullet q, \quad \text{then} \quad x \bullet p, \tag{6}$$

we can easily get contraposition. But such a result would be rejected by those who want to dismiss contraposition on the basis of considerations on the conditional, and also by some paraconsistentists. Different philosophical parties (classicists, intuitionists, paraconsistentists, etc.) have opposed views on what negation is, whereas the aim here is to provide an intuitive depiction on which all parties can agree; this is why I find formalization useful only to a certain extent.

4 Minimal LNC

Independently of the possible additional characterization, NOT has some nice features. First, is not explicitly defined via the concept *truth*. As Graham Priest has pointed out to me (in communication), this may not prevent truth from jumping in again. I have been forced to admit that, given some (albeit debatable) metaphysical assumptions, we may need propositional quantification to spell out the details of NOT. And such quantification is inter-definable with truth. But what NOT is explicitly referred to is the concept *exclusion*, whose primitiveness has been argued for above.

Secondly, NOT has a strong pre-theoretical appeal as an *exclusion-expressing* tool: it allows us to rule out things by claiming that something incompatible with them is going on. This is what at least some of the items we qualify as negation should help us to do.

Finally, dialetheists grasp the notion of exclusion. They ask us to stop using "not" or "true" as exclusion-expressing devices, because "not-A" is insufficient by itself to rule out A, and "A is true" is insufficient by itself to rule out that A is also false. But Priest's account of acceptance and rejection shows that the dialetheist believes in the impossibility of some couples of facts', or states of affairs', simultaneously obtaining; or, equivalently, that she assumes that some things materially exclude some others: x's simultaneously catching and missing the bus, for instance; and, of course, x's simultaneously accepting and rejecting the same A, this being, as we have seen, a basic step in Priest's answer to the Exclusion Problem. NOT is supposed to work even in a framework in which nothing is ruled out on logical grounds alone, because it is not merely logically, i.e. formally, but metaphysically ("materially") founded. The dialetheist may have a vacuous notion of logical, formal incompatibility. But she does have a notion of material incompatibility.

Now for the final step: express the LNC via NOT. Take Aristotle's traditional formulation of the LNC, in Book Γ of the *Metaphysics*, and just put in it our NOT. The formulation can be taken as a definition of "the impossible":

> For the same thing to hold good and NOT hold good simultaneously of the same thing and in the same respect is impossible.[18] (7)

"P_1 does NOT hold good of x" should be a short form for "to x belongs some property P_2, which is materially incompatible with P_1." This does not seem to be questionable by the dialetheist anymore, provided she has understood NOT — and to understand NOT is to understand exclusion. If the dialetheist refuses to subscribe to the characterization of NOT via the intuitive notion of exclusion, she seems to actually end up as unable to express the exclusion of any position (is she trying to exclude exclusion?). And a dialetheism without the LNC stated in terms of NOT looks very much like a trivialism (I totally agree with Dutilh Novaes, who presses a point very similar to this one in her essay).[19] Such a LNC, to use Aristotle's words, is "a principle which every one must have who knows anything about being."[20]

Does this reply to Dutilh Novaes' challenge for (strong) paraconsistency, namely, that of defining "P-contradictions, that is, contradictions that are so threatening to a theory that they really compromise rational inference-making within it"?[21] To some extent, yes — if the steps of the argumentation proposed above work. But the success for the version of the LNC phrased in terms of NOT is very limited: for that LNC simply rules out the simultaneous obtaining of reciprocally exclusionary states of affairs. The question remains open of *which* are the exclusionary states of affairs (or properties, etc.). And now the discussion between dialetheists and anti-dialetheists can develop with a significant decrease in issues of question-begging and clashes of intuitions. What is incompatible with what? Given two properties P_1 and P_2, the question whether they are exclusive can involve broadly empirical matters, difficult analyses of our conceptual toolkit and/or of our use of ordinary language expressions. Some cases may be easy to resolve; but others may produce battles of intuitions: are *young* and *old* actually exclusive? *Blue* and *green*? *True* and *false*? *Circular* and *square*? I have claimed that material exclusion is based on the content of facts, concepts, or properties; but how do we know what the content of a concept

---

[18] See Aristotle *Met.* 1005b 18–21 (Aristotle, 1984).

[19] See (Dutilh Novaes, 2007, p. 487)).

[20] Arist. *Met.* 1005b 14–15 (Aristotle, 1984).

[21] (Dutilh Novaes, 2007, p. 489).

is, or which are the actual fields of applications of a property? The formal characterization of NOT, of course, does not entail special commitments on which are the specific properties, or concepts, or states of affairs, between which it holds. Such commitments are fallible. We can come to believe that some properties, or concepts, or states of affairs, are incompatible, and then find out that they are not. Would this entail explosion, that is, anything being derivable, and trivialism? Well, not: the standard strategy in this case is simply to retract our previous assumption that they were.

So the dialetheist who has no troubles with our minimal (7) can still object to other formulations of the LNC, e.g., because they are phrased in terms of truth and falsity: those who rule out that any sentence could be both *true* and *false* take truth and falsity as exclusionary concepts; the dialetheist has qualms on this, and perhaps counterexamples to offer (say, the Liar sentences). But the issue addressed here is whether *all* concepts (or properties, etc.) are like that; and the dialetheist agrees that some concepts (or properties, etc.) do rule out each other. This is the shared, basic intuition NOT appeals to.

Francesco Berto
Department of Philosophy and Theory of Sciences, University of Venice-Ca'Foscari
Dorsoduro 3484 D, 30123 Venice, Italy
bertofra@unive.it

References

Aristotle. (1984). Metaphysics. In J. Barnes (Ed.), *The Complete Works of Aristotle* (Vol. 2). Princeton, N. J.: Princeton University Press.

Batens, D. (1990). Against global paraconsistency. *Studies in Soviet Thought*, *39*, 209–229.

Berto, F. (2006). Meaning, metaphysics, and contradiction. *American Philosophical Quarterly*, *43*, 283–297.

Berto, F. (2007). *How to Sell a Contradiction. The Logic and Philosophy of Inconsistency.* London: College Publications.

Berto, F. (2008). Adynaton and material exclusion. *Australasian Journal of Philosophy*, *86*, 165–190.

Berto, F., & Priest, G. (2008). Dialetheism. In *The Stanford Encyclopedia of Philosophy.* Stanford, CA: CSLI. Available from http://plato.stanford.edu/entries/dialetheism/

Brandom, R. (1994). *Making it Explicit.* Cambridge, MA: Harvard University Press.

Dunn, J. (1996). Generalized ortho negation. In H. Wansing (Ed.), (pp. 3–26). Berlin–New York: De Gruyter.

Dutilh Novaes, C. (2007). Contradiction: The real philosophical challenge for paraconsistent logic. In J. Béziau, W. Carnielli, & D. Gabbay (Eds.), *Handbook of Paraconsistency* (pp. 477–492). London: College Publications.

Grim, P. (2004). What is a contradiction? In G. Priest, J. Beall, & B. Armour-Garb (Eds.), (pp. 49–72). Oxford: Clarendon Press.

Harman, G. (1986). *Change in view.* Cambridge: MIT Press.

Parsons, T. (1990). True contradictions. *Canadian Journal of Philosophy, 20*, 335–354.

Peacocke, C. (1987). Understanding logical constants: a realist's account. *Proceedings of the British Academy, 73*, 153–200.

Price, H. (1990). Why 'Not'? *Mind, 99*, 221–238.

Priest, G. (1979). The logic of paradox. *Journal of Philosophical Logic, 8*, 219–241.

Priest, G. (1987). *In contradiction: a study of the transconsistent.* Dordrecht: Martinus Nijhoff. (2nd, extended, edn, Oxford: Oxford University Press, 2006.)

Priest, G. (1989). Reductio ad absurdum et modus tollendo ponens. In G. Priest, R. Routley, & N. J. (Eds.), *Paraconsistent Logic. Essays on the Inconsistent* (pp. 613–626). München: Philosophia Verlag.

Priest, G. (2006). *Doubt Truth to be a Liar.* Oxford: Oxford University Press.

Priest, G., Beall, J., & Armour-Garb, B. (Eds.). (2004). *The Law of Non-Contradiction. New Philosophical Essays.* Oxford: Clarendon Press.

Shapiro, S. (2004). Simple truth, contradiction, and consistency. In G. Priest, J. Beall, & B. Armour-Garb (Eds.), (pp. 336–354). Oxford: Clarendon Press.

Smiley, T. (1993). Can contradictions be true? I. *Proceedings of the Aristotelian Society, 67*, 17–34.

Tennant, N. (2004). An anti-realist critique of dialetheism. In G. Priest, J. Beall, & B. Armour-Garb (Eds.), (pp. 355–384). Oxford: Clarendon Press.

Wansing, H. (Ed.). (1996). *Negation. A Notion in Focus.* Berlin–New York: De Gruyter.

Wansing, H. (2001). Negation. In L. Goble (Ed.), *The Blackwell Guide to Philosophical Logic* (pp. 415–436). Oxford: Blackwell.

Medieval *Obligationes* as a Regimentation of 'the Game of Giving and Asking for Reasons'

Catarina Dutilh Novaes*

1 Introduction

Medieval *obligationes* disputations were a highly regimented form of oral disputation opposing two participants, respondent and opponent, and where inferential relations between sentences took precedence over their truth or falsity. In (Dutilh Novaes, 2005), (Dutilh Novaes, 2006) and (Dutilh Novaes, 2007, Ch. 3) I presented an interpretation of *obligationes* as logical games of consistency maintenance; this interpretation had many advantages, in particular that of capturing the goal-oriented, rule-governed nature of this kind of disputation by means of the game analogy. It also explained several of its features that remained otherwise mysterious in alternative interpretations, such as the role of impertinent sentences and why, while there is always a winning strategy for respondent, the game remains hard to play. However, the logical game interpretation did not provide a full account of the deontic aspect of *obligationes* — of what being *obliged* to a certain statement really consists in — beyond the general (and superficial) commitment towards playing (and winning) a game. After all, the very name invokes normativity, so an interpretation of *obligationes* that does not fully account for the deontic component seems to be missing a crucial aspect of the general spirit of the enterprise. In order to amend this shortcoming in my previous analysis I here present an extension of the game-interpretation based on the notion of 'the game of giving and asking for reasons' — henceforth, GOGAR[1] — presented in Chapter 3 of R. Brandom's *Making it Explicit* (Brandom, 1994) as constituting the ultimate basis for social linguistic practices. The basic

* Thanks to Edgar Andrade-Lotero and Ole Thomassen Hjortland for comments on an earlier draft of the paper.

[1] Following J. MacFarlane's terminology, cf. `http://johnmacfarlane.net/gogar.html`.

idea is that *obligationes* can be seen as a regimentation of some of the core aspects of GOGAR.

What is to be gained from a comparison between *obligationes* and GOGAR? From the point of view of the latter, the comparison can shed light on its general logical structure: if *obligationes* really are a regimentation of GOGAR, then they can certainly contribute to making its structure explicit (which is of course another crucial element of Brandom's general enterprise). Indeed, an *obligatio* is something of a *Sprachspiel* for GOGAR, a simplified model whereby some of GOGAR's properties can be made manifest. As for *obligationes*, what can be gained from the comparison, besides the emphasis on its fundamentally deontic nature, is a better understanding of its general purpose. At first sight, this highly regimented form of disputation, where truth does not seem to have any major role to play, may seem like sterile scholastic logical gymnastics. But if it is put in the context of GOGAR — which (presumably) captures the essence of our social, linguistic and rational behaviors — then its significance would appear to go well beyond the (mere) development of the ability to recognize inferential relations and to maintain consistency.

2 GOGAR

A crucial element of the philosophical system presented by Brandom in *Making it Explicit* (and further expanded in several of his subsequent writings) is the model of language use that he refers to as 'the game of giving and asking for reasons'. Brandom insists that language use and language meaningfulness can only be understood in the context of social practices articulating information exchange and actions — linguistic speech-acts (typically, the making of a claim) as well as non-linguistic actions.

In fact, GOGAR should account for what makes us social, linguistic and rational animals. As Brandom construes it, GOGAR is fundamentally a normative game in that the *propriety* of the moves to be undertaken by the participants is at the central stage. It is, however, not a transcendental kind of normativity, requiring an almighty judge outside the game to keep track of the correctness of the moves undertaken; rather, the participants themselves are in charge of evaluating whether the moves undertaken are appropriate. It is a "deontic scorekeeping model of discursive practice". In GOGAR, we are all players (speakers) *and* scorekeepers concomitantly; we undertake moves and keep track of everybody's moves (including our own) at the same time. The focus on (giving and asking for) reasons is an important aspect of how the model captures the concept of rationality: we are responsible for the claims we make, and thus must be prepared to provide *reasons*[2] for them when challenged. Underlying this fact is the idea of a logical articulation of contents such that some contents count as appropriate reasons for other contents.

In principle, as a general model of language use, GOGAR should encompass all different kinds of speech-acts: assertions, questions, but also promises, orders, expressions of doubt etc. However, for Brandom there is one fundamental kind of speech-act in the game, namely that of making an assertion.[3] An assertion is both something that can count as a reason (a justification) for another assertion and something that may constitute a challenge — typically, when a speaker S makes an assertion incompatible with something previously said by T — and thus provoke the need for further reasons (T must defend the original assertion): hence, giving *and* asking for reasons.

The need to defend one's assertions threatened by challenges through further reasons indicates that one is somehow responsible for one's assertions. This is indeed the case according to the GOGAR model, and this fact is accounted for by the absolutely crucial concept of doxastic *commitment.* Just as a promise creates the commitment to fulfill what has been promised, the making of an assertion creates the commitment to defend it, i.e., to have had good reasons to make it. This is because one often relies on the information conveyed by an assertion made by another person in order to assess a particular situation and then act upon the assessment; but if false information is transmitted, then the assessment will probably be mistaken, and the action in question will probably not have the desired outcome; it may even have deleterious consequences for the agent. In such cases, it is fair to say that the person having conveyed the incorrect information is responsible for the infelicitous outcome, just as a reckless driver is responsible for the accidents he/she (directly or indirectly) causes. If somebody shouts 'fire!' as a prank in a completely full stadium, for example, this will probably cause considerable mayhem, and the infelicitous joker will be held accountable for all the damage caused. So given the potential *practical consequences* of an assertion, it is not surprising at all that liability should be involved in the making of an assertion.

For Brandom, the commitment to the content[4] of an assertion in fact goes beyond the assertion itself: one is also committed to everything that follows from the original assertion, i.e., everything that can be inferred from it. The

[2] Etymologically, rationality comes from *ratio*, 'reason' in Latin.

[3] "The fundamental sort of move in the game of giving and asking for reasons is making a *claim* — producing a performance that is propositionally contentful in that it can be the offering of a reason, and reasons can be demanded for it." (Brandom, 1994, p. 141).

[4] It is not entirely clear to me though whether Brandom sees commitments as having contents or sentences or claims as their objects, but it seems to me that contents would be the most appropriate objects of commitments.

inferential relations between assertions are a primitive element of Brandom's system (codified in terms of *material* inferences, not formal ones); as he sometimes says, they are "unexplained explainers" (Brandom, 1994, p. 133). Material inferences are painstakingly discussed in (Brandom, 1994, Ch. 2), but for our purposes what is important is to realize that commitment to a content transfers over to other contents by means of inferential relations.

But besides being committed to contents, there is another primitive deontic status that a speaker may or may not enjoy with respect to contents: *entitlement*. From the point of view of a scorekeeper,[5] for a speaker S to be entitled to asserting a given content amounts to S being in the position to offer grounds that justify belief in the content, and thus the making of the corresponding assertion; this deontic status is attributed when the speaker has good (enough) reasons to believe the content to be the case. Brandom remarks that "commitment and entitlement correspond to the traditional deontic primitives of obligation and permission" (Brandom, 1994, p. 160); he rejects this terminology because he wishes to avoid the stigmata of norms associated with hierarchy and commands (as noted above, the scorekeeping is done horizontally by all participants). But ultimately, a commitment is indeed an obligation, and an entitlement is indeed a permission, and thus being committed to a content amounts to being *obliged* to it in exactly the same sense of being obliged during an *obligatio* disputation (as we shall see): one has a duty towards a certain content, which transfers over to all the contents that follow from it.

From the two primitive concepts of commitment and entitlement, Brandom derives the equally important concept of incompatibility: content p being incompatible with content q amounts to commitment to p precluding entitlement to q. It is not so much that it is factually impossible for one to be committed to p while believing oneself to be entitled to q; this can occur, just as one can make conflicting promises and hold inconsistent beliefs. But again, this is a matter of deontic scorekeeping: from the point of view of the scorekeepers, if a speaker is committed to p there is a whole series of contents q, t, etc. to which the speaker in question is simply not entitled as long as he maintains his commitment to p. But if he nevertheless insists in being committed to p and entitled to q at the same time, then he is simply making a bad move within GOGAR.

Brandom correctly notices that incompatibility, as much as entailment, is essentially a relation between sets of contents, not between contents them-

[5] The deontic statuses of commitment and entitlement are always perspectival, i.e. defined by the deontic attitudes of (self-)attributing commitments and entitlements of each scorekeeper. "Such statuses are creatures of the practical attitudes of the members of a linguistic community — they are instituted by practices governing the taking and treating of individuals *as* committed." (Brandom, 1994, p. 142).

selves.[6] Take a set of three contents, e.g., those expressed by the sentences 'Every man is running', 'Socrates is a man' and 'Socrates is not running'. Commitment to either one of the two first contents alone does nor preclude entitlement to the third content, but commitment to both of them does preclude entitlement to the third, just as commitment to the first two contents simultaneously entails commitment to the content 'Socrates is running'. This aspect will be significant for the comparison with *obligationes* later on, as it hints at the fundamentally dynamic nature of the GOGAR model: every new assertion made requires the recalibration of everybody's deontic statuses by the scorekeepers — of the asserter, in particular, but in fact of everybody else as well, as GOGAR also accounts for inter-personal transmission of entitlement by testimony. In other words, a speaker's deontic status — her commitments and entitlements — is modified every time an assertion is made, more saliently but not exclusively by the speaker herself.

Indeed, there seem to be four main sources of entitlement according to the GOGAR model.

1. Interpersonal, intracontent deferential entitlement: Speaker 1 is entitled to (asserting) content p because speaker 2, a reliable source, asserted p.
2. Intrapersonal, intercontent inferential entitlement: Speaker 1 is entitled to (asserting) q because she is entitled to (asserting) p and p entails q.
3. Perception: Speaker 1 is entitled to (asserting) p because she has had a (reliable) perceptual experience corresponding to p.
4. Default entitlement: 'free moves', the contents entitlement to which is shared by all speakers insofar as these contents constitute common knowledge — everybody knows it, and everybody knows that everybody knows it.

A final point I wish to address in my brief presentation of GOGAR is the notion of inference, more specifically material inference. Brandom criticizes the formalist view of inference, according to which every valid inference is an instance of a formally valid schema; rather, the inferential relations that are the primitive elements of his inferential semantics are of a conceptual nature, while also firmly embedded in practices: "Inferring is a kind of doing."(Brandom, 1994, p. 91) The focus on the notion of material inference also echoes important features of *obligationes*, as in the latter

[6] (Brandom, 2008, Lect. 5). The cases of relations involving single contents can be seen as limit-cases, relating singleton sets.

framework the relation of 'following' (*sequitur*) in question is not restricted to formally valid schemata.[7]

3 Medieval *obligationes*

An *obligatio* disputation has two participants, Opponent and Respondent. In the case of *positio*, the most common and widely discussed form of *obligationes*, the game starts with Opponent putting forward a sentence, usually called the *positum*, which Respondent must accept for the sake of the disputation, unless it is contradictory in itself. Opponent then puts forward other sentences (the *proposita*), one at a time, which Respondent must either grant, deny or doubt on the basis of inferential relations with the previously accepted or denied sentences — or, in case there are none (and these are called impertinent[8] sentences) on the basis of the common knowledge shared by those who are present. In other words, if Respondent fails to recognize inferential relations or if he does not respond to an impertinent sentence according to its truth-value within common knowledge, then he responds badly. Respondent 'loses the game' if he concedes a contradictory set of propositions. The disputation ends if and when Respondent grants a contradiction, or else when Opponent says '*cedat tempus*', 'time is up'. Opponent and possibly a larger panel of masters present at the disputation are in charge of keeping track of Respondent's replies and of evaluating them once the disputation is over.

An *obligatio* disputation can be represented by the following tuple:

$$Ob = \langle K_C, \Phi, \Gamma, R(\phi_n) \rangle$$

K_C is the state of common knowledge of those present at the disputation. Φ is an ordered set of sentences, namely the very sentences put forward during the disputation. Γ is an ordered set of sets of sentences, which are formed by Respondent's responses to the various ϕ_n. Finally, $R(\phi_n)$ is a function from sentences to the values 1, 0, and ?, corresponding to the rules Respondent must apply to reply to each ϕ_n.

The rules for the *positum* are

- $R(\phi_0) = 0$ iff $\phi_0 \Vdash \bot$,
- $R(\phi_0) = 1$ iff $\phi_0 \nVdash \bot$.

[7] Indeed, the terminology of formal vs. material consequences, from which the terminology used by Brandom (directly borrowed from Sellars) ultimately derives, was consolidated in the 14th century; see (Dutilh Novaes, 2008).

[8] Throughout the text, I will use the terms 'pertinent' and 'impertinent', the literal translations of the Latin terms '*pertinens*' and '*impertinens*'. But notice that they are often translated as 'relevant' and 'irrelevant', for example in the translation of Burley's treatise (Burley, 1988).

The rules for the *proposita* are

- Pertinent propositions: $\Gamma_{n-1} \Vdash \phi_n$ or $\Gamma_{n-1} \Vdash \neg\phi_n$;
  - If $\Gamma_{n-1} \Vdash \phi_n$ then $R(\phi_n) = 1$;
  - If $\Gamma_{n-1} \Vdash \neg\phi_n$ then $R(\phi_n) = 0$;
- Impertinent propositions: $\Gamma_{n-1} \nVdash \phi_n$ and $\Gamma_{n-1} \nVdash \neg\phi_n$;
  - If $K_C \Vdash \phi_n$ then $R(\phi_n) = 1$;
  - If $K_C \Vdash \neg\phi_n$ then $R(\phi_n) = 0$;
  - If $K_C \nVdash \phi_n$ and $K_C \nVdash \neg\phi_n$ then $R(\phi_n) =?$.

As the disputation progresses, different sets of sentences are formed at each round, namely the sets formed by the sentences that Respondent has granted and the contradictories of the sentences he has denied. These sets Γ_n can be seen as models of the successive stages of deontic statuses of Respondent with respect to the commitments undertaken by him at each reply. The sets Γ_n are defined as follows:

- If $R(\phi_n) = 1$ then $\Gamma_n = \Gamma_{n-1} \cup \{\phi_n\}$;
- If $R(\phi_n) = 0$ then $\Gamma_n = \Gamma_{n-1} \cup \{\neg\phi_n\}$;
- If $R(\phi_n) =?$ then $\Gamma_n = \Gamma_{n-1}$.

For reasons of space, I shall keep my presentation of *obligationes* very brief. The interested reader is urged to consult the vast primary and secondary literature on the topic,[9] but further aspects of the framework will be discussed in the next comparative sections as well.

4 Comparison

In this section I undertake a systematic comparison of the two frameworks. The emphasis will be laid on similarities, but I will also mention some important dissimilarities. Essentially, what is at stake during an *obligatio* disputation is the ability to appreciate the (logical and practical) consequences of the commitments undertaken by Respondent. Responded is committed (i.e. obligated) to the sentences he grants as well as to the contradictories of the sentences he denies. The deontic status of entitlement plays a less prominent role within *obligationes*, as the point really is to explore what one is obligated to once one obligates oneself to the *positum*. Besides this

[9] My own previous work (Dutilh Novaes, 2005), (Dutilh Novaes, 2006), (Dutilh Novaes, 2007) can serve as a starting point.

general and fundamental point of similarity, there are several specific similarities between GOGAR's and obligational concepts:[10]

The key role of inferential relations In both models, (intra-personal, inter-content) transfer of commitment takes place through inferential relations, not restricted to formally valid inferences. By means of the transfer of commitment, Respondent is obligated to everything that follows from what he has granted/denied so far, as well as to the contradictories of what is incompatible (*repugnans*) with what he has granted/denied so far. Indeed, the notion of 'repugnant' sentences corresponds precisely to Brandom's notion of incompatibility.

The relation of inference relates sets of sentences/contents Both frameworks correctly treat the relation of inference (and the corresponding transfer of commitment) as relating sets of sentences to sets of sentences (although usually the consequent set is a singleton). Indeed, within rational discursive practices, what counts are not so much the inferential relations between individual sentences/contents; as a matter of fact, we are usually committed to a wide range of sentences/contents. It is the interaction between these different commitments that counts to define our further commitments: what very often happens is that commitment to p alone or to q alone does not commit the speaker to t, but joint commitment to p and to q does. In the obligational framework, every *propositum* that is granted or denied modifies Respondent's commitments.

The dynamic nature of both models A corollary of the previous point is that both models are dynamic, i.e., temporality is an important factor. In (Dutilh Novaes, 2005), I have explored in detail the dynamic nature of *obligationes*, and GOGAR is dynamic in very much the same way. Both models deal with phenomena that take place in successive steps, and each step is to some extent determined by the previous steps (a feature that is accurately captured by the game metaphor). In both cases, the order of occurrence of these steps is crucial. For example, if the *positum* of an *obligatio* is 'Every man is running', and the next step is 'You are running', this *propositum* must be denied as impertinent and false (since nothing has been said about Respondent being a man so far). However, if after the same *positum* 'You are a man' is proposed and accepted (as impertinent and true), and afterward 'You are running' is proposed, then the latter should

[10] For reasons of space, I here treat only the most salient points of similarity. Notice though that there are others, for example the role played by pragmatic elements in both cases.

be accepted as following from what has been granted so far, contrary to the first scenario.

Impertinent propositions and default entitlement Even though *obligationes* deal essentially with commitments and less so with entitlements, one specific kind of entitlement is nevertheless present in the framework. While Respondent's replies to pertinent sentences are fully determined by his previous commitments, there are no commitments concerning impertinent sentences (as this is exactly what they are: thus far uncommitted-to contents). What must determine his replies to impertinent sentences are exactly the uncontroversial entitlements shared by all those who are present at the disputation. These include circumstantial information (such as being in Paris or being in Rome), as well as very general common knowledge, for example that the Pope is a man. In other words, Respondent is entitled to accepting, denying or doubting a sentence on the basis of his factual knowledge concerning them; these are Brandom's 'free moves', with the same social dimension insofar as it concerns common knowledge.

Scorekeeping Within GOGAR, scorekeeping is something of a metaphor rather than a reality — nobody explicitly writes down the commitments and entitlements of other speakers. Scorekeeping is rather something done tacitly, and usually one is not even really aware of doing it. But within *obligationes*, scorekeeping is for real. This is exactly what I mean when I say that the latter is a regimented model of the former: some implicit, tacit elements of GOGAR are made explicit and tangible within *obligationes.* Indeed, those present at the disputation (in particular Opponent) explicitly keep score of Respondent's successive deontic statuses of commitments during a disputation; when he then fails to recognize a previously taken commitment, he responds badly and loses the game. Moreover, once the disputation is over, Respondent's performance is explicitly evaluated by a panel of Masters present at the occasion.

Caveats While the resemblance between the two frameworks is overwhelming, there are of course important points of dissimilarity. More specifically, and as noted before, *obligationes* is a less encompassing model, treating only a subclass of the phenomena captured by GOGAR.

- *Obligationes* only account for the commitments and entitlements of one speaker, namely Respondent.
- Asking for reasons is not part of an *obligatio*: Opponent cannot challenge Respondent, except by saying '*cedat tempus*' if Respondent grants a contradiction.

- *Obligationes* offer no extensive treatment of the different kinds of entitlements and of the mechanisms of transfer of entitlement.

- GOGAR is meant to be a model of the very meaningfulness of language — i.e., the relations of commitment-preserving entailment and entitlement-preserving entailment define the meaning of utterances — whereas *obligationes* operate with a language that is meaningful from the start.

For these reasons, an *obligatio* is best seen as a simplified model of how a speaker must behave towards assertions. This simplification may on the one hand entail loss of generality, but on the other hand it may offer a viewpoint from which some properties of our social discursive practices are made manifest and can thus more easily be studied.

5 What is gained through the comparison?

For obligationes

The deontic nature of obligationes Ever since scholars of medieval philosophy became interested in *obligationes* halfway the 20th century, the very name of this form of disputation was a source of puzzlement. In what sense exactly did such a disputation consist in an obligation? *Who* was obliged, and *what* was he obliged to? Although some modern analyses did emphasize its deontic nature (see (Knuuttila & Yrjonsuuri, 1988)), it is fair to say that the deontic component was essentially overlooked in most of them (including my own game interpretation). On a personal note, I can say that I only fully understood how thoroughly deontic the *obligationes* framework really was against the background of GOGAR, and in particular by means of the concept of commitment.

Recall that I have accounted for the notion of commitment to a statement/content in terms of the *practical consequences* that the reliance on its truth can have for other people's lives, insofar as they assume the statement to be true unless they have good reasons not to (Brandom's 'default entitlement' and Lewis' 'convention of truthfulness and trust') and insofar as they make practical decisions on the basis on their reliance on its truth. Of course, given the somewhat 'artificial' setting of an *obligatio* disputation, no practical consequences are to be expected. Nevertheless, the basic idea seems to be that commitment — obligation — transfers over by means of inferential relations: if respondent is committed to ϕ_n and ϕ_n implies ϕ_m, then respondent is also committed to ϕ_m. Now, since respondent is always committed to at least one statement, the *positum*, this first commitment sets the whole wheel of commitments in motion. So an *obligatio* is not only about logical relations between sentences and consistency maintenance;

more importantly, it is about the deontic statuses of commitments and entitlements and the (intrapersonal, inter-content) mechanisms of inheritance of these statuses.

The general purpose of obligationes More pervasive and significant than the puzzlement caused by the term *obligationes* itself is the still widespread perplexity of scholars concerning the very purpose of such disputations: after all, what's the point? What are *obligationes* about? They are not about truth, as more straightforward forms of disputation are, given that the *positum* is generally, and conspicuously, a possible but *false* sentence. Many of the modern interpretations have sought to establish a rationale for *obligationes* — a logic of conditionals, a framework for belief revision —, but the shortcomings of each of these interpretations only contributed to the growing frustration related to the apparent elusiveness of the 'point' of *obligationes*. It couldn't possibly be a mere form of testing a student's skills, i.e. "schoolboy's exercise", as suggested in the early secondary literature of the 1960's. If there is no real purpose to it beyond the intricate logical structure of the framework, then it might be merely sterile scholastic logical gymnastics after all, just as most of the techniques of scholasticism according to the standard post-scholastic (i.e. Renaissance) criticism.

But when put in the context of GOGAR, *obligationes* seem to provide a model of what it means to act and talk rationally, i.e. to take part in (mainly, but not exclusively) *discursive* social practices. Thus viewed, *obligationes* could also most certainly fulfill an important pedagogical task, namely that of teaching a student how to argue *rationally* — i.e., how to argue mindful of one's entitlements and commitments, of the reasons (grounds) for endorsing or rejecting statements, and of the need to defend one's own commitments — but its importance clearly goes beyond merely pedagogical purposes. Interestingly, throughout the later Middle Ages the format of *obligationes* was extensively adopted for scientific investigations, precisely because it provides a good model for rational argumentation. In a wide variety of contexts (ranging from logic to theology, from ethics to physics), one encounters extensive use of the *obligationes* vocabulary and concepts in the presentation of arguments. Thus seen, the framework is far from being a futile logical exercise: rather, it presents a regimentation of some crucial aspects of what it is to argue and act rationally, of which GOGAR is also a (more encompassing) model.

For the game of giving and asking for reasons

Underlying logical structure While in terms of the 'bigger picture', it is mostly *obligationes* that can benefit from the comparison, on the level of (logical) detail GOGAR has much to learn from *obligationes*. Ever since the publication of (Brandom, 1994), Brandom has been refining the logical structure underlying GOGAR in particular and his inferentialist semantics in general, especially through the development of what he calls 'incompatibility semantics'. Nevertheless, and despite the powerfulness of the modern logical techniques often employed by Brandom and his collaborators, these fairly recent developments are still somewhat overshadowed by the centuries of research (involving a very large number of logicians) on the logic of *obligationes*.[11] Indeed, the (primary and secondary) literature on the topic contains sophisticated analyses of the logical and pragmatic properties of the framework, which are (presumably) applicable to GOGAR so that the comparison can contribute to making GOGAR's logical structure explicit.

For example, I have proved elsewhere (Dutilh Novaes, 2005) that the class of models satisfying Γ_n becomes smaller in the next step of the game only if ϕ_{n+1} is impertinent; if ϕ_{n+1} is pertinent, then the class of models satisfying Γ_n is the same as the class of models satisfying Γ_{n+1}, even though Γ_n and Γ_{n+1} are not the same.[12] This result can be interpreted in terms of GOGAR in the following manner: when a speaker makes an assertion p which actually follows from any sentence or set of sentences previously asserted by him, then his set of commitments is thereby *not* augmented. In other words, his deontic status remains the same, as he was *de facto* already committed to p. Mixing the two vocabularies, one can say that a speaker's deontic status is modified only if he asserts an *impertinent sentence*; assertions of *pertinent* sentences have no effect whatsoever in this sense. Now, this is just one example of how, given that the *obligationes* framework is a more regimented form of the rational, discursive practices also modeled by GOGAR, such logical properties are more easily investigated against the background of the former rather than the latter.

Strategic perspective When speaking of 'the *game* of giving and asking for reasons', Brandom seems to be taking seriously the analogy between the rational discursive practices presumably captured by GOGAR and actual games. It is undoubtedly also a reference to Wittgenstein's language-games, but the question immediately arises: how much of a *game* is GOGAR, actually? To the best of my knowledge, Brandom does not further explore the comparison to games, just as he does not discuss specific game-theoretic properties of GOGAR; this seems to me, however, to be a promising line of investigation. Two important game-theoretical properties that come to mind are the goal(s) to be attained within a certain game, i.e. the expected

[11] *Obligationes* were one of the main topics of investigation in the late medieval Latin tradition, as attested by the very large number of surviving texts ranging from the 12th to the 15th Century.

[12] Assuming, of course, that Respondent has replied according to the rules.

outcome, and the possible strategies to play the game (usually, one is interested in maximizing the payoff, i.e. in the ratio of best possible outcome vs. the most economical strategy). Based on these two concepts, it would seem that GOGAR is in fact a family of games, not a single game, as each particular game of the GOGAR family has its own goals. Most of them are cooperative games, where participants have a common goal rather than that of beating the opponent, e.g., dialogues where people exchange information and coordinate their actions. Nevertheless, there are of course numerous situations of discursive practices where the point really is to beat the opponent, such as, e.g., in a court of law. In each case, different strategies must be employed: in the case of cooperative games, Gricean maxims may be seen as a good account of strategies to maximize understanding between the parties; in the case of competitive games, however, a completely different strategy must be used, one where deceit, for instance, may even have some role to play.

Obligationes is obviously a competitive game: if Respondent grants a contradiction, he loses the game; but if he is able to maintain consistency, he beats Opponent. And even though the medieval authors themselves did not account for *obligationes* in terms of games (nor did they have knowledge of the specific game-theoretical concepts just discussed), medieval treatises on *obligationes* are filled with strategic advice for Respondent on how to perform well during an *obligatio*. These treatises present not only rules defining the legitimate moves within the disputation but also practical, strategic advice.[13] Some of the strategic rules presented in Burley's treatise are: " One must pay particular attention to the order [of the propositions]" (Burley, 1988, p. 385); "When a possible proposition has been posited, it is not absurd to grant something impossible *per accidens*" (Burley, 1988, p. 389); "When a false contingent proposition is posited, one can prove any false proposition that is compossible with it" (Burley, 1988, p. 391).

The point here is that the strategic perspective present in these *obligationes* treatises can very likely be transposed to the GOGAR framework to produce interesting results. In the case of GOGAR games where the different speakers are truly opposed to one another and the point is really to beat the opponent, then the strategic tips from the *obligationes* treatises can be used straightforwardly. But even in the case of cooperative games of GOGAR, the obligational strategies may still be quite helpful, as they essentially describe procedures that may enable one to maintain consistency — certainly a desirable outcome in the context of rational discursive practices. The heart of the matter is that GOGAR does not emphasize the player-perspective: rather, Brandom's description of GOGAR is that of

[13] "It is important to know that there are some rules that constitute the practice of this art and others that pertain to its being practiced well." (Burley, 1988, p. 379)

the theorist standing outside the game and offering a model to explain the use(s) and meaningfulness of language. In this sense, the player-perspective offered by *obligationes* may come as an interesting complement.

The role of doubting Brandom presents GOGAR as having only one quintessential kind of move, i.e. making a claim. We have seen that challenging is also an important move, but a challenge is made by means of the *assertion* of an incompatible content. In contrast, *obligationes* feature three main kinds of moves for Respondent: granting, denying and doubting. Granting obviously corresponds to asserting, and in a sense, denying is also a kind of assertion within the *obligationes* framework, namely the assertion of the contradictory of the denied sentence. I have also pointed out that challenging is not a legitimate move for either Respondent or Opponent, a fact that is related to the regimented and simplified nature of *obligationes* as a model of rational discursive practices. But GOGAR claims to be much more encompassing than *obligationes* does, so while it seems reasonable for *obligationes* to leave some important elements out, the same does not hold of GOGAR. Now, GOGAR has no resources to deal with the phenomenon of not being sure, of recognizing that one does not dispose of sufficient grounds to assert a content or its contradictory (knowing that you don't know), whereas this seems to be a very important element of our rational discursive practices. In contrast, by having doubting as one of the legitimate moves for Respondent, the obligational framework fares batter in this respect.

It might be argued that doubting is not relevant for GOGAR insofar as it has no impact on a speaker's deontic status, as it is simply the lack of commitment or entitlement; not so. A particular rule presented in Kilvington's treatment of *obligationes* in his *Sophismata* (Kilvington, 1990, sophism 48) shows that doubting can indeed alter a speaker's deontic status. The rule is the following: if 'p implies q' is a good consequence, and if Respondent has doubted p, then he must not deny q, i.e., he is not entitled to $\neg q$. This is because, in a valid consequence, if the consequent is (known to be) false, then the antecedent will also be (known to be) false, so if Respondent has doubted the antecedent, he must either doubt or grant the consequent. This is just an example of the intricacies of the logic of doubting and of how doubting can indeed have an impact on one's deontic status. The obligational literature is filled with many more of such examples, in particular in the treatments of *dubitatio*,[14] one of the forms of an obligational dis-

[14] In a *dubitatio*, the first sentence (the *obligatum*) is not a *positum*, it is a *dubium*, a sentence which Respondent must doubt for the sake of the disputation just as he accepts the *positum* in a *positio*; he must then see what follows (in terms of his commitments and entitlements) from having doubted the first sentence.

putation along with *positio* (which is in some sense the 'standard' form of *obligationes*, and the one discussed in this text so far). Thus, I suggest that GOGAR should pay more attention to speech-acts other than assertions as also having an impact on a speaker's deontic status — doubting in particular, as shown within the obligational framework.

6 Conclusion

In this brief comparative analysis of GOGAR and medieval *obligationes* I hope to have indicated how fruitful a systematic comparison between the two frameworks can be. For reasons of space I have here merely sketched such a comparison, and a more thorough analysis shall remain a topic for future work.

Catarina Dutilh Novaes
Department of Philosophy and ILLC, University of Amsterdam
Nieuwe Doelenstraat 15, 1012 CP Amsterdam, The Netherlands
`c.dutilhnovaes@uva.nl`
`http://staff.science.uva.nl/~dutilh/`

References

Brandom, R. (1994). *Making it Explicit.* Cambridge, MA: Harvard University Press.

Brandom, R. (2008). *Between saying and doing.* Oxford: Oxford University Press.

Burley, W. (1988). Obligations (selection). In N. Kretzmann & E. Stump (Eds.), *The Cambridge Translations of Medieval Philosophical Texts: Logic and the Philosophy of Language* (pp. 369–412). Cambridge: Cambridge University Press.

Dutilh Novaes, C. (2005). Medieval *obligationes* as logical games of consistency maintenance. *Synthese*, *145*(3), 371–395.

Dutilh Novaes, C. (2006). Roger Swyneshed's *obligationes*: A logical game of inference recognition? *Synthese*, *151*(1), 125–153.

Dutilh Novaes, C. (2007). *Formalizing medieval logical theories.* Berlin: Springer.

Dutilh Novaes, C. (2008). Logic in the 14th Century after Ockham. In D. Gabbay & J. Woods (Eds.), *Handbook of the history of logic* (Vol. 2: Mediaeval and Renaissance Logic, pp. 433–504). Amsterdam: Elsevier.

Kilvington, R. (1990). *Sophismata.* Cambridge: Cambridge University Press. (English translation, historical introduction and philosophical commentary by N. Kretzmann and B. E. Kretzmann.)

Knuuttila, S., & Yrjonsuuri, M. (1988). Norms and actions in obligational disputations. In O. Pluta (Ed.), *Die Philosophie im 14. und 15. Jahrhundert* (pp. 191–202). Amsterdam: B. R. Güner.

Truth Value Intervals, Bets, and Dialogue Games

Christian G. Fermüller*

Fuzzy logics in Zadeh's 'narrow sense' (Zadeh, 1988), i.e., truth functional logics referring to the real closed unit interval $[0, 1]$ as set of truth values, are often motivated as logics for 'reasoning with imprecise notions and propositions' (see, e.g., (Hájek, 1998)). However the relation between these logics and theories of vagueness, as discussed in a prolific discourse in analytic philosophy (Keefe & Rosanna, 2000), (Keefe & Smith, 1987), (Shapiro, 2006) is highly contentious. We will not directly engage in this debate here but rather pick out so-called interval based fuzzy logics as an instructive example to study

1. how such logics are usually motivated informally,
2. what problems may arise from these motivations, and
3. how betting and dialogue games may be used to analyze these logics with respect to more general principles and models of reasoning.

The main technical result[1] of this work consists in a characterization of an important interval based logic, considered, e.g., in (Esteva, Garcia-Calvés, & Godo, 1994), in terms of a dialogue *cum* betting game, that generalizes Robin Giles's game based characterization of Łukasiewicz logic (Giles, 1974), (Giles, 1977). However, our aim is to address foundational problems with certain models of reasoning with imprecise information. We hope to show that the traditional paradigm of dialogue games as a possible foundation of logic (going back, at least, to (Lorenzen, 1960)) combined with bets as 'test cases' for rationality in the face of uncertainty might help to sort out some of the relevant conceptual issues. This is intended to highlight a particular meeting place of logic, games, and decision theory at the foundation of a field often called 'approximate reasoning' (see, e.g., (Zadeh, 1975)).

* This work is supported by FWF project I143–G15.

[1] Due to limited space, we state propositions without proofs.

1 T-norm based fuzzy logics and bilattices

Petr Hájek, in the preface of his influential monograph on mathematical fuzzy logic (Hájek, 1998) asserts:

> The aim is to show that fuzzy logic as a logic of imprecise (vague) propositions does have well developed formal foundations and that most things usually named 'fuzzy inference' can be naturally understood as logical deduction. (Hájek, 1998, p. viii)

As the qualification 'vague', added in parenthesis to 'imprecise', betrays, some terminological and, arguably, also conceptional problems may be located already in this introductory statement. These problems relate to the fact that fuzzy logic is often subsumed under the general headings of 'uncertainty' and 'approximate reasoning'. In any case, Hajek goes on to introduce a family of formal logics, based on the following design choices (compare also (Hájek, 2002)):

1. The set of degrees of truth (truth values) is represented by the real unit interval $[0, 1]$. The usual order relation $\leq$ models comparison of truth degrees; 0 represents absolute falsity, and 1 represents absolute truth.

2. The truth value of a compound statement shall only depend on the truth values of its subformulas. In other words: the logics are truth functional.

3. The truth function for (strong) conjunction (&) should be a continuous, commutative, associative, and monotonically increasing function $* : [0,1]^2 \rightarrow [0,1]$, where $0 * x = 0$ and $1 * x = x$. In other words: $*$ is a continuous t-norm.

4. The residuum $\Rightarrow_*$ of the t-norm $*$ — i.e., the unique function $\Rightarrow_*$: $[0,1]^2 \rightarrow [0,1]$ satisfying $x \Rightarrow_* y = \sup\{z \mid x * z \leq y\}$ — serves as the truth function for implication.

5. The truth function for negation is $\lambda x[x \Rightarrow_* 0]$.

Probably the best known logic arising in this way is Łukasiewicz logic **L** (Łukasiewicz, 1920), where the t-norm $*_{\mathbf{L}}$ that serves as truth function for & is defined as $x *_{\mathbf{L}} y = \max(0, x + y - 1)$. Its residuum $\Rightarrow_{\mathbf{L}}$ is given by $x \Rightarrow_{\mathbf{L}} y = \min(1, 1 - x + y)$. A popular alternative choice for conjunction takes the *minimum* as its truth function. Besides 'strong conjunction' (&), also this latter 'weak (min) conjunction' ($\wedge$) can be defined in all t-norm based logics by $A \wedge B \stackrel{\text{def}}{=} A\&(A \rightarrow B)$. Maximum as truth function for disjunction ($\vee$) is always definable from $*$ and $\Rightarrow_*$, too.

Other important logics, like Gödel logic **G**, and Product logic **P**, can be obtained in the same way, but we will confine attention to **L**, here. At this point we like to mention that a rich, deep, and still growing subfield of mathematical logic, documented in hundreds of papers and a number of books (beyond (Hájek, 1998)) is triggered by this approach. Consequently it became evident that degree based fuzzy logics are neither a 'poor man's substitute for probabilistic reasoning' nor a trivial generalization of finite-valued logics.

A number of researchers have pointed out that, while modelling degrees of truth by values in $[0, 1]$ might be a justifiable choice in principle, it is hardly realistic to assume that there are procedures that allow us to assign concrete values to concrete (interpreted) atomic propositions in a coherent and principled manner. While this problem might be ignored as long as we are only interested in an abstract characterization of logical consequence in contexts of graded truth, it is deemed desirable to refine the model by incorporating 'imprecision due to possible incompleteness of the available information' (Esteva et al., 1994) about truth values. Accordingly, it is suggested to replace single values $x \in [0, 1]$ by whole *intervals* $[a, b] \subseteq [0, 1]$ of truth values as the basic semantic unit assigned to propositions. The 'natural truth ordering' $\leq$ can be generalized to intervals in different ways. Following (Esteva et al., 1994) we arrive at these definitions:

Weak truth ordering: $[a_1, b_1] \leq^* [a_2, b_2]$ iff $a_1 \leq a_2$ and $b_1 \leq b_2$

Strong truth ordering: $[a_1, b_1] \prec [a_2, b_2]$ iff $b_1 \leq a_2$ or $[a_1, b_1] = [a_2, b_2]$

On the other hand, set inclusion ($\subseteq$) is called *imprecision ordering* in this context. The set of closed subintervals $\mathrm{Int}_{[0,1]}$ of $[0, 1]$ is augmented by the empty interval $\varnothing$ to yield so-called *enriched bilattice* structures $\langle \mathrm{Int}_{[0,1]}, \leq^*, 0, 1, \varnothing, L, N^* \rangle$ as well as $\langle \mathrm{Int}_{[0,1]}, \prec, 0, 1, \varnothing, L, N^* \rangle$, where L is the standard lattice on $[0, 1]$, with minimum and maximum as operators, and N^* is the extension of the negation operator N to intervals; in our particular case $N^*([a, b]) = [1 - b, 1 - a]$ and $N^*(\varnothing) = \varnothing$.

Quite a number of papers have been devoted to the study of logics based on such interval generated bilattices. Let us just mention that the Ghent school of Kerre, Deschrijver, Cornelis, and colleagues has produced an impressive amount of work on interval bilattice based logics (see, e.g., (Cornelis, Deschrijver, & Kerre, 2006)).

While it is straightforward to generalize both types of conjunction (t-norm and minimum) as well as disjunction (maximum) from $[0, 1]$ to $\mathrm{Int}_{[0,1]}$ by applying the operators point-wise, it seems less clear how the 'right' generalization of the truth function for implication should look like. In (Cornelis, Arieli, Deschrijver, & Kerre, 2007), (Cornelis, Deschrijver, & Kerre, 2004) $[a, b] \Rightarrow^*_C [c, d] \stackrel{\mathrm{def}}{=} [\min(a \Rightarrow c, b \Rightarrow d), b \Rightarrow d]$ is studied, but in

(Esteva et al., 1994) the authors suggest $[a,b] \Rightarrow_E^* [c,d] \stackrel{\text{def}}{=} [b \Rightarrow c, a \Rightarrow d]$. As has been pointed out in (Hájek, n.d.) there seems to be a kind of trade off involved here. While $\Rightarrow_C^*$ preserves a lot of algebraic structure — in particular it yields a *residuated* lattice which contains the underlying lattice over $[0,1]$ as a substructure — the function $\Rightarrow_E^*$ is not a residuum, but leads to the following desirable preservation property that is missing for $\Rightarrow_C^*$. If $\mathcal{M}_2$ is a precisiation of $\mathcal{M}_1$ (meaning: for each propositional variable p, $\mathcal{M}_2$ assigns a subinterval of the interval assigned to p by $\mathcal{M}_1$), than any formula satisfied by $\mathcal{M}_1$ is also satisfied by $\mathcal{M}_2$.[2] Below, we will try to show that a game based approach might justify the preference of $\Rightarrow_E^*$ over $\Rightarrow_C^*$ against a background that takes the challenge of deriving formal semantics from first principles about logical reasoning more seriously than the mentioned literature on 'interval logics'.

2 Worries about truth functionality

It is interesting to note that both, (Esteva et al., 1994) and (Cornelis et al., 2007), (Cornelis et al., 2004), refer to Ginsberg (Ginsberg, 1988), who explicitly introduced bilattices following ideas of (Belnap, 1977). Most prominently Ginsberg considers $\mathcal{B} = \langle\{0, \top, \bot, 1\}, \leq_t, \leq_k, \neg\rangle$ as endowed with the following intended meaning:

- 0 and 1 represent (classical) falsity and truth, respectively, $\top$ represents 'inconsistent information' and $\bot$ represents 'no information'. The idea here is that truth values are assigned after receiving relevant information from different sources. Accordingly $\top$ is identified with the information set $\{0,1\}$, $\bot$ with $\varnothing$ and the classical truth values with their singleton sets.

- $\leq_t$, defined by $0 \leq_t \top/\bot \leq 1$, is the '*truth ordering*'.

- $\leq_k$, defined by $\bot \leq_t 0/1 \leq 1$, is the '*knowledge ordering*'.

- Negation is defined by $\neg(0) = 1$, $\neg(1) = 0$, $\neg(\top) = \top$, $\neg(\bot) = \bot$.

While the four 'truth values' of $\mathcal{B}$ may justifiably be understood to represent different states of knowledge about propositions, it is very questionable to try to define corresponding 'truth functions' for connectives other than negation. Indeed, it is surprising to see how many authors[3] followed (Belnap, 1977) in defending a four valued, truth functional logic based on $\mathcal{B}$. It should be clear that, in the underlying classical setting that is taken for granted by Belnap, the formula $A \wedge \neg A$ can only be false (0), independently of the

[2] Here, a formula is defined to be satisfied if it evaluates to the degenerate interval $[1,1]$.
[3] Dozens of papers have been written about Belnap's 4-valued logic.

kind of information, if any, we have about the truth of A. On the other hand, if we neither have information about A nor about B, then $B \wedge \neg A$ could be true as well as false, and therefore $\perp$ should be assigned not only to A, B, and $\neg A$, but also to $B \wedge \neg A$ (in contrast to $A \wedge \neg A$). This simple argument illustrates that knowledge does not propagate compositionally — a well known fact that, however, has been ignored repeatedly in the literature. (For a recent, forceful reminder on the incoherency of the intended semantics for Belnap's logic we refer to (Dubois, n.d.).)

In our context this warning about the limits of truth functionality is relevant at two separate levels. First, it implies that 'degrees of truth' for compound statements cannot be interpreted *epistemically* while upholding truth functionality. Indeed, most fuzzy logicians correctly emphasize that the concept of degrees of truth is *orthogonal* to the concept of degrees of belief. While truth functions for degrees of truth can be motivated and justified in various ways — below we will review a game based approach — degrees of belief simply don't propagate compositionally and call for other types of logical models (e.g., 'possible worlds'). Second, concerning the concept of *intervals* of degrees of truth, one should recognize that it is incoherent to insist on both at the same time:

1. *truth functions* for all connectives, lifted from $[0,1]$ to $\mathrm{Int}_{[0,1]}$, and
2. the interpretation of an interval $[a,b] \subseteq [0,1]$ assigned to a (compound) proposition F as representing a situation where our best *knowledge* about the (definite) degree of truth $d \in [0,1]$ of F is that $a \leq d \leq b$.

Given the mathematical elegance of 1, that results, among other desirable properties, in a low computational complexity of the involved logics,[4] one should look for alternatives to 2. Godo and Esteva[5] have pointed out that, if we insist on 2 just for *atomic* propositions, then at least we can assert that the corresponding 'real', but unknown truth degree of any *composite* proposition F cannot lie outside the interval assigned to F according to the truth functions considered in (Esteva et al., 1994) (described above). However, these bounds are not optimal, in general. As we will see in Section 5, taking clues from Giles's game based semantic for **L** (Giles, 1974), (Giles, 1977), a tighter characterization emerges if we dismiss the idea that intervals represent sets of unknown, but *definite* truth degrees.

[4] It is easy to see that coNP-completeness of testing validity for **L** (and many other t-norm based logics) carries over to the interval based logics described above.

[5] Private communication.

3 Revisiting Giles's game for L

Giles's analysis (Giles, 1974), (Giles, 1977) of approximate reasoning originally referred to the phenomenon of 'dispersion' in the context of physical theories. Later (Giles, 1976) explicitly applied the same concept to the problem of providing 'tangible meanings' to (composite) fuzzy propositions.[6] For this purpose he introduces a game that consists of two independent components:

Betting for positive results of experiments.

Two players — say: *me* and *you* — agree to pay 1€ to the opponent player for every false statement they assert. By $[p_1, \ldots, p_m \| q_1, \ldots, q_n]$ we denote an *elementary state* of the game, where I assert each of the q_i in the multiset $\{q_1, \ldots, q_n\}$ of atomic statements (represented by propositional variables), and you assert each atomic statement $p_i \in \{p_1, \ldots, p_m\}$.

Each propositional variable q refers to an experiment E_q with binary (yes/no) result. The statement q can be read as 'E_q yields a positive result'. Things get interesting as the experiments may show dispersion; i.e., the same experiment may yield different results when repeated. However, the results are not completely arbitrary: for every run of the game, a fixed *risk value* $\langle q \rangle^r \in [0, 1]$ is associated with q, denoting the probability that E_q yields a negative result. For the special atomic formula $\bot$ (*falsum*) we define $\langle \bot \rangle^r = 1$. The risk associated with a multiset $\{p_1, \ldots, p_m\}$ of atomic formulas is defined as $\langle p_1, \ldots, p_m \rangle^r = \sum_{i=1}^{m} \langle p_i \rangle^r$. It specifies the expected amount of money (in €) that has to be paid according to the above agreement. The risk $\langle \rangle^r$ associated with the empty multiset is 0. The risk associated with an elementary state $[p_1, \ldots, p_m \| q_1, \ldots, q_n]$ is calculated from my point of view. Therefore the condition $\langle p_1, \ldots, p_m \rangle^r \geq \langle q_1, \ldots, q_n \rangle^r$ expresses that I do not expect (in the probability theoretic sense) any loss (but possibly some gain) when we bet on the truth of the involved atomic statements as stipulated above.

[6] E.g., Giles suggests to specify the semantics of the fuzzy predicate 'breakable' by assigning an experiment like 'dropping the relevant object from a certain height to see if it breaks'. The expected dispersiveness of such an experiment represent the 'fuzziness' of the corresponding predicate. An arguably even better example of a dispersive experiment in the intended context might consist in asking an arbitrarily chosen competent speaker for a yes/no answer to questions like 'Is Chris tall?' or 'Is Shakira famous?' for which truth may cogently be taken as a matter of degree.

A dialogue game for the reduction of composite formulas.

Giles follows ideas of Paul Lorenzen that date back already to the 1950s (see, e.g., (Lorenzen, 1960)) and constrains the meaning of logical connectives by reference to rules of a dialogue game that proceeds by systematically reducing arguments about composite formulas to arguments about their subformulas.

The dialogue rule for implication can be stated as follows:

$R_{\rightarrow}$ If I assert $A \rightarrow B$ then, whenever you choose to attack this statement by asserting A, I have to assert also B. (And vice versa, i.e., for the roles of me and you switched.)

This rule reflects the idea that the meaning of implication is specified by the principle that an assertion of 'if A, then B' ($A \rightarrow B$) obliges one to assert B, if A is granted.[7]

In the following we only state the rules for 'me'; the rules for 'you' are perfectly symmetric. For disjunction we stipulate:

$R_{\vee}$ If I assert $A_1 \vee A_2$ then I have to assert also A_i for some $i \in \{1, 2\}$ that I myself may choose.

The rule for (weak) conjunction is dual:

$R_{\wedge}$ If I assert $A_1 \wedge A_2$ then I have to assert also A_i for any $i \in \{1, 2\}$ that you may choose.

One might ask whether asserting a conjunction shouldn't oblige one to assert *both* disjuncts. Indeed, for strong conjunction[8] we have

$R_{\&}$ If I assert $A_1 \& A_2$ then I have to assert either both A_1 and A_2, or just $\perp$.

The possibility of asserting $\perp$ instead of the attacked conjunction reflects Giles's 'principle of hedged loss': one never has to risk more than 1€ for each assertion. Asserting $\perp$ is equivalent to (certainly) paying 1€.

In contrast to dialogue games for intuitionistic logic (Lorenzen, 1960), (Felscher, 1985) or fragments of linear logic, no special regulation on the succession of moves in a dialogue is required here. Moreover, we assume that each assertion is attacked at most once: this is reflected by the removal of the occurrence of the formula F from the multiset of formulas asserted by a player, as soon as it has been attacked, or whenever the other player has indicated that she will not attack this occurrence of F during the whole

[7] Note that, since $\neg F$ is defined as $F \rightarrow \perp$, according to $R_{\rightarrow}$ and the above definition of risk, the risk involved in asserting $\neg p$ is $1 - \langle p \rangle^r$.

[8] Giles did not consider strong conjunction. The rule is from (Fermüller & Kosik, 2006).

run of the dialogue game. Every run thus ends in finitely many steps in an elementary state $[p_1, \ldots, p_m \| q_1, \ldots, q_n]$. Given an assignment $\langle\cdot\rangle^r$ of risk values to all p_i and q_i we say that I *win* the corresponding run of the game if I do not have to expect any loss in average, i.e., if $\langle p_1, \ldots, p_m\rangle^r \geq \langle q_1, \ldots, q_n\rangle^r$.

As an almost trivial example consider the game where I initially assert $p \to q$ for some atomic formulas p and q; i.e., the initial state is $[\|p \to q]$. In response, you can either assert p in order to force me to assert q, or explicitly refuse to attack $p \to q$. In the first case, the game ends in the elementary state $[p\|q]$; in the second case it ends in state $[\|]$. If an assignment $\langle\cdot\rangle^r$ of risk values gives $\langle p\rangle^r \geq \langle q\rangle^r$, then I win, whatever move you choose to make. In other words: I have a winning strategy for $p \to q$ in all assignments of risk values where $\langle p\rangle^r \geq \langle q\rangle^r$.

Note that winning, as defined here, does not guarantee that I do not loose money. I have a winning strategy for $p \to p$, resulting either in state $[\|]$ or in state $[p\|p]$ depending on your (the opponents) choice. In the second case, although the winning condition is clearly satisfied, I will actually loose 1€, if the execution of the experiment E_p associated with *your* assertion of p happens to yield a positive result, but the execution of the same experiment associated with *my* assertion of p yields a negative result. It is only guaranteed that my *expected loss* is non-positive. ('Expectation', here, refers to standard probability theory. Under a frequentist interpretation of probability we may think of it as *average loss*, resulting from unlimited repetitions of the corresponding experiments.)

To state Giles's main result, recall that a valuation v for Łukasiewicz logic $\mathbf{L}$ is a function assigning values $\in [0, 1]$ to the propositional variables and 0 to $\bot$, extended to composite formulas using the truth functions $*_\mathbf{L}$, max, min, and $\Rightarrow_\mathbf{L}$, for strong and weak conjunction, disjunction and implication, respectively.

Theorem 1 ((Giles, 1974), (Fermüller & Kosik, 2006)). *Each assignment $\langle\cdot\rangle^r$ of risk values to atomic formulas occurring in a formula F induces a valuation v for $\mathbf{L}$ such that $v(F) = x$ if and only if my optimal strategy for F results in an expected loss of $(1 - x)$€.*

Corollary 1. *F is valid in $\mathbf{L}$ if and only if for all assignments of risk values to atomic formulas occurring in F I have a winning strategy for F.*

4 Playing under partial knowledge

It is important to realize that Giles's game model for reasoning about vague (i.e., here, unstable) propositions implies that *each occurrence* of the same atomic proposition in a composite statement may be evaluated differently at the level of results of associated executions of binary experiments. This feature induces truth functionality: the value for $p \vee \neg p$ is *not* the probability that experiment E_p either yields a positive or a negative result, which is 1 by definition; it rather is $1-x$, where $x = \min(\langle p \rangle^r, 1 - \langle p \rangle^r)$ is my expected loss (in €) after having decided to bet either for a positive or for a negative result of an execution of E_p (whatever carries less risk for me).

The players only know the individual success probabilities[9] of the relevant experiments. Alternatively, one may disregard individual results of binary experiments altogether and simply identify the assigned probabilities with 'degrees of truth'. In this variant the 'pay-off' just corresponds to these truth values, and Giles's game turns into a kind of Hintikka style evaluation game for **L**.

How does all this bear on the mentioned problems of interpretation for interval based fuzzy logics? Remember that both, (Esteva et al., 1994) and (Cornelis et al., 2007, 2006, 2004) seem to suggest that an interval of truth values $[a, b]$ represents 'imprecise knowledge' about the 'real truth value' c, in the sense that only $c \in [a, b]$ is known. For the betting and dialogue game semantic this suggests that the players (or at least player 'I') now have to choose their moves in light of corresponding 'imprecise' (partial) knowledge about the success probabilities of the associated experiments. However, while this may result in an interesting variant of the Giles's game, its relation to the truth functional semantics suggested for logics based on $\mathrm{Int}_{[0,1]}$ and **L**-connectives is dubious.

The following simple example illustrates this issue. Suppose the interval $v^*(p) = [v_1^*(p), v_2^*(p)]$ assigned to the propositional variable p is $[0, 1]$, reflecting that we have no knowledge at all about the 'real truth value' of the proposition represented by p. According to the truth functions presented in Section 1, the formula $p \vee \neg p$ evaluates also to $[0, 1]$, since $v^*(\neg p) = [1 - v_2^*(p), 1 - v_1^*(p)] = [0, 1]$ and hence $v(p \vee \neg p) = [\max(0, 0), \max(1, 1)] = [0, 1]$. Sticking with the 'imprecise knowledge' interpretation, the resulting interval should reflect my knowledge about my expected loss if I play according to an optimal strategy. However, while $1 - v_2^*(p \vee \neg p) = 0$ is the correct lower bound on my expected loss after performing the relevant instance of E_p, to require that $1 - v_1^*(p \vee \neg p) = 1$ is the *best upper bound* for the loss that I have to expect when playing the game is problematic. When playing a mixed strategy that results in my assertion of either p or of $\neg p$ with equal probability, then my resulting *expected* loss is 0.5€, not 1€.

We introduce some notation to assist precise statements about the relation between the interval based semantics of (Esteva et al., 1994) and Giles's

[9] These might well be purely *subjective probabilities* that may differ for the two players. To prove Theorem 1 one only has to assume that I can act on assigned probabilities that determine 'my expectation' of loss.

game. Let v^* be an interval assignment, i.e. an assignment of closed intervals $\subseteq [0, 1]$ to the propositional variables PV. Then $v^*_{\mathbf{L}}$ denotes the extension of v^* from PV to arbitrary formulas via the truth functions $\Rightarrow^*_E$ for implication and the point-wise generalizations of max, min, and $*_{\mathbf{L}}$ for disjunction, weak conjunction, and strong conjunction, respectively. Call any assignment v of reals $\in [0, 1]$ *compatible* with v^* if $v(p) \in v^*(p)$ for all $p \in$ PV. The corresponding risk value assignment $\langle\cdot\rangle^r_v$, defined by $\langle p\rangle^r_v = 1 - v(p)$, is also called compatible with v^*.

Proposition 1. *If, given an interval assignment v^*, the formula F evaluates to $v^*_{\mathbf{L}}(F) = [a, b]$ then the following holds:*

- *For the game in Section 3, played on F: All (pure) strategies for me that are optimal with respect to some fixed risk value assignment $\langle\cdot\rangle^r_v$ compatible with v^* result in an expected loss of at most $(1 - a)$ €, but at least $(1 - b)$ €.*

Note that in the above statement my expected loss refers to a risk value assignment $\langle\cdot\rangle^r_v$ that is fixed before the dialogue game begins. I will play optimally with respect to this assignment. Since the corresponding *expected* pay-off is all that matters here, we technically still have a game of perfect information and therefore no generality is lost by restricting attention to pure strategies. The bounds given by $v^*_{\mathbf{L}}$ for my expected loss are not optimal in general. In other words, the inverse direction of Proposition 1 does not hold. To see this, consider again the interval assignment $v^*(p) = [0, 1]$ resulting in $v^*_{\mathbf{L}}(p \vee \neg p) = [0, 1]$. Obviously, I cannot loose more than 1€, even if I play badly, but my *expected* loss under any fixed risk value assignment $\langle\cdot\rangle^r_v$ is never greater than 0.5€ if I play *optimally* with respect to $\langle\cdot\rangle^r_v$.

On the other hand, sticking with our example '$p \vee \neg p$', one can observe that the best upper bound for my loss is indeed 1€ if I do *not know* the relevant risk values and I still have to stick with some pure strategy. This is because the chosen strategy might suggest to assert p even if, unknown to me, the experiment E_p always has a negative result. In other words, the bounds 1 and 0 are optimal now and coincide with the limits of $v^*_{\mathbf{L}}(p \vee \neg p)$. However, in general, this scenario — playing a pure strategy referring to risk values that need not coincide with the risk values used to calculate the expected pay-off — may lead to an expected loss outside the interval corresponding to $v^*_{\mathbf{L}}$. For a simple example consider $p \vee q$, where $v^*(p) = [0.4, 0.4]$, i.e., the players know that the expected loss associated with an assertion of p is 0.6€, and $v^*(q) = [0, 1]$, i.e., the risk associated with asserting q can be any value between 1 and 0. We have $v^*_{\mathbf{L}}(p \vee q) = [\max(0, 0.4), \max(0.4, 1)] = [0.4, 1]$. Under the assumption that $\langle q\rangle^r_v = 0$, which is compatible with $v^*(q)$, my best strategy calls for asserting q in consequence of asserting $p \vee q$. But if the state $[\|q]$ is evaluated using the risk value $\langle q\rangle_v^r = 1$, which is also compatible with $v^*(q)$, then I have to expect a sure loss of 1€, although $1-1=0$ is outside $[0.4, 1]$.

5 Cautious and bold betting on unstable propositions

We suggest that a more convincing justification of the formal semantics of (Esteva et al., 1994) arises from the following alternative game based model of reasoning under imprecise knowledge. Like above, let v^* be an assignment of intervals $\subseteq [0,1]$ to the propositional variables. Again, we leave the dialogue part of Giles's game unchanged. But in reference to the partial information represented by v^*, we assign two different success probabilities to each experiment E_q corresponding to a propositional variable q, reflecting whether q is asserted by me or by you and consider best case and worst case scenarios (from my point of view) concerning the resulting expected pay-off. More precisely, my expected loss for the final state $[p_1, \ldots, p_m\|q_1, \ldots, q_n]$ when evaluated v^*-*cautiously* is given by $\sum_{i=1}^{n} \langle q_i\rangle_h^r - \sum_{i=1}^{m} \langle p_i\rangle_l^r$, but when evaluated v^*-*boldly* it is given by $\sum_{i=1}^{n} \langle q_i\rangle_l^r - \sum_{i=1}^{m} \langle p_i\rangle_h^r$, where the risk values $\langle q\rangle_h^r$ and $\langle q\rangle_l^r$ are determined by the limits of the interval $v^*(q) = [a,b]$ as follows: $\langle q\rangle_h^r = 1-a$ and $\langle q\rangle_l^r = 1-b$.

Proposition 2. *Given an interval assignment v^*, the following statements are equivalent:*

(i) *Formula F evaluates to $v_L^*(F) = [a,b]$.*

(ii) *For the dialogue game in Section 3, played of F: if elementary states are evaluated v^*-cautiously then the minimal expected loss I can achieve by an optimal strategy is $(1-b)$€; if elementary states are evaluated v^*-boldly then my optimal expected loss is $(1-a)$€.*

6 Conclusion

We have been motivated by various problems that arise from insisting on truth functionality for a particular type of fuzzy logic intended to capture reasoning under 'imprecise knowledge'. Most importantly for the current purpose, we have employed a dialogue *cum* betting game approach to model logical inference in a context of 'dispersive experiments' for testing the truth of atomic assertions. This analysis not only leads to different characterizations of an important interval based fuzzy logic, but relates concerns about properties of fuzzy logics to reflections on rationality *qua* playing optimally in adequate games for 'approximate reasoning'.

Christian G. Fermüller
Institut für Computersprachen, TU Wien
Favoritenstraße 9–11, A–1040 Vienna, Austria
chrisf@logic.at
http://www.logic.at/staff/chrisf/

References

Belnap, N. D. (1977). A useful four–valued logic. In G. Epstein & J. M. Dunn (Eds.), *Modern uses of multiple–valued logic* (pp. 8–37).

Ciabattoni, A., Fermüller, C. G., & Metcalfe, G. (2005). Uniform rules and dialogue games for fuzzy logics. In (pp. 496–510). Springer Verlag.

Cornelis, C., Arieli, O., Deschrijver, G., & Kerre, E. E. (2007). Uncertainty modeling by bilattice–based squares and triangles. *IEEE Transactions on fuzzy Systems*, *15*(2), 161–175.

Cornelis, C., Deschrijver, G., & Kerre, E. E. (2004). Implication in intuitionistic and interval–valued fuzzy set theory: construction, classification, application. *Intl. J. of Approximate Reasoning*, *35*, 55–95.

Cornelis, C., Deschrijver, G., & Kerre, E. E. (2006). Advances and challenges in interval–valued fuzzy logic. *Fuzzy Sets and Systems*, *157*(5), 622–627.

Dubois, D. (n.d.). *On ignorance and contradiction considered as truth-values.* (To appear in the *Logic Journal of the IGPL.*)

Esteva, F., Garcia-Calvés, P., & Godo, L. (1994, March). Enriched interval bilattices: An approach to deal with uncertainty and imprecision. *International Journal of Uncertainty, Fuzziness and Knowledge-Based Systems*, *1*, 37–54.

Felscher, W. (1985). Dialogues, strategies, and intuitionistic provability. *Annals of Pure and Applied Logic*, *28*, 217–254.

Fermüller, C. G. (2003). Theories of vagueness versus fuzzy logic: Can logicians learn from philosophers? *Neural Network World Journal*, *13*(5), 455–466.

Fermüller, C. G. (2004, October). *Revisiting Giles's game: Reconciling fuzzy logic and supervaluation.* (To appear in *Logic, Games and Philosophy: Foundational Perspectives*, Prague Colloquium October 2004.)

Fermüller, C. G. (2007). Exploring dialogue games as foundation of fuzzy logic. In M. Štěpnička et al. (Eds.), *New dimensions in fuzzy logic and related technologies. Proceedings of the* 5th *EUSFLAT conference* (Vol. I, pp. 437–444).

Fermüller, C. G., & Kosik, R. (2006). *Combining supervaluation and degree based reasoning under vagueness* (No. 4246). Springer Verlag.

Giles, R. (1974). A non-classical logic for physics. *Studia Logica*, *33*(4), 399–417.

Giles, R. (1976). Łukasiewicz logic and fuzzy set theory. *International Journal of Man-Machine Studies*, *8*(3), 313–327.

Giles, R. (1977). A non-classical logic for physics. In R. Wojcicki & G. Malinkowski (Eds.), *Selected papers on Łukasiewicz sentential calculi* (pp. 13–51). Ossolineum: Polish Academy of Sciences.

Ginsberg, M. L. (1988). Multivalued logics: a uniform approach to reasoning in artificial intelligence. *Computational Intelligence*, *4*(3), 265–316.

Hájek, P. (n.d.). *On a fuzzy logic with imprecise truth value.* (Unpublished manuscript.)

Hájek, P. (1998). *Metamathematics of fuzzy logic.* Dordrecht: Kluwer.

Hájek, P. (2002). Why fuzzy logic? In D. Jacquette (Ed.), *A companion to philosophical logic* (pp. 595–606). Oxford: Blackwell.

Keefe, R., & Rosanna. (2000). *Theories of vagueness.* Cambridge: Cambridge University Press.

Keefe, R., & Smith, P. (Eds.). (1987). *Vagueness: A reader.* Cambridge, MA–London: MIT Press.

Lorenzen, P. (1960). Logik und Agon. In *Atti congr. internat. di filosofia* (Vol. 4, pp. 187–194). Sansoni, Firenze.

Łukasiewicz, J. (1920). O logice tròjwartościowej. *Ruch Filozoficzny*, *5*, 169–171.

Shapiro, S. (2006). *Vagueness in context.* Oxford: Oxford University Press.

Zadeh, L. A. (1975). Fuzzy logic and approximate reasoning. *Synthese*, *30*(3–4), 407–428.

Zadeh, L. A. (1988). Fuzzy logic. *IEEE: Computer*, *21*(4), 83–93.

Procedural Semantics for Mathematical Constants

Bjørn Jespersen   Marie Duží*

1 Introduction

Consider numerical constants like '1' and 'π'. What is their semantics? We are going to argue in favour of a realist procedural semantics, according to which sense and denotation are correlated as procedure and product. So it is obvious that our procedural semantics bears similarities to Moschovakis's as based on algorithm and value. We are in opposition to Kripke's unrealistic realist contention that the semantics of 'π' consists in nothing other than 'π' rigidly denoting π. Yes, 'π' does denote π — indeed, 'π' qualifies as a strongly rigid designator of π, cf. (Kripke, 1980, p. 48) — but there is substantially more to the semantics of 'π' than merely the denotation relation. In this paper we focus on 'π', since our general top-down strategy is to develop a semantics for the hardest (or a very hard) case and then generalise downwards to increasingly less hard cases from there.

In outline, our procedural semantics says that 'π' expresses as its sense a procedure whose product is π. The procedure is, as a matter of mathematical convention, a *definition* of π and the product is, as a matter of mathematical fact, the (transcendental) *number* so defined. For comparison, '1' expresses as its sense the procedure consisting in applying the successor function to zero once and denotes whatever (natural) number emerges as the product of this procedure.

The upside of a procedural semantics for 'π' is that to *understand*, as a reader or hearer, and exercise *linguistic competence*, as a writer or speaker, one must merely understand a particular numerical definition and need not know which number it defines. Procedural semantics, whether realist or idealist, construes sense as an *itinerario mentis* abstracting from the itinerary's destination. Making the denotation of a numerical constant irrelevant to

* This work is supported by Grant No. 401/07/0451, *Semantisation of Pragmatics*, of the Grant Agency of the Czech Republic.

understanding and linguistic competence is not pressing in the case of '1', but it is so in the case of 'π'. The downside, however, is that at least two equivalent, but obviously distinct, definitions of π are vying for the role as *the* sense of 'π'. One is the *ratio of a circle's area and its radius squared*; the other is *the ratio of a circle's circumference to its diameter.* They are equivalent, because the same number is harpooned by both definitions. But the procedures are conceptually different, so they should not both be assigned to 'π' as its sense on pain of installing homonymy. This kind of predicament has become historically famous. Says Frege,

> Solange nur die Bedeutung dieselbe bleibt, lassen sich diese Schwankungen des Sinnes ertragen, wiewohl auch sie in dem Lehrgebäude einer beweisenden Wissenschaft zu vermeiden sind und in einer vollkommenen Sprache nicht vorkommen dürften. (Frege, 1892, n. 2, p. 42)

We shall suggest a solution to this predicament. The crust of the solution is to relegate each definition of π to individual *conceptual systems.* Since an interpreted sign such as 'π' is a pair whose elements are a character (in this case the Greek letter 'π') and a sense, there will be as many such pairs as there are conceptual systems defining π. Disambiguation of 'π'-involving discourse will consist in making explicit which particular π-defining system should supply the sense of a token of the character 'π'.

A related predicament, which we shall also address, is whether 'π' is best construed as a *name* for π or as a shorthand for a *definite description.* If a name, the sense of 'π' will, in our semantics, be a primitive procedure consisting in the instruction to obtain, or access, π in one step. The procedure will not tell us *how* to obtain π, but only *that* π is to be obtained. This does not sit well with π being something as complicated as a transcendental number. But it does sit well with 'π' being itself a primitive, or simple, character not disclosing any information about its denotation. So at least on a literal analysis, according to which syntactic and semantic structures are by and large isomorphic, 'π' should be paired off with a non-compound sense. If 'π' is a definite description (in disguise), the sense of 'π' will, in our semantics, be a compound procedure consisting in the instruction to manipulate various mathematical operations and concepts in order to define a number. Only the problem, as we just pointed out, is, *which* procedure? Is it the instruction to calculate the ratio of a circle's area and its radius squared, or is it the instruction to calculate the ratio of a circle's circumference and its diameter, or is it some yet other instruction? Whichever it may be, though, the grammatical constant 'π' will be *synonymous* with the definite description 'the ratio...' chosen. The problem of homonymy does not rear its head in case the sense of 'π' is a primitive procedure, for then

'π' is only *equivalent* (co-denoting) with a particular definition. In fact, since all the variants of definitions co-denote the same number, 'π' will be equivalent with all such descriptions.

Our underlying semantic schema is depicted in the following figure.

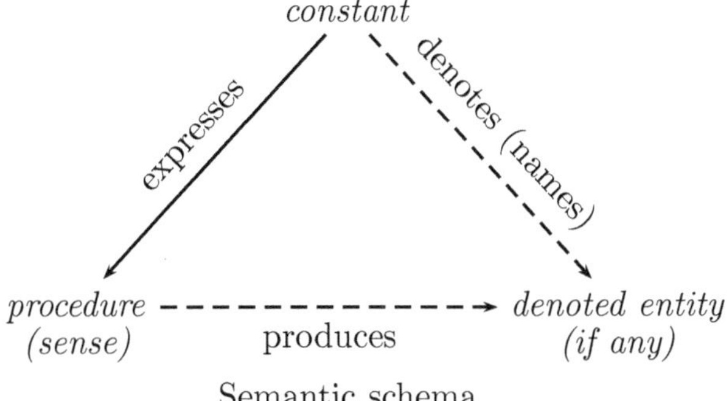

Semantic schema

The relation *a priori* of expressing as obtaining between constant and sense exhausts the pure semantics of the constant. As soon as a procedure is explicitly given, its product (if any) is implicitly given, for the relation from procedure to product is an internal one: a procedure can have at most one product, and that product is invariant. The procedure will produce its product independently of any algorithm; this is why the relation between procedure and product is an internal one. But for epistemological reasons we will need some way or other of calculating its product to learn what it is, so we need a π-calculating algorithm to show us what number satisfies whatever π-defining condition. Such an algorithm will, *ipso facto*, reveal to us what the denotation of 'π' is. The number 3.14159... which is π is itself no player in the pure semantics of 'π'. π is just whatever number rolls out as the value of the given procedure. The number 3.14159... is itself of little mathematical interest and of no semantic import. The *properties* of π, by contrast, are of great interest; e.g., whether π is normal in some base; and establishing that π is transcendental (and not just rational) was a major mathematical achievement.

An algorithm may appear in one of two capacities. Either it is an intermediary between the definition and the number so defined: then the algorithm (whichever it is) is no player in the pure semantics of 'π'. Or an algorithm is the very sense of 'π': then the algorithm is a player in the pure semantics of 'π'. Our procedural semantics allows that a π-calculating algorithm may itself be elevated to playing the role of sense of 'π'. In such a case 'π' will have as its sense one particular way of calculating π. An algorithm is a particular kind of procedure and can as such figure as a linguistic sense relative to a procedural semantics.

In the former case, if the definition is a *condition* then the algorithm will calculate the *satisfier* of the condition. Full competence with respect to the definition *the ratio*... will yield knowledge of a condition to be satisfied by a real number, but will not yield knowledge of which number satisfies it. So the definition is, strictly speaking, a definition of something for a number to be; namely, the ratio of two geometric proportions. If the sense defines π as the ratio between the area of a circle and its radius squared, a matching algorithm must calculate this ratio. Full linguistic competence with respect to 'π' neither presupposes, nor need involve, knowledge of how to calculate π. What competence consists in depends on whether the sense of 'π' is a primitive or compound procedure. If primitive, competence requires knowing which transcendental real 3.14159... is π. If compound, competence requires understanding the concept *the ratio of*, as well as either the concepts *the area of, the radius of, the square of*, or the concepts *the circumference of* and *the diameter of*, together with knowledge of how to mathematically manipulate them. A school child will understand such a complex procedure; it takes a professional mathematician to develop and comprehend a π-calculating algorithm. The task facing the mathematician is to come up with an algorithm equivalent with the definition defining the given ratio.

In the latter case, where an algorithm is the sense of 'π', full linguistic competence with respect to 'π' is to understand a definition of π and, again, not of the number so defined. But since the algorithm is now not an intermediary between definition and number, linguistic competence will be harder to come by, since the sense of 'π' is now likely to involve much more complicated mathematical notions than just, say, those of ratio, area, and circumference, such as the limit of an infinite series.

2 Beyond Benacerraf

Assume that the truth-condition of "...π..." requires π to exist as an independent, abstract entity. Assume, further, that we can have no epistemic access to entities that we can have no causal interaction with. Then next stop is Benacerraf's dilemma as formulated for π: we do not know what number is π; yet we want to dub π 'π' in order to talk about π in "...π...". So how is 'π' to be introduced into mathematese? Moreover, now that 'π' has actually been introduced into standard mathematical vocabulary and been in use for three hundred years, what would a realist (as opposed to constructivist or otherwise idealist) construal of its semantics look like?

We propose placing our procedural semantics within the general Fregean programme of explicating sense (*Sinn*) as the *mode of presentation* (*Art des Gegebenseins*) of the entity (*Bedeutung*) that a sense determines. Muskens correctly points out that "The idea was provided with extensive philosophical justification in Tichý [(1988)]" and that "[Tichý's] notion of senses as *constructions* essentially captures the same idea." (Tichý, 2004, p. 474) Going with this Fregean programme, however, raises a batch of questions deserving and demanding to be answered. Just how finely are senses sliced? What is the ontological status of a sense? What does a sense 'look like'; in particular, what is its structure? And how does a sense *determine* something?

We agree with Moschovakis' conception of sense ('referential intension', in his vernacular) as 'an (abstract, idealized, not necessarily implementable) algorithm which computes the denotation of [a term]' (Moschovakis, 2006, p. 27); see also (Moschovakis, 1994).[1]

Moschovakis outlines his conception thus:

> The starting point... [is] the insight that a correct understanding of programming languages should explain the relation between a program and the algorithm it expresses, so that the basic interpretation scheme for a programming language is of the form
>
> $$\text{program } P \rightarrow \text{algorithm}(P) \rightarrow \text{den}(P).$$
>
> It is not hard to work out the mathematical theory of a suitably abstract notion of algorithm which makes this work; and once this is done, then it is hard to miss the similarity with the basic Fregean scheme for the interpretation of a natural language,
>
> $$\text{term } A \rightarrow \text{meaning}(A) \rightarrow \text{den}(A).$$
>
> This suggested at least a formal analogy between algorithms and meanings which seemed worth investigating, and proved after some work to be more than formal: when we view natural language with a programmer's eye, it seems almost obvious that we can represent the meaning of a term A by the algorithm which is expressed by A and which computes its denotation. (Moschovakis, 2006, p. 42)

In modern jargon, TIL belongs to the paradigm of *structured meaning*. However, Tichý does not reduce structure to set-theoretic sequences, as

[1] Moschovakis' notion of algorithm borders on being too permissive, since algorithms are normally understood to be effective. (See (Cleland, 2002) for discussion.) Tichý separates algorithms from constructions: "The notion of construction is... correlative not with the notion of algorithm itself but with what is known as a particular algorithmic *computation*, the sequence of steps prescribed by the algorithm when it is applied to a particular input. But not every construction is an algorithmic computation. An algorithmic computation is a sequence of effective *steps*, steps which consist in subjecting a manageable object... to a feasible operation. A construction, on the other hand, may involve steps which are not of this sort." (Moschovakis, 1994, p. 526), (Moschovakis, 2006, p. 613)

do Kaplan and Cresswell.[2] Nor does Tichý fail to explain how the sense of a molecular term is determined by the senses of its atoms and their syntactic arrangement (as Moschovakis objects to 'structural' approaches in (Moschovakis, 2006, p. 27)).

In general, a procedure is a structure encompassing one or more steps that individually detail how to determine a product and jointly detail how to determine the product of the procedure that they are sub-procedures of. (This holds even for one-step procedures.) Structures are needed as molecular units in which to organise atomic sub-procedures in a particular order. A compound structure constitutes a hierarchy of sub-procedures. The philosophical idea informing our procedural semantics is that since senses are procedures, any two senses are identical just when they are, roughly speaking, procedurally indistinguishable. (We shall individuate senses in terms of *procedural isomorphism*; see below.) Intuitively, any two procedures are identical just when they are instructions to do the same to the same things in the same order.

3 Logical foundations

TIL constructions are procedures. Constructions divide into atomic and compound, according as they encompass one or more steps. The atomic ones are *Variable* and *Trivialization*; the compound ones, *Composition* and *Closure*.[3] A variable x constructs an object relative to a valuation function pairing variables and entities off, such that x constructs the value assigned to it. The Trivialization ${}^0X$ constructs the entity X (which may be whatever sort of entity found in the ontology of TIL). A Composition is the procedure of applying a function at one of its arguments to obtain the value (if any) at that argument; the functional value is the product of that procedure. A Closure is the procedure of arranging objects $x_1, \ldots, x_n$ and y as functional arguments and values, respectively; the resulting function is the product of that procedure. If the sense of 'π' is simple, its sense is the Trivialization of π: ${}^0\pi$. If complex, it is a Composition. In either case the product of the respective procedure is the same transcendental number.

[2] Kaplan may well have been the one to reintroduce the notion of structured meaning into mainstream analytic philosophy of language. See (Kaplan, 1978), written in 1970; but see also (Lewis, 1972). (Cresswell, 1985) has become the standard point of reference. All three agree that structure, especially a structured proposition, is (or can be modelled as) an ordered n-tuple. This won't do, though, since sequences underdetermine structure and so cannot solve Russell's old problem of propositional unity.

[3] And four others — *Execution, Double Execution, Tuple, Projection* — that we do not need here.

Here follows an outline of the logical backbone of our procedural semantics for 'π'. TIL constructions, as well as the entities they construct, all receive a logical (as opposed to linguistic) *type*.

Definition 1 (Type of order 1).
Let B be a base, where a base is a collection of pair-wise disjoint, non-empty sets. Then:

(i) Every member of B is an elementary *type of order* 1 *over* B.

(ii) Let α, $\beta_1, \ldots, \beta_m$ ($m > 0$) be types of order 1 over B. Then the collection $(\alpha\beta_1 \ldots \beta_m)$ of all m-ary partial mappings from $\beta_1 \times \cdots \times \beta_m$ into α is a functional *type* of order 1 over B.

(iii) Nothing is a *type of order* 1 *over* B unless it so follows from (i) and (ii).

Definition 2 (Construction).

(i) The *Variable* x is a construction that constructs an object O of the respective type dependently on a valuation v; it v-constructs O.

(ii) *Trivialization*: Where X is an object whatsoever (an extension, an intension or a *construction*), ${}^0X$ is the *construction Trivialization*. It constructs X without any change.

(iii) The *Composition* $[XY_1 \ldots Y_m]$ is the following *construction*.
If X *v-constructs* a function f of a type $(\alpha\beta_1 \ldots \beta_m)$, and $Y_1 \ldots Y_m$ *v-construct* entities $B_1, \ldots, B_m$ of types $\beta_1, \ldots, \beta_m$, respectively, then the *Composition* $[XY_1 \ldots Y_m]$ *v-constructs* the value (an entity, if any, of type α) of f on the tuple-argument $\langle B_1, \ldots, B_m\rangle$. Otherwise the *Composition* $[XY_1 \ldots Y_m]$ does not *v-construct* anything and so is *v-improper*.

(iv) The *Closure* $[\lambda x_1 \ldots x_m Y]$ is the following *construction*.
Let $x_1, x_2, \ldots, x_m$ be pairwise distinct variables v-constructing entities of types $\beta_1, \ldots, \beta_m$ and Y a construction v-constructing an α-entity. Then $[\lambda x_1 \ldots x_m Y]$ is the *construction λ-Closure* (or *Closure*). It *v-constructs* the following function f of type $(\alpha\beta_1 \ldots \beta_m)$. Let $v(B_1/x_1, \ldots, B_m/x_m)$ be a valuation identical with v at least up to assigning objects $B_1, \ldots, B_m$ of types $\beta_1, \ldots, \beta_m$, respectively, to variables $x_1, \ldots, x_m$. If Y is $v(B_1/x_1, \ldots, B_m/x_m)$-improper (see (iii)), then f is undefined on $\langle B_1, \ldots, B_m\rangle$. Otherwise the value of f on $\langle B_1, \ldots, B_m\rangle$ is the entity of type α $v(B_1/x_1, \ldots, B_m/x_m)$-constructed by Y.

(v) Nothing is a *construction*, unless it so follows from (i) through (iv).

Definition 3 (Ramified hierarchy of types).
Let B be a base. Then:
T_1 (*types of order* 1): defined by Definition 1.
C_n (*constructions of order* n)

(i) Let x be a variable ranging over a type of order n. Then x is a *construction of order* n *over* B.

(ii) Let X be a member of a type of order n. Then ${}^0X$, ${}^1X$, ${}^2X$ are *constructions of order* n *over* B.

(iii) Let $X, X_1, \ldots, X_m$ $(m > 0)$ be constructions of order n over B. Then $[XX_1 \ldots X_m]$ is a *construction of order* n *over* B.

(iv) Let $x_1, \ldots, x_m, X$ $(m > 0)$ be constructions of order n over B. Then $[\lambda x_1 \ldots x_m X]$ is a *construction of order* n *over* B.

(v) Nothing is a construction of order n over B unless it so follows from C_n (i)–(iv).

T_{n+1} (*types of order* $n+1$)
Let $*_n$ be the collection of all constructions of order n over B.

(i) $*_n$ and every type of order n are types of order $n+1$.

(ii) If $0 < m$ and $\alpha, \beta_1, \ldots, \beta_m$ are types of order $n+1$ over B, then $(\alpha\beta_1 \ldots \beta_m)$ (see T_1 (ii)) is a type of order $n+1$ over B.

(iii) Nothing is a type of order $n+1$ over B unless it so follows from (i) and (ii).

The ontological status of a construction is an objective, abstract, structured procedure residing in a Platonic realm. Constructions are not inherently linguistic senses, for they exist prior to and independently of language. But they may be made, via linguistic convention, to serve as linguistic senses. That is, in true realist fashion, TIL considers language a *code*.[4] Programmatically stated, our semantics for 'π' complements the ontology for π put forward in (Brown, 1990).

A construction *determines* what it constructs by *constructing* it. So the logic of determination consists in the constructional descent from a procedure to its product, as specified for each particular kind of construction in Definition 2. Constructions are too finely individuated to figure as linguistic senses, since some of the procedural differences they embody are logically insignificant and are not encoded linguistically. Most obviously, two α-equivalent constructions like $\lambda x\,[{}^0{>}\,x\,{}^0 0]$ and $\lambda y\,[{}^0{>}\,y\,{}^0 0]$ are just that —

[4] See (Tichý, 1988, pp. 228ff.).

two constructions of the class of positive numbers and not one; yet the difference between the λ-bound variables x and y is procedurally irrelevant. The solution to the granularity problem consists in forming equivalence classes of *procedurally isomorphic* constructions and privileging a member of each such class as *the* procedural sense of a given unambiguous term or expression. Technically, the quest is for a suitable degree of extensionality in the λ-calculus. Needless to say, it remains an open research question exactly what the desirable calibration of linguistic senses should be, but our current thesis is that procedures, and hence senses, should be identified up to α- and η-equivalence.[5]

4 Kripke's 'π' and ours

Central to Kripke's denotational semantics is the distinction between *fixing the reference* and *giving the meaning/a synonym*.[6] One of Kripke's illustrations is this:

> ['π'] is not being used as *short* for the phrase 'the ratio of the circumference of a circle to its diameter' [...] It is used as a *name* for a real number, which in this case is necessarily the ratio of the circumference of a circle to its diameter. (Kripke, 1980, p. 60)

Kripke's semantics for 'π' is simple (simplistic, as it turns out):

$$\text{`}\pi\text{'} \xrightarrow[\text{rigidly designates}]{} \pi$$

The description 'the ratio...' serves to single out the unique ratio shared by all circles, after which that number is baptised 'π'. The description is subsequently kicked off and so does not form part of the semantics proper of Kripke's 'π'. This is problematic. Nobody knows of some one particular real that it is π. So nobody knows of some one particular real that it is the reference of 'π'. So it is obscure what linguistic competence with respect to 'π' would consist in. Note that it is not an option to say that 'π' designates whatever real is the ratio of a circle's circumference to its diameter, for this uniqueness condition forms no part of Kripke's semantics for 'π'.[7] Kripke's introduction of 'π' is impeccable, and his 'π' does denote π. But we cannot use his 'π' to denote π, nor can we understand anyone else's use of 'π', since

[5] See (Duží, Jespersen, & Materna, ms. 1, § 2.2) or (Jespersen, ms). For discussion of Frege's quest for the right calibration of *Sinn*, see (Sundholm, 1994) and (Penco, 2003).

[6] — a distinction anticipated at least by (Geach, 1969).

[7] The Kripkean can have recourse to some causal theory of reference in the case of words for empirical entities like tigers, lemons and gold. But Benacerraf's first horn blocks this avenue. We surmise that Kripkean rigid designation cannot possibly be extended to numerical constants and other terms denoting abstract entities.

we cannot know which particular transcendental number is π. In short, Kripke's 'π' has been severed radically from any humanly possible linguistic practice, so it is inoperative.

In the idiom of procedural semantics, Kripke focuses entirely on the product at the expense of the *procedure*. As a matter of mathematical fact, 3.14159... is π, but why introduce a non-descriptive name when that name severs the link between condition/procedure and satisfier/product? It seems that on Kripke's semantics it will be a discovery, and not a convention, that π is the ratio of a circle's circumference to its diameter. If so, it also seems that Kripke's 'π' misconstrues mathematical practice.

Some π-producing procedure must figure in the semantics of 'π'; only how? TIL faces a dilemma of its own, as we saw above. On the one hand, a literal analysis of 'π' would dictate that the sense of 'π' be ${}^0\pi$, yielding the schema

$$\text{`}\pi\text{'} \xrightarrow[\text{expresses}]{} {}^0\pi \xrightarrow[\text{constructs}]{} \pi$$

The advantage of this construal is that what looks like a constant *is* a constant (and not a definite description masquerading as one). However, this is too close to Kripke's 'π' for comfort. We would be reinstating the problem that the semantics of 'π' pairs no mathematical condition off with 'π'. To master 'π', ${}^0\pi$ would suffice. The Trivialization merely instructs us to construct π and not also how to construct it.

On the other hand, not least epistemic concerns dictate that the sense of 'π' ought to be an *ontological definition* of π, yielding the schema

$$\text{`}\pi\text{'} \xrightarrow[\text{expresses}]{} [\iota x[\forall y[x = [{}^0\text{Ratio}\,[\ldots y \ldots][\ldots y \ldots]]]]] \xrightarrow[\text{constructs}]{} \pi$$

(By 'ontological definition, we mean a compound construction (here, a Composition) that, in this case, constructs the number π, thereby laying down what π is. An ontological definition contrasts with a linguistic definition, which introduces a new term as synonymous with an existing term.) This makes 'π' a shorthand term synonymous with 'the ratio...', and its sense is an ontological definition of π. The advantage of this construal is that it pairs a mathematical condition off with 'π'; but again, which? There is no criterion to help decide which of the possible ontological definitions should be *the* sense of 'π'. It would be arbitrary to select one and assign it as sense; but assigning them all introduces homonymy.

It would seem evident that a language-user needs to know at least one definition of π in order to use and understand 'π'. If we go with the Trivialization-based analysis of 'π', the first step toward enhancing it is to make the logico-semantic fact that ${}^0\pi$ is *equivalent* with $[\iota x[\forall y[x =$

$[{}^0\text{Ratio}\,[\ldots y\ldots][\ldots y\ldots]]]]]$ part of the semantics of 'π'. ${}^0\pi$ is indifferent to how π is constructed by this or that compound construction, so as far as the equivalence relation goes, any compound π-construction is as good as any other. 'π' may be *introduced* as equivalent with

$$[\iota x \forall y[x = [{}^0 \mathit{Ratio}\,[{}^0 \mathit{Area}\,y][{}^0 \mathit{Square}\,[{}^0 \mathit{Radius}\,y]]]]] \tag{1}$$

or

$$[\iota x \forall y[x = [{}^0 \mathit{Ratio}\,[{}^0 \mathit{Circumference}\,y][{}^0 \mathit{Diameter}\,y]]]], \tag{2}$$

or any other compound π-constructing construction. *Understanding* is another matter. One thing is to understand (1); another thing is to understand (2). One may well know that 'π' is equivalent to this Composition without knowing, *ipso facto*, that it is equivalent to that Composition.

5 Realistic realism?

Both causal theory of reference and denotational semantics are neither here nor there as a theory of terms for abstract entities such as numbers. We are putting forward a procedural semantics as a rival theory in order not to get gored by Benacerraf's horns or turning linguistic competence with mathematical constants into an enigma. We suggest, in the final analysis, that the semantics of 'π' ought to be that it is shorthand for, and therefore synonymous with, a definite description expressing a definition of π and denoting the number so defined. But for each definition$_n$ of π there is going to be a pair $\langle$'π', definition$_n(\pi)\rangle$. So how do we handle the resulting homonymy? *Schwankungen des Sinnes* are neither here nor there in a regimented language such as mathematese. Our solution revolves around *conceptual systems*.

By 'conceptual system' we mean a set of constructions that is fully determined by the chosen set of *simple concepts*. Simple concepts are Trivialisations of nonconstructional entities of order 1. The compound concepts of a conceptual system are then all the compound constructions that are formed according to the rules of Definition 2 (plus perhaps involving additional constructions) using simple concepts and variables. The exact definition of *conceptual system* can be found in (Materna, 2004).

Relative to a particular conceptual system, a pair $\langle$'π', definition$_n(\pi)\rangle$ is an unambiguous assignment of exactly one definition of π to 'π', provided the conceptual system is independent, i.e., its set of simple concepts is minimal. Consequently, 'π' is not ambiguous, for this character must always be given together with a particular definition of π culled from a particular conceptual system. The appearance of ambiguity arises only when two or more conceptual systems are invoked in the course of a discourse in which tokens of 'π' occur.

The upshot of our solution is that there are several π-denoting constants sharing the same first element, 'π'. So when two mathematicians are both deploying tokens of 'π', there is a risk of them talking at cross purposes, until and unless they compare notes and, in case of invoking different conceptual systems, come to agree on the same definition of π in the interest of synonymy. Yet the mathematical results they may have π individually obtained with respect to π are bound to be equivalent, for any two definitions of π are bound to converge in the same number. After all, the problem was always to do with *Schwankungen des Sinnes* and never *Schwankungen der Bedeutung*.[8]

Bjørn Jespersen
Section of Philosophy, Delft University of Technology
The Netherlands
b.t.f.jespersen@tudelft.nl

Marie Duží
Department of Computer Science, VSB–Technical University Ostrava
17. listopadu 15, 708 33 Ostrava, Czech Republic
marie.duzi@vsb.cz
http://www.cs.vsb.cz/duzi/

References

Benacerraf, P. (1973). Mathematical truth. *Journal of Philosophy*, *70*, 661–679.

Brown, J. R. (1990). π in the sky. In A. D. Irvine (Ed.), *Physicalism in mathematics* (pp. 95–120). Dordrecht: Kluwer.

Cleland, C. E. (2002). On effective procedures. *Minds and Machines*, *12*, 159–179.

Cresswell, M. J. (1975). Hyperintensional logic. *Studia Logica*, *34*, 25–38.

Cresswell, M. J. (1985). *Structured meanings.* Cambridge, MA: MIT Press.

Duží, M., Jespersen, B., & Materna, P. (ms. 1). *Procedural semantics for hyperintensional logic: Foundations and applications of TIL.* (Forthcoming in Springer.)

[8] Versions of this paper were read by Marie at the Joint Paris/Arché Workshop Philosophy of Mathematics and Abstract Entities, ENS, Paris, February 28th–March 1st, 2008, and by Bjørn at LOGICA '08, Hejnice, June 16th–20th, 2008, at Department of Fuzzy Modelling, TU Ostrava, June 25th, 2003, at Department of Philosophy, University of Milan, April 22nd, 2008, and at Department of Philosophy, University of Padua, May 5th, 2008. Portions of the present paper have been lifted from (Duží et al., ms. 1, § 3.2.1). This paper is an abridged version of (Duží, Jespersen, & Materna, ms. 2).

Duží, M., Jespersen, B., & Materna, P. (ms. 2). 'π' in the sky. (Forthcoming in *Acts of knowledge: history, philosophy and logic. A Festschrift for Göran Sundholm*, London: College Publications Tribute Series, 2009.)

Frege, G. (1892). *Über Sinn und Bedeutung* (G. Patzig, Ed.). Göttingen: Vandenhoeck und Ruprecht. (Reprint 1986.)

Geach, P. T. (1969). The perils of Pauline. *Review of Metaphysics*, *23*, 287–300.

Jespersen, B. (ms). Hyperintensions and procedural isomorphism: Alternative (1/2). (Forthcoming in K. Kijania-Placek (ed.). London: College Publications.)

Kaplan, D. (1978). Dthat. In P. Cole (Ed.), *Syntax and semantics.* New York: Academic Press.

Kripke, S. (1980). *Naming and necessity.* Oxford: Blackwell.

Lewis, D. (1972). General semantics. In D. Davidson & G. Harman (Eds.), *Semantics of natural language* (pp. 169–218). Dordrecht: Reidel.

Materna, P. (2004). *Conceptual systems.* Berlin: Logos–Verlag.

Moschovakis, Y. (1994). Sense and denotation as algorithm and value. Berlin: Springer–Verlag.

Moschovakis, Y. (2006). A logical calculus of meaning and synonymy. *Linguistics and Philosophy*, *29*, 27–89.

Muskens, R. (2005). Sense and the computation of reference. *Linguistics and Philosophy*, *28*, 473–504.

Penco, G. (2003). Frege: two theses, two senses. *History and Philosophy of Logic*, *24*, 87–109.

Sundholm, B. G. (1994). Proof-theoretical semantics and Fregean identity criteria for propositions. *The Monist*, *77*, 294–314.

Sundholm, B. G. (2000). Virtues and vices of interpreted classical formalisms. In T. Childers & J. Palomäki (Eds.), *Between words and worlds: A Festschrift for Pavel Materna.* Prague: Filosofia, Czech Academy of Sciences.

Tichý, P. (1986). Constructions. *Philosophy of Science*, *53*, 514–534. (Reprinted in (Tichý, 2004).)

Tichý, P. (1988). *The foundations of Frege's logic.* Berlin: de Gruyter.

Tichý, P. (2004). *Collected papers in logic and philosophy* (V. Svoboda, B. Jespersen, & C. Cheyne, Eds.). Prague and Dunedin: Filosofia, Czech Academy of Sciences and University of Otago Press.

Neighborhood Incompatibility Semantics for Modal Logic

Kohei Kishida*

This paper introduces neighborhood semantics for propositional modal logic into the framework of Brandom's (2008) incompatibility semantics. Neighborhood semantics for modal logic, as it is conventionally studied, can be considered to be a kind of possible-world semantics, in the sense that a system of neighborhoods codifies a generalized accessibility relation among points of the space, or *worlds*, at which the *truth values* of propositions are evaluated. Such a semantics features the representational notions of truth and possible world as its basic primitive constituents. Brandom's incompatibility semantics, in contrast, is founded upon the inferential notions of *incoherence* and *incompatibility* of sentences. The chief goal of this paper is to show that this inferentialist framework of incompatibility semantics can also adopt the notion of neighborhood to interpret modality, as the core idea of neighborhood semantics works independently of the representational notions.

1 Incompatibility semantics: a quick review

Here we quickly review the basic definitions and facts in incompatibility semantics that are relevant to this paper; see (Brandom, 2008) for a full exposition of the semantics.

We write $\mathcal{L}$ both for a given sentential language and for the set of its sentences. Let Inc be any subset of $\mathcal{PL}$, the powerset of $\mathcal{L}$, that is closed upward in terms of $\subseteq$, i.e., if $X \in \text{Inc}$ and $X \subseteq Y$ then $Y \in \text{Inc}$. We say X is *incoherent* if $X \in \text{Inc}$; then Inc being $\subseteq$-upward closed means that adding more sentences to an incoherent set X of sentences never cures the incoherence. We also say Y is *incompatible with* X if $X \cup Y \in \text{Inc}$, and

*The author would like to thank Alp Aker, Robert Brandom, Jaroslav Peregrin, José Martínez Fernández, and especially Nuel Belnap, for insightful comments and helpful suggestions.

write $\mathcal{I}(X)$ for the collection of $Y \subseteq \mathcal{PL}$ incompatible with X, i.e.

$$\mathcal{I}(X) = \{Y \subseteq \mathcal{L} \mid X \cup Y \in \mathrm{Inc}\}.$$

When $p \in \mathcal{L}$, we write $\mathcal{I}(p)$ for $\mathcal{I}(\{p\})$.

The entailment relation, $p \vDash q$, is defined by $\mathcal{I}(q) \subseteq \mathcal{I}(p)$, i.e., that if X is incompatible with q then it is incompatible with p. In general, $X \vDash Y$, i.e., the conjunction of X entailing the disjunction of Y, is defined by

$$X \vDash Y \qquad \Longleftrightarrow \qquad \bigcap_{p \in Y} \mathcal{I}(p) \subseteq \mathcal{I}(X),$$

that is, anything that rules out all $p \in Y$ rules out X. Applying this to the case $Y = \varnothing$ in particular, with $\bigcap_{p \in \varnothing} \mathcal{I}(p) = \mathcal{PL}$, we have the following, which agrees with what is usually meant by $X \vDash \varnothing$:

$$X \vDash \varnothing \qquad \Longleftrightarrow \qquad \mathcal{PL} \subseteq \mathcal{I}(X) \qquad \Longleftrightarrow \qquad X \in \mathrm{Inc}\,.$$

When $\mathcal{L}$ has a negation operator $\neg$, we assume that $\mathcal{I}$ satisfies

$$X \in \mathcal{I}(\neg p) \qquad \Longleftrightarrow \qquad X \vDash p \qquad \text{for every} \quad p \in \mathcal{L};$$

i.e., $\neg p$ is incompatible with all and only X that entail p. Then we have $\mathcal{I}(\neg\neg p) = \mathcal{I}(p)$, and hence $X \in \mathcal{I}(p) \iff X \vDash \neg p$. Also, when $\mathcal{L}$ has a disjunction operator $\vee$, $\mathcal{I}$ is assumed to satisfy $\mathcal{I}(p \vee q) = \mathcal{I}(p) \cap \mathcal{I}(q)$; i.e., X implies neither p nor q is the case if and only if it denies both p and q.[1] A pair $(\mathcal{L}, \mathcal{I})$ of such $\mathcal{L}$ and $\mathcal{I}$ is called an *incompatibility frame.*

2 Neighborhood semantics in the possible-world framework

To introduce neighborhood incompatibility semantics, it is helpful to first review the neighborhood semantics as conventionally studied in the possible-world framework and to then draw a formal analogy.

Let us recall that possible-world semantics interprets a modal language $\mathcal{L}$ with a set $\mathbb{W}$ of possible worlds by assigning to each sentence $p \in \mathcal{L}$ a subset $[\![p]\!]$ of $\mathbb{W}$, sometimes called a proposition. Then that a world $w \in \mathbb{W}$ lies in the interpretation $[\![p]\!]$ of p means p is true at w, and hence p semantically entails q if and only if $[\![p]\!] \subseteq [\![q]\!]$. So, for example, the T axiom $\Box p \vdash p \vdash \Diamond p$ of modal logic corresponds to $[\![\Box p]\!] \subseteq [\![p]\!] \subseteq [\![\Diamond p]\!]$. This suggests, from the point of view of topology, that the $\Box$ operator corresponds to a generalized interior operation, while $\Diamond$ corresponds to a generalized closure, defined on a system of neighborhoods on $\mathbb{W}$ as follows.

[1] The official definition in (Brandom, 2008) first defines $\mathcal{I}(p \wedge q) = \mathcal{I}(\{p, q\})$ and then defines $p \vee q$ as $\neg(\neg p \wedge \neg q)$, which still entails $\mathcal{I}(p \vee q) = \mathcal{I}(p) \cap \mathcal{I}(q)$.

Each world w, or *point* of the *space* $\mathbb{W}$, is assigned a collection of subsets of the space, called the *neighborhoods* of w; we write $\mathcal{N}_w$ for this collection. Then, given $A \subseteq X$, a point w is in the *interior* (in the generalized sense) of A if and only if it has a neighborhood U contained in A to witness that w is "well inside" A. And w is in the *closure* of A if and only if all of its neighborhoods intersect A, or in other words, if and only if w has no neighborhood U disjoint from A to witness that w is "well outside" A. Now, given an interpretation $[\![p]\!] \subseteq X$ of p, its interior and closure interpret $\Box p$ and $\Diamond p$, respectively. More formally, we have the following ($\Box_{\text{pw}}$) and ($\Diamond_{\text{pw}}$) (the subscript "pw" is just to connote "the possible-world case"):[2]

$$w \in [\![\Box p]\!] \iff \exists U \in \mathcal{N}_w \cdot U \subseteq [\![p]\!], \qquad (\Box_{\text{pw}})$$

$$w \in [\![\Diamond p]\!] \iff \forall U \in \mathcal{N}_w \cdot U \cap [\![p]\!] \neq \varnothing. \qquad (\Diamond_{\text{pw}})$$

Let us note two things here. First, this neighborhood setting subsumes Kripke semantics. That is because Kripke semantics is obtained from this neighborhood semantics by further assuming that each world w has exactly one neighborhood R_w, called the worlds "accessible from w". Then ($\Box_{\text{pw}}$) and ($\Diamond_{\text{pw}}$) boil down to the following conditions, which are clearly equivalent to the usual truth conditions for $\Box p$ and $\Diamond p$:

$$w \in [\![\Box p]\!] \iff R_w \subseteq [\![p]\!],$$

$$w \in [\![\Diamond p]\!] \iff R_w \cap [\![p]\!] \neq \varnothing.$$

Second, this neighborhood semantics has $\Box$ and $\Diamond$ dual to each other, i.e., $\Diamond$ is just $\neg\Box\neg$, while $\Box$ is just $\neg\Diamond\neg$, because the condition $[\![\neg p]\!] = \mathbb{W} \setminus [\![p]\!]$ (i.e., that negation is interpreted by complement in $\mathbb{W}$) implies

$$\begin{aligned} w \in [\![\Diamond p]\!] &\iff \forall U \in \mathcal{N}_w \cdot U \cap [\![p]\!] \neq \varnothing && \text{by } (\Diamond_{\text{pw}}) \\ &\iff \neg\exists U \in \mathcal{N}_w \cdot U \subseteq \mathbb{W} \setminus [\![p]\!] = [\![\neg p]\!] \\ &\iff w \notin [\![\Box\neg p]\!] && \text{by } (\Box_{\text{pw}}) \\ &\iff w \in \mathbb{W} \setminus [\![\Box\neg p]\!] = [\![\neg\Box\neg p]\!]. \end{aligned}$$

[2] This definition of interior and closure is less general than the standard version in neighborhood semantics (see, e.g., (Chellas, 1980)). Indeed, the former is equivalent to the latter with the assumption that all families of neighborhoods are $\subseteq$-upward closed. The reason I adopt the less general definition in this paper is a *philosophical* one that, when imported to the incompatibility framework, it renders available to us the counterfactual-robustness interpretation of neighborhoods, which we will see in Section 3. There is no *technical* reason we could not adopt and import the standard, fully general formulation to the incompatibility framework.

3 Neighborhood semantics in the incompatibility framework

To import the idea from the possible-world framework to the incompatibility framework, let us compare the two frameworks at the ground (i.e. non-modal) level.

First, while the possible-world semantics interprets sentences p with subsets $[\![p]\!]$ of $\mathbb{W}$ containing worlds w, the incompatibility semantics interprets them with subsets $\mathcal{I}(p)$ of $\mathcal{PL}$ containing sets X of sentences. This comparison suggests that we consider $\mathcal{PL}$ rather than $\mathbb{W}$ to be the space, and $X \in \mathcal{PL}$ rather than $w \in \mathbb{W}$ to be points in this space.

Then recall that, because that a point X of this space $\mathcal{PL}$ lies in a semantic interpretant $\mathcal{I}(p)$ means incompatibility, entailment $p \vDash q$ corresponds to the reverse inclusion $\mathcal{I}(p) \supseteq \mathcal{I}(q)$ in the incompatibility framework, rather than the inclusion $[\![p]\!] \subseteq [\![q]\!]$ in the possible-world case. For example, the T axiom $\Box p \vdash p \vdash \Diamond p$ corresponds to $\mathcal{I}(\Diamond p) \subseteq \mathcal{I}(p) \subseteq \mathcal{I}(\Box p)$. This suggests that, in the incompatibility framework, $\Diamond$ should be interpreted by interior rather than closure, and $\Box$ by closure rather than interior.

Therefore, the neighborhood incompatibility semantics should simply replace $\mathbb{W}$ with $\mathcal{PL}$, and switch $\Diamond$ and $\Box$ in the interpretation. This idea can be put as follows. A *neighborhood incompatibility frame* is a triple $(\mathcal{L}, \mathcal{I}, \mathcal{N})$ consisting of:

- A (sentential) language $\mathcal{L}$ with modal operators;
- A map $\mathcal{I}$ such that $(\mathcal{L}, \mathcal{I})$ is an incompatibility frame on $\mathcal{L}$, treating the non-modal operators of $\mathcal{L}$ properly (in the manner reviewed in Section 1);
- A *neighborhood function* $\mathcal{N} : \mathcal{PL} \to \mathcal{PPPL}$ whose interior and closure operations interpret $\Diamond$ and $\Box$, i.e., that satisfies the following:

$$X \in \mathcal{I}(\Diamond p) \iff \exists U \in \mathcal{N}_X \cdot U \subseteq \mathcal{I}(p), \qquad (\Diamond)$$

$$X \in \mathcal{I}(\Box p) \iff \forall U \in \mathcal{N}_X \cdot U \cap \mathcal{I}(p) \neq \varnothing. \qquad (\Box)$$

We define $\vDash$ as before: $(\mathcal{L}, \mathcal{I}, \mathcal{N})$ has $X \vDash Y$ if and only if $(\mathcal{L}, \mathcal{I})$ has $\bigcap_{p \in Y} \mathcal{I}(p) \subseteq \mathcal{I}(X)$.

As this definition is purely formal, we need to explain what, conceptually, is going on here. First, fix a point $X \in \mathcal{PL}$. Then X is a set of sentences, e.g., $p \notin X$, $q \in X$,... When $U \subseteq \mathcal{PL}$ is a neighborhood of X, namely $U \in \mathcal{N}_X$, it contains other points Y, Z,... each of which is a set of sentences, e.g., $p \in Y$, $q \notin Y$,... Then Y might be obtained by adding p to X, dropping q from X, and so on, and similarly for Z. Hence we can consider Y or Z to be modifying X with counterfactual hypotheses; for example, when we are at the information state X, Y is just X with the counterfactual supposition "If p were known true, but q were unknown, and so on". Then each neighborhood $U \subseteq \mathcal{PL}$ of X is an "admissible" way of grouping together such counterfactual hypotheses on X, where the "admissibility" is formally expressed by U lying in $\mathcal{N}_X$.

Now, under this interpretation of neighborhoods $U \in \mathcal{N}_X$, $U \subseteq \mathcal{I}(p)$ (i.e., $\forall Y \in U[Y \in \mathcal{I}(p)]$) means "Whatever counterfactual hypothesis Y (within the range of U) we may make on X, it would still be incompatible with p". In short, $U \in \mathcal{N}_X$ such that $U \subseteq \mathcal{I}(p)$ witnesses the incompatibility of X with p is *counterfactually robust.* Accordingly, ($\Diamond$) states that X is incompatible with possibly-p if and only if some neighborhood U of X witnesses the counterfactual robustness of X being incompatible with p. On the other hand, $U \in \mathcal{N}_X$ such that $U \cap \mathcal{I}(p) = \varnothing$ (i.e., $\forall Y \in U[Y \notin \mathcal{I}(p)]$) witnesses that the compatibility of X with p is counterfactually robust, and hence ($\Box$) states that X is incompatible with necessarily-p if and only if the counterfactual robustness of X being compatible with p is never witnessed. Or, rewriting ($\Box$) in terms of entailment with $X \vDash \neg p \iff X \in \mathcal{I}(p)$, we have

$$\begin{aligned} X \vDash \neg\Box p \quad &\iff \quad X \in \mathcal{I}(\Box p) \iff \quad \forall U \in \mathcal{N}_X \cdot U \cap \mathcal{I}(p) \neq \varnothing \\ &\iff \quad \forall U \in \mathcal{N}_X \exists Y \in U \cdot Y \in \mathcal{I}(p) \\ &\iff \quad \forall U \in \mathcal{N}_X \exists Y \in U \cdot Y \vDash \neg p; \end{aligned}$$

that is, X entails not-necessarily-p if and only if every neighborhood U of X contains a counterfactual hypothesis Y that entails not-p.

4 Logic for neighborhood incompatibility semantics

This section reviews what rules and axioms are valid or invalid in neighborhood incompatibility semantics. By saying that an axiom (scheme) $X \vdash Y$ is valid in a neighborhood incompatibility frame $(\mathcal{L}, \mathcal{I}, \mathcal{N})$, we mean that $(\mathcal{L}, \mathcal{I}, \mathcal{N})$ has $X \vDash Y$, i.e. $\bigcap_{p \in Y} \mathcal{I}(p) \subseteq \mathcal{I}(X)$ (for all instances of the scheme).[3]
Also, by saying that a rule (scheme)

$$\frac{X_0 \vdash Y_0}{X_1 \vdash Y_1}$$

is valid in $(\mathcal{L}, \mathcal{I}, \mathcal{N})$, we mean that if $(\mathcal{L}, \mathcal{I}, \mathcal{N})$ has $X_0 \vDash Y_0$ then it has $X_1 \vDash Y_1$ (for all instances of the scheme).

[3] We use $\vDash$ and $\vdash$ differently as follows: $X \vDash Y$ is a statement that X entails Y (in a given frame); in contrast, $X \vdash Y$ is a sequent rather than a statement.

First, the $\Diamond$ operator preserves the order of entailment $\models$; that is, the rule M below is valid in all neighborhood incompatibility frames.[4]

$$\frac{p \vdash q}{\Diamond p \vdash \Diamond q} \qquad \text{M}$$

M is valid because $p \models q$ (i.e., $\mathcal{I}(q) \subseteq \mathcal{I}(p)$) implies $\Diamond p \models \Diamond q$ (i.e., $\mathcal{I}(\Diamond q) \subseteq \mathcal{I}(\Diamond p)$) in any frame. To show this, assume $\mathcal{I}(q) \subseteq \mathcal{I}(p)$ and $X \in \mathcal{I}(\Diamond q)$. Then, by ($\Diamond$), X has some $U \in \mathcal{N}_X$ such that $U \subseteq \mathcal{I}(q) \subseteq \mathcal{I}(p)$, which means, again by ($\Diamond$), that $X \in \mathcal{I}(\Diamond p)$. A similar argument shows that $\Box$ also preserves entailment, because any neighborhood intersecting $\mathcal{I}(q) \subseteq \mathcal{I}(p)$ intersects $\mathcal{I}(p)$ as well.

Apart from these preservation rules, neighborhood semantics is so general as to provide counterexamples to many rules and axioms that are valid in other (most notably Kripke's relational) semantics. For example, $\Diamond$ may not preserve incoherence; that is, even when p is incoherent (viz., $\mathcal{I}(p) = \mathcal{PL}$), $\Diamond p$ may be coherent (viz., $\mathcal{I}(\Diamond p) \neq \mathcal{PL}$), thereby failing the rule

$$\frac{p \vdash}{\Diamond p \vdash}. \qquad \text{N}_\Diamond$$

This fails in a pathological frame of neighborhoods where some $X \in \mathcal{PL}$ has $\mathcal{N}_X = \varnothing$ (and hence, by ($\Diamond$), $X \notin \mathcal{I}(\Diamond q)$ for any q). Also, the rule

$$\frac{p \vdash}{\Box p \vdash} \qquad \text{N}_\Box$$

can fail in another kind of pathological frame where some $X \in \mathcal{PL}$ has $\varnothing \in \mathcal{N}_X$ (and hence, by ($\Box$), $X \notin \mathcal{I}(\Box q)$ for any q). In fact, $\text{N}_\Diamond$ and $\text{N}_\Box$ are valid in the frames without these pathologies.

The facts described so far apply to neighborhood semantics in general, not only in the incompatibility framework but also in the conventional possible-world framework. One major divergence of the former from the latter is the following point regarding completeness. Note that every frame satisfies one of the following (1)–(3):

$$\mathcal{N}_\varnothing \neq \varnothing \text{ but } \varnothing \notin \mathcal{N}_\varnothing, \tag{1}$$

$$\varnothing \in \mathcal{N}_\varnothing, \tag{2}$$

$$\mathcal{N}_\varnothing = \varnothing. \tag{3}$$

The modal logic **MN** that is obtained by adding M, $\text{N}_\Diamond$, $\text{N}_\Box$ to classical logic is sound and complete with respect to the frames satisfying (1), in the

[4] This rule can be avoided if we adopt the standard, more general definition of interior; see Footnote 2. The rule that is valid instead of M in the more general formulation is to infer $\Diamond p \vdash \Diamond q$ from both $p \vdash q$ and $q \vdash p$.

sense that a sequent or an inference from sequents to another is valid in all those frames if (soundness) and only if (completeness) it is a theorem or a derivable rule of **MN**. Also, let $\mathbf{MI}_\Diamond$ and $\mathbf{MI}_\Box$ be the logics obtained by adding M as well as the following axioms $\mathrm{I}_\Diamond$ and $\mathrm{I}_\Box$, respectively, to classical logic:

$$\Diamond p \vdash, \qquad \mathrm{I}_\Diamond$$

$$\Box p \vdash . \qquad \mathrm{I}_\Box$$

Then $\mathbf{MI}_\Diamond$ and $\mathbf{MI}_\Box$ are sound and complete with respect to the frames satisfying (2) and (3), respectively. Moreover, we need to formulate the completeness of an axiom or rule in a manner that depends on these three classes of frames, as follows. We say an axiom or rule A is complete with respect to a semantic condition C if the logics obtained by adding A to **MN**, $\mathbf{MI}_\Diamond$, $\mathbf{MI}_\Box$, respectively, are complete with respect to the classes of frames satisfying C as well as (1), (2), (3), respectively. On the other hand, we can define the soundness of A with respect to C in the usual manner.

The most notable divergence of the neighborhood incompatibility semantics from the conventional neighborhood semantics is that the duality of $\Diamond$ and $\Box$ may fail in the former, even though the non-modal base of the logic is classical. Recall that the proof (at the end of Section 2) of the duality in the possible-world framework was based on the interpretation of negation, $\neg$, in terms of complement in the space $\mathbb{W}$, so that $[\![p]\!] \cap [\![\neg p]\!] = \varnothing$ and $[\![p]\!] \cup [\![\neg p]\!] = \mathbb{W}$. The incompatibility framework interprets $\neg$ differently, which is why the duality fails in the framework. Even though a full construction of counterexamples requires too many details to cover here, a rough but heuristic description can be given as follows. In the incompatibility framework, $\mathcal{I}(p)$ and $\mathcal{I}(\neg p)$ normally intersect with each other; indeed, $\mathcal{I}(p) \cap \mathcal{I}(\neg p)$ equals the set Inc of incoherent sets of sentences. So, a coherent point X may have a nonempty $U \subseteq \mathrm{Inc} = \mathcal{I}(p) \cap \mathcal{I}(\neg p)$ as its only neighborhood. Then $U \cap \mathcal{I}(p) \neq \varnothing$ and hence $X \in \mathcal{I}(\Box p)$ by ($\Box$) (because $\mathcal{N}_X = \{U\}$), but at the same time $U \subseteq \mathcal{I}(\neg p)$; this means $X \in \mathcal{I}(\Diamond \neg p)$ by ($\Diamond$), which then implies $X \notin \mathcal{I}(\neg \Diamond \neg p)$ since X is coherent, that is, $X \notin \mathrm{Inc} = \mathcal{I}(\Diamond \neg p) \cap \mathcal{I}(\neg \Diamond \neg p)$. So this X witnesses $\mathcal{I}(\Box p) \not\subseteq \mathcal{I}(\neg \Diamond \neg p)$, i.e., $\neg \Diamond \neg p \nvDash \Box p$. The upshot is that this entailment fails when neighborhoods are too strong (i.e., when points in them lie in both $\mathcal{I}(p)$ and $\mathcal{I}(\neg p)$), which cannot happen in the classical possible-world framework (i.e., no point ever lies in both $[\![p]\!]$ and $[\![\neg p]\!]$). Quite expectably, the other direction $\Box p \vDash \neg \Diamond \neg p$ of the duality fails when neighborhoods are too weak (i.e., when points in them lie in neither $\mathcal{I}(p)$ nor $\mathcal{I}(\neg p)$), which cannot happen in the classical possible-world framework (i.e., every point has to lie in either $[\![p]\!]$ or $[\![\neg p]\!]$).

While there are certain semantic conditions to the effect that neighborhoods are not too strong, or not too weak, with respect to which either of

$\neg\Diamond\neg p \vdash \Box p$ and $\Box p \vdash \neg\Diamond\neg p$ is sound and complete, the more important point is that neighborhood semantics is more expressive in the incompatibility framework than in the classical possible-world framework. This is not merely in the sense that the duality axioms are invalid, but the semantics separates the $\Diamond$ and $\Box$ versions of many axioms; for example,

$$p \vdash \Diamond p, \qquad \mathrm{T}_\Diamond$$

$$\Box p \vdash p. \qquad \mathrm{T}_\Box$$

Even though these axioms are treated just as equivalent in possible-world semantics, they correspond to two different semantic conditions on neighborhood incompatibility frames.

To lay out these conditions, we need to introduce a preorder (i.e. a reflexive and transitive relation) $\precsim$ on $\mathcal{PL}$ defined as follows:

$$X \precsim Y \iff \mathcal{I}(X) \subseteq \mathcal{I}(Y).$$

So, $X \precsim Y$ roughly means X is weaker than Y, or that Y conjunctively entails X conjunctively. Then $\precsim$ generalizes $\subseteq$ in the sense that $X \subseteq Y$ entails $X \precsim Y$,[5] and moreover every semantic interpretant $\mathcal{I}(p)$ is closed upward in terms of $\precsim$. Then, given any set $U \subseteq \mathcal{PL}$ of points, we write $\uparrow U$ and $\downarrow U$ for the $\precsim$-upward and $\precsim$-downward closures of $U \subseteq \mathcal{PL}$, respectively, i.e.,

$$\uparrow U = \{Y \in \mathcal{PL} \mid X \precsim Y \text{ for some } X \in U\},$$
$$\downarrow U = \{X \in \mathcal{PL} \mid X \precsim Y \text{ for some } Y \in U\}.$$

In the possible-world framework, the T axiom $p \vdash \Diamond p$ (or $\Box p \vdash p$) corresponds to the semantic condition that $w \in U$ for all $U \in \mathcal{N}_w$. In contrast, in the incompatibility framework, using the notions defined above we can modify this condition to obtain the following two versions:

$$X \in \uparrow U \qquad \text{for all } U \in \mathcal{N}_X, \tag{4}$$

$$X \in \downarrow U \qquad \text{for all } U \in \mathcal{N}_X. \tag{5}$$

Then (4) and (5) have the axioms $\mathrm{T}_\Diamond$ and $\mathrm{T}_\Box$, respectively, sound and complete.

Here we only show the soundness. That $\mathrm{T}_\Diamond$ is sound with respect to (4) means that (4) entails $\mathcal{I}(\Diamond p) \subseteq \mathcal{I}(p)$. To show this, let us assume $X \in \mathcal{I}(\Diamond p)$. It means some $U \in \mathcal{N}_X$ is included in $\mathcal{I}(p)$. Then, because $\mathcal{I}(p)$ is $\precsim$-upward closed, $\uparrow U$ is still included in $\mathcal{I}(p)$. Hence (4) implies $X \in \uparrow U \subseteq \mathcal{I}(p)$, thereby establishing the soundness. We can similarly

[5] There are more senses in which we can say $\precsim$ generalizes $\subseteq$. See (Kishida, n.d.-b).

show the soundness of $\mathrm{T}_\Box$ with respect to (5), i.e., (5) entailing $\mathcal{I}(p) \subseteq \mathcal{I}(\Box p)$, by assuming $X \notin \mathcal{I}(\Box p)$. This means some $U \in \mathcal{N}_X$ is disjoint from $\mathcal{I}(p)$. Then, again because $\mathcal{I}(p)$ is $\precsim$-upward closed, $\downarrow U$ is still disjoint from $\mathcal{I}(p)$. Hence (5) implies $X \in \downarrow U$ and therefore $X \notin \mathcal{I}(p)$, establishing the soundness.

Upward and downward closures differentiate the $\Diamond$ and $\Box$ versions in the case of the T axioms, and they do so in the following case as well. Consider the condition:

$$\text{If } U \in \mathcal{N}_X \text{, there is } V \in \mathcal{N}_X \text{ such that every } Y \in V \text{ has some } U_Y \in \mathcal{N}_Y \text{ with } U_Y \subseteq U. \quad (6)$$

This means that every neighborhood U of X has another V of X such that U is (a superset of) a neighborhood of every point in V; in short, (6) says that each neighborhood is a neighborhood of a neighborhood. Now replace the last U in (6) with $\uparrow U$, to obtain:

$$\text{If } U \in \mathcal{N}_X \text{, there is } V \in \mathcal{N}_X \text{ such that every } Y \in V \text{ has some } U_Y \in \mathcal{N}_Y \text{ with } U_Y \subseteq \uparrow U. \quad (7)$$

Then the following $\Diamond$ version of the S4 axiom is sound and complete with respect to (7):

$$\Diamond\Diamond p \vdash \Diamond p. \qquad \mathrm{S4}_\Diamond$$

To show the soundness, i.e., that (7) entails $\mathcal{I}(\Diamond p) \subseteq \mathcal{I}(\Diamond\Diamond p)$, assume $X \in \mathcal{I}(\Diamond p)$. This means $\mathcal{I}(p)$ includes some $U \in \mathcal{N}_X$ and $\uparrow U$. Then (7) yields $V \in \mathcal{N}_X$ such that every $Y \in V$ has some $U_Y \subseteq \uparrow U \subseteq \mathcal{I}(p)$, i.e. $Y \in \mathcal{I}(\Diamond p)$, which means $V \subseteq \mathcal{I}(\Diamond p)$. Therefore V witnesses $X \in \mathcal{I}(\Diamond\Diamond p)$.

Replacing $\uparrow U$ with $\downarrow U$ in (7), we can show by essentially the same idea that $\mathrm{S4}_\Box$ is sound:

$$\Box p \vdash \Box\Box p. \qquad \mathrm{S4}_\Box$$

Here is, however, some asymmetry between $\Diamond$ and $\Box$. Even though $\mathrm{S4}_\Box$ is sound with respect to the condition (7) with $\downarrow U$ in place of $\uparrow U$, it is not complete. To achieve the completeness, we need to weaken the condition a little bit, to obtain:

$$\begin{aligned}&\text{If } \mathcal{N}_\varnothing \neq \varnothing \text{, then the following holds for every } X \in \mathcal{PL}\text{:}\\ &\quad \text{if } U \in \mathcal{N}_X \text{, there is } V \in \mathcal{N}_X \text{ such that}\\ &\qquad \text{every } Y \in V \text{ has some } U_Y \in \mathcal{N}_Y \text{ with } U_Y \subseteq \downarrow U.\end{aligned}$$

In the following case the asymmetry between $\Diamond$ and $\Box$ is even bigger. Consider the following axioms (which are dual to each other in the conventional framework):

$$\Diamond(p \vee q) \vdash \Diamond p \vee \Diamond q, \qquad \mathrm{C}_\Diamond$$

$$\Box p \wedge \Box q \vdash \Box(p \wedge q). \qquad \mathrm{C}_\Box$$

$\mathrm{C}_\Diamond$ is sound and complete with respect to:

$$U_0, U_1 \in \mathcal{N}_X \Longrightarrow \text{there is } U_2 \in \mathcal{N}_X \text{ such that } U_2 \subseteq \uparrow U_0 \text{ and } U_2 \subseteq \uparrow U_1. \tag{8}$$

For the soundness (i.e., (8) entailing $\mathcal{I}(\Diamond p \vee \Diamond q) \subseteq \mathcal{I}(\Diamond(p \vee q))$), assume $X \in \mathcal{I}(\Diamond p \vee \Diamond q) = \mathcal{I}(\Diamond p) \cap \mathcal{I}(\Diamond q)$. This means X has some neighborhoods $U_0 \subseteq \mathcal{I}(p)$ and $U_1 \subseteq \mathcal{I}(q)$, which, as always, entails $\uparrow U_0 \subseteq \mathcal{I}(p)$ and $\uparrow U_1 \subseteq \mathcal{I}(q)$. Now, (8) yields $U_2 \in \mathcal{N}_X$ such that $U_2 \subseteq \uparrow U_0 \subseteq \mathcal{I}(p)$ and $U_2 \subseteq \uparrow U_1 \subseteq \mathcal{I}(q)$, i.e. $U_2 \subseteq \mathcal{I}(p) \cap \mathcal{I}(q) = \mathcal{I}(p \vee q)$. Hence U_2 witnesses $X \in \mathcal{I}(\Diamond(p \vee q))$. In this way, $\mathrm{C}_\Diamond$ is sound with respect to (8), and in fact complete. Nevertheless, the announced asymmetry between $\Diamond$ and $\Box$ is that it is an open problem even what condition has $\mathrm{C}_\Box$ sound and complete, because it does not seem to work to replace $\uparrow U$ with $\downarrow U$.[6] It is also interesting to see how $\Diamond$ and $\Box$ interact with each other. The axiom

$$\Box p \vdash \Diamond p \qquad \mathrm{D}$$

is sound and complete with respect to the condition that $U \cap V \neq \varnothing$ for all $U, V \in \mathcal{N}_X$. It is easy to show D to be sound because if $X \in \mathcal{I}(\Diamond p)$, i.e., if a $U \in \mathcal{N}_X$ has $U \subseteq \mathcal{I}(p)$, then the condition says every $V \in \mathcal{N}_X$ intersects U, thereby intersecting $\mathcal{I}(p)$, i.e. $X \in \mathcal{I}(\Box p)$.

It is an open problem with respect to what condition the following axiom B is complete:

$$p \vdash \Box\Diamond p; \qquad \mathrm{B}$$

but there is a condition with respect to which B is sound:

$$\begin{aligned}&\text{Given a collection } \{\, X_i \in \mathcal{PL} \mid i \in I \,\} \text{ of any size and } Y \in \mathcal{PL},\\ &\qquad\text{if each } X_i \text{ has some } U_i \in \mathcal{N}_{X_i} \text{ with } Y \notin U_i,\\ &\qquad\qquad\text{then a } V \in \mathcal{N}_Y \text{ has } X_i \notin V \text{ for all } X_i.\end{aligned} \tag{9}$$

To show the soundness (i.e., (9) entailing $\mathcal{I}(\Box\Diamond p) \subseteq \mathcal{I}(p)$), suppose $Y \notin \mathcal{I}(p)$. Write $\mathcal{I}(\Diamond p) = \{X_i \in \mathcal{PL} \mid i \in I\}$; then each X_i lies in $\mathcal{I}(\Diamond p)$, i.e., X_i has some $U_i \in \mathcal{N}_{X_i}$ such that $U_i \subseteq \mathcal{I}(p)$ and hence $Y \notin U_i$. Then (9) yields $V \in \mathcal{N}_Y$ such that $X_i \notin V$ for all X_i, i.e. $V \cap \mathcal{I}(\Diamond p) = \varnothing$; therefore $Y \notin \mathcal{I}(\Box\Diamond p)$.

[6] The difficulty arises partly from the lack of understanding of what $\mathcal{I}(\Box p \wedge \Box q) = \mathcal{I}(\{\Box p, \Box q\})$ looks like, in contrast to $\mathcal{I}(\Diamond p \vee \Diamond q)$ understood simply as $\mathcal{I}(\Diamond p) \cap \mathcal{I}(\Diamond q)$.

5 Conclusion

We have shown that the idea of neighborhood semantics to interpret modal operators with interior and closure operations can be straightforwardly imported — quite independently of the notions of truth and possible worlds — to the framework of incompatibility semantics. The philosophical advantage of putting the notion of neighborhood in this framework is that the connection between neighborhoods and modality can be directly and naturally interpreted in terms of the idea of counterfactual robustness. We have also shown the technical merit of the semantics that we can separate the behaviors of $\Diamond$ and $\Box$ while keeping the non-modal base of the logic classical, which, combined with the counterfactual-robustness interpretation, will enable us to apply modal logic to an even wider range of cases.

Kohei Kishida
Department of Philosophy, University of Pittsburgh
1001 Cathedral or Learning, Pittsburgh, PA 15260 U.S.A.
kok6@pitt.edu

References

Brandom, R. (2008). Incompatibility, modal semantics, and intrinsic logic. In *Between saying and doing: Towards an analytic pragmatism* (chap. 5). Oxford: Oxford University Press.

Chellas, B. (1980). Modal logic: An introduction. Cambridge-New York: Cambridge University Press.

Kishida, K. (n.d.-a). *Neighborhood incompatibility semantics and completeness.* (Draft.)

Kishida, K. (n.d.-b). *Possible–world representation of incompatibility.* (Draft.)

What do Gödel Theorems Tell us about Hilbert's Solvability Thesis?

Vojtěch Kolman*

When dealing with the foundational questions of elementary arithmetic, we find ourselves standing in the shadow of Gödel, just as our predecessors stood in the shadow of Kant, to the extent that we tend to see Gödel's famous incompleteness theorems as a new Critique of Pure Reason. In its most exuberant form (common particularly among the so-called working mathematicians) this amounts to claiming that

> human reason has encountered its limits by proving that there are truths which are humanly unprovable ("inaccessible") and that it is impossible for our mind to prove its own consistency.[1]

This attitude is not only at variance with the (Kantian) doubts about the possibility of proving the unprovability in an absolute sense, but, more specifically and famously, with the so-called Hilbert program of solving every mathematical problem by axiomatic means. In his *Parisian address*,[2] Hilbert not only phrased the conjecture that all questions which human mind asks must be answerable (the so-called axiom of solvability)[3] but supplemented it, as a kind of challenge, with a list of ten and later of twenty-three problems of prime interest, including the *Second Problem* of the consistency (and completeness) of arithmetical axioms.

In Hilbert's later writings, particularly in his *Königsberg address*,[4] the solvability argument takes a more subtle form. Introducing the finite mode

* Work on this paper has been supported in part by grant No. 401/06/0387 of the Grant Agency of the Czech Republic and in part by the research project MSM 0021620839 of the Ministry of Education of the Czech Republic.

[1] (Gödel, 1995, p. 310) himself phrased it like this: "there exist absolutely unsolvable diophantine problems [...], where the epithet 'absolutely' means that they would be undecidable, not just within some particular axiomatic system, but by any mathematical proof the human mind can conceive."

[2] See (Hilbert, 1900).

[3] See (Hilbert, 1900, p. 297).

[4] See (Hilbert, 1930).

of thought (*finite Einstellung*)[5] as a new kind of Kantian intuition, Hilbert argues that the harmony between *nature* (experience) and *thought* (theory) must lie exactly in the transcendental fact they are both finite.[6] As a consequence, the seeming infinity of human knowledge (particularly in the realm of mathematics) must have finite roots which are to be identified with a finite (or finitely describable) system of rules and axioms, and finite deductions from them.[7] Hence, "we must know, we shall know."[8] Obviously, this is a *transcendental deduction* of its own kind, namely of inferentialism or broader axiomatism from finitism, starting with the words: in the beginning was a sign.[9]

Gödel (1931), so we are usually told, put an end to Hilbert's optimism by proving that the Second Problem is essentially unsolvable. This verdict is sometimes supported by the seemingly analogous case of Hilbert's *First Problem*, the Continuum Hypothesis, which, partially also due to Gödel, was proved to be undecidable on the basis of currently accepted axioms. In this paper I would like to present Gödel's theorems not as a direct refutation of Hilbert's axiom but only as an impulse to phrase it with more caution, in such a way that the Continuum Hypothesis is no longer regarded as a *real* problem. I will draw on two rather different sources, both, however, connected to Hilbert's philosophy, namely

- the late metamathematical views of Zermelo and
- Lorenzen's post-Hilbertian program of operative mathematics.

This will lead me to a closer analysis of the distinction between proof and truth which does not endorse one of them at the expense of the other, as Lorenzen, the *constructivist*, and Zermelo, the *Platonist*, still tend to do.

1

First, let us discuss the possibility of proving the unsolvability of something. There is a general pattern: if someone comes along with a positive solution to a given problem, one can check to see that it does the required work. But if it is to be shown that the problem is unsolvable, one has to give a precise delimitation of the methods that can be employed. This brings us to the difference between *method* in the *broader* (general) and in the *narrower* (limited) sense.

---

[5] See (Hilbert, 1930, p. 385) and also (Hilbert, 1926, p. 161).
[6] See (Hilbert, 1930, pp. 380–381).
[7] See (Hilbert, 1930, p. 379) and also (Hilbert, 1918).
[8] (Hilbert, 1930, p. 387).
[9] See (Hilbert, 1922, p. 163).

To illustrate the point let us take some famous geometrical problems like the trisection of an angle or the quadrature of the circle. Due to the methods of modern algebra we positively know that these problems are unsolvable by straightedge and compass. However, we also know that the ancient mathematicians (Hippias, Archimedes) already solved them by extended — so-called mechanical — means (quadratix, spiral)[10] where the epithet "mechanical" means mainly that they were devised *ad hoc*. Similarly, if I give you — meaning somebody sufficiently educated in predicate logic — a formula, I am quite sure you will be able to decide, in a finite number of steps, whether it is a tautology or not. What you might not be able to do, however, is to solve the problem with pre-chosen schematic methods such as with *one particular* Turing machine.

Now, as may be expected, a similar observation applies to Gödel's theorems, only this time it is the provability itself the limits of which beg the question. Gödel showed that for any axiomatic system of arithmetic there will always be an individual sentence that is undecidable by it. The gist of his argument lies in the fact that this unprovable sentence of arithmetic (informally *saying* "I am unprovable") is unprovable because it is true (it *is* unprovable), its truth being proven as a part of the argument. So, the whole argument works only because it employs two different concepts of *proof*, the first being that of *Principia Mathematica* (or Peano arithmetic) and the second being the broader one in which the argument is clinched.

Zermelo, in his unjustly infamous correspondence with Gödel, was probably the first person to make this observation. Setting himself the natural question, "What does one understand by a proof?", his answer went like this:

> In general, a proof is understood as a system of propositions that, when accepting the premises, yields the validity of the assertion as being *reasonable*. And there remains only the question of what may be "reasonable". In any case — as you are showing yourself — not only the propositions of some finitary scheme that, also in your case, may always be extended. So, in this respect, we are of the same opinion, however, I *a priori* accept a more general scheme that does not need to be extended. And in this system, really all propositions are decidable.[11]

What needs to be explained now is the nature of the difference between proof in the narrower and broader sense, or between the *proof* and *truth*, and the sense in which the second one is "decidable", or better: complete and unextendable, as Zermelo claims.

[10] See, e.g., (Heath, 1931).
[11] See (Gödel, 2003, p. 431).

2

The analogous differences between the general and narrower *construability* or *decidability* is less problematic since the *ad hoc* constructive or decision methods (like quadratix or spiral) are still bound to some humanly feasible means, and so quite naturally counted as constructions and algorithms. The traditional problem of arithmetic is its very relationship to the empirical world, as (already before Kant) expressed in the claim it is a science of analytical nature. Hence, the whole issue of the difference between the truth and proof can be boiled down to a single question:

what is arithmetical truth outside of a specific axiomatic system?

It is exactly the lack of any explicit answer to this question that leads to the Platonist account of arithmetical truth. The usual model-theoretical exposition operating with an unexplained concept of standard model ("$2 + 2 = 4$" is true if and only if $2 + 2 = 4$) confirms this image, particularly when it starts to invoke our "intuitions".

However, to understand sentences like "$2 + 2 = 4$" and "$23 + 4 < (6 \times 3) + 2$" you need no more mathematics than that provided by a good secondary education. This is to say that they are not true or false, at least not in the first place, *because* they are deducible in Peano arithmetic, or happen to inexplicably hold in the standard model, but *because* they are transformable into the simpler forms of "$4 = 4$" and "$27 < 20$" where only knowledge of the sequence $1, 2, 3, 4, \ldots$ and the ability to compare symbols is needed. This is the basis of the operativist account of arithmetical truth as developed by Lorenzen in his *Einführung in die operative Logik und Arithmetik* (Lorenzen, 1955), in opposition to the usual standards of Frege that consider such justifications prescientific. According to Lorenzen,[12] the ultimate foundation of arithmetic (including higher analysis) lies exactly in these prescientific practices of counting and operating with *symbols*. They can be made explicit in synthetic (recursive) definitions like

$$\Rightarrow |, \qquad\qquad \Rightarrow x + | = x|$$
$$x \Rightarrow x|, \qquad\qquad x + y = z \Rightarrow x + y| = z|$$

$$\Rightarrow | \times x = x \qquad\qquad \Rightarrow | < x|$$
$$x \times y = p,\ p + y = q \Rightarrow x| \times y = q \qquad\qquad x < y \Rightarrow x < y|$$

introducing (in unary form) the number series, the operations $+$, $\times$ and the relation $<$ respectively. The (true) arithmetical sentences are then defined as

[12] See (Lorenzen, 1974, pp. 199–200).

the consequences of these definitions, prospectively within the broader frame of game- or proof-theoretical semantics (Lorenzen's dialogical games).[13]

As for Gödel's results, Lorenzen[14] claims that instead of being about arithmetic, as completely given by its operative definition, they merely tell us something about Peano's formalism in its particular shape of a first-order scheme within the language containing 0, s, $+$ and $\times$. So, coming from the other side, Lorenzen arrived at the same basic difference as Zermelo. It is also in accord both with Lorenzen's later views, as developed in his *Metamathematik* (1962), and with Zermelo's late project of infinitist logic,[15] to rephrase this difference in *inferentialist* terms as the distinction between two different kinds of consequence: strongly effective or *full-formal* $\vdash$ and the more liberal or *semi-formal* $\models$.[16] Now, simplifying heavily:

Full-formal arithmetic, like the arithmetic of Peano, is arithmetic *in the narrower sense*, and deals with schematically or mechanically given and controllable axioms and rules. Semi-formal arithmetic or *the* arithmetic *proper* employs — in accord with the infinite nature of the number sequence $1, 2, 3, \ldots$ — rules with infinitely many premises, particularly the (ω)-rule

$$A(1), A(2), A(3), \text{ etc.} \quad \Rightarrow \quad (\forall x)(Ax). \tag{ω}$$

As an arithmetical rule it is transparent and sound enough (or "reasonable", as Zermelo would say), as long as one interprets the "etc." correctly. In fact, Tarski's idea of semantics[17] employs this kind of rules systematically, with the (ω)-rule as a special case of the more general

$$A(N) \text{ for all substituents } N \quad \Rightarrow \quad (\forall x)A(x). \tag{$\forall$}$$

This rule is then nothing else than the well-known part of the so-called *semantic definition of truth.* Hence, the significance of semi-formalism is to make us think of semantic definitions as special (more generously conceived) systems of rules (proof systems) which — starting with some elementary sentences — evaluate the complex ones by *exactly one* of *two* truth values. The most important point to notice is that the semi-formal rules are called semantic not because they are infinite but because they, unlike Peano's formalism, work with a uniquely determined range of quantification.

As a consequence, arithmetical truth need not be guaranteed by God or by intuition, but, as (Zermelo, 1932, p. 87) put it, simply by the fact that the broader concept of "mathematical proof is nothing other than a system of propositions which is well-founded by quantification." Zermelo's

[13] See (Lorenzen & Lorenz, 1978).
[14] See (Lorenzen, 1974, p. 21–22).
[15] See (Zermelo, 1932).
[16] Both distinctions are due to (Schütte, 1960).
[17] See especially (Tarski, 1936).

claim that all the sentences are decided by his "more general scheme", i.e., completely and correctly evaluated by arithmetical semi-formalism, can be "proved" by an easy meta-induction like this:

1. Elementary arithmetical sentences ($M = P$, $M < N$) are evaluated unambiguously as true or false only on the basis of calculations with numerals.

2. Tarski's definition provides for the evaluation of more complex sentences, particularly because: either for every term N from $1, 2, 3, \ldots$, the sentence $A(N)$ is true and hence $(\forall x)A(x)$ is true, or there is N from $1, 2, 3, \ldots$ such that $A(N)$ is false, and $(\forall x)A(x)$ is false, *tertium non datur.*

It is a known fact that the intuitionists and some constructivists (including Lorenzen,[18] but not, e.g., Weyl[19]) question the completeness of this evaluation, arguing that the existence of concrete strategies for proving or refuting every $A(N)$ doesn't entail the existence of a general strategy for $A(x)$. To give a familiar example: there is no problem in demonstrating whether, for any given even number M, it is the sum of two primes. However, the truth value of the general judgment that every even number is the sum of two primes (Goldbach Conjecture) is still unknown, 250 years after the problem was first posed. Hence, it is possible that we have proofs for all the sentences $A(N)$ without knowing it, i.e., without having the general strategy of how to prove a proposition concerning them all.

Consequently, a decision must be made whether the infinite vehicles of truth and judgment such as $(\forall)$ or (ω) should be referred to as rules

1. only in the case when we positively know that all their premises are true, i.e., when we have at our disposal some general strategy for proving all of them at once, or

2. more liberally, if we know somehow that all their premises are positively true or false. The general distinction between the constructive and classical methods in arithmetic is based on this.

3

Now, if one leaves, like, e.g., Lorenzen and Bishop, the concept of effective procedure or proof to a large extent open and does not tie it, like, e.g., Goodstein and Markov, to the concept of the Turing machine,[20] there is still

[18] See, e.g., (Lorenzen, 1968, p. 83).
[19] See (Weyl, 1921, p. 156).
[20] For further discussion of these differences see, e.g., (Bridges & Richman, 1987).

room for an effective, yet liberal enough *semantics* (semi-formal system) and a strongly effective or 'mechanical' syntax or *axiomatics* (full-formal system). Hence, the constructivist reading does not necessarily wipe out the differences between the proof and truth, as, e.g., Brouwer's mentalism or Wittgensteins's verificationism seem to. As a result, one can officially differentiate not only between full-formal $\vdash$ and semi-formal $\models$ consequence, but also between semi-formal consequence in a stricter (constructive) sense and in the more liberal (classical) sense. All these differences stem from (Gödel, 1931) for the following reason:

Gödel's theorem affects only the full-formal systems, because their schematic nature makes it possible to devise a general meta-strategy for constructing true arithmetical sentences not provable in them. The unprovable sentence of Gödel is of the so-called Goldbach type, i.e., it is of the form $(\forall x)A(x)$ where $A(x)$ is a decidable property of numbers. Now, Gödel's argument shows that this decision is done already by Peano axioms in the sense that all the instances $A(N)$ are deducible and, hence, set as true. So, with Gödel's proof we have a general strategy for proving all the premises $A(N)$ at once, which makes the critical unprovable sentence $(\forall x)A(x)$ constructively true, i.e. provable by means of the (ω)-rule interpreted constructively.

> Lorenzen (1974, p. 222) put it like this:[21]
> ω-incompleteness [...] demonstrates that not all constructively true propositions are logically deducible from the axioms. This should come as no surprise. A universal proposition $(\forall x)A(x)$ is constructively true when $A(N)$ for all N is true. But in order *logically* to deduce the universal proposition $(\forall x)A(x)$, we must first deduce $A(x)$ with a free variable x. So we should have expected ω-incompleteness. But Peano arithmetic is ω-complete if we restrict ourselves to addition. The point of Gödel's proof was to demonstrate that Peano arithmetic with only addition and multiplication (without the higher forms of inductive definition) already shows the ω-incompleteness that was to be expected in general.

It is of real significance here that it was none other than (Hilbert, 1931) who — probably still unaware of Gödel's result[22] — employed the (ω)-rule as a means of improving his old project of founding arithmetic on axiomatic grounds. So, our claim that Gödel's theorems did not destroy but refine Hilbert's optimism in the suggested semi-formal way is sound also from a historical perspective. And using the concept of semi-formalism again, we can extend this optimism yet further by claiming that full-formal systems

[21] Translation by K. R. Pavlovic in (Lorenzen, 1987, p. 240–241).

[22] See Bernays' remarks in (Hilbert, 1935, p. 215) but also Feferman's commentary in (Gödel, 1986, pp. 209–210).

such as Peano and Robinson arithmetic are consistent simply because their axioms are provable in the arithmetical semi-formalism and, moreover, even in its constructive variant. This, in fact, is the usual model-theoretic argument:

> if a theory is inconsistent, then it does not have a model,

in a relative setting:

> if Peano arithmetic is inconsistent, then so is the arithmetical semi-formalism.

In the first case the consequent is precluded "by fiat". In the second case one does not need to use such tricks, because it was actually proved that the rules of semi-formalism do not evaluate arithmetical sentences incorrectly.

4

Now, should we perhaps follow Zermelo further and discard the narrower concept of proof totally by saying that everything true is provable? While the danger of the first extreme lies in the fact that the narrower, limited methods can and eventually will fail because of their limitedness, the shortcoming of Zermelo's alternative is that it is safe to the point of becoming totally idle. The problems of set theory are a particularly good example of such a situation. Let me illustrate it very briefly with the help of the concept of continuum.[23]

Continuum has had an intricate historical development, from the Pythagorean definition of proportion by means of a reciprocal subtraction, through the Euclidian theory of points constructible by means of a ruler and compass, to the Cartesian idea of numbers as roots of polynomials. By grasping real numbers as *arbitrary* (Cauchy) sequences, rather than as sequences that are in some sense law-like, Cantor believed himself to have won the whole game by simple "fiat". But this was no more substantiated than it would have been for the Greeks to define real numbers as points constructible by whatever means, or for us now to say that everything true is provable. Obviously, this would dispose of problems like the quadrature of the circle, the axiomatizability of arithmetic, or the "Entscheidungsproblem", but it would also dispose of the whole of mathematics — insofar as it is understood as an enterprise of solving problems somehow related to human lives rather than as a pure science indulged in for its own sake. Hence, the reason for retaining and developing the difference between the broader (and vaguer) and the narrower (more limited) sphere of methods lies in the fact that it

[23] For a detailed account see (Kolman, n.d.).

mirrors the general process of *explaining* something complicated through something less complicated.

Set theory runs into problems because of its failure to keep these differences apart. Set theorists believe, on the one hand, that the Continuum Hypothesis is either true or false whether we know it or not, but, on the other hand, the only specific idea they can give us about its standard model is one loosely connected to Zermelo's full-formalism, by which it is, however, undecidable, i.e. neither true nor false. So, because the only criterion of truth is the incomplete and possibly inconsistent full-formalism, we must face the possibility that the status of questions like "how big is the continuum?" may be similar to that of questions like "how many hairs does Othello have?", not because we do not yet know the answer, but because no answer is available. This deficit does not make such questions human-independent, but only deeply fictitious, the reason for which, again, is not that they are still undecided (such a decision is not difficult to make, e.g., by endorsing $V = L$) but because nothing really important hinges on them.

My conclusion may resemble the position of (Feferman, 1998, p. 7), according to whom the Continuum Hypothesis, unlike Hilbert's Second Problem, "does not constitute a *genuine* definite mathematical problem," because it is an "inherently vague or indefinite one, as are propositions of higher set theory more generally." I have attempted, however, to be more specific about where the difference between set theory and arithmetic comes from. The so-called iterative hierarchy, described in a pseudo-constructive manner by Zermelo's axioms, is not a model in the same sense in which the standard model of arithmetic is, because the concept of subset is left unexplained, along with the range of quantification and the respective ($\forall$)-rule.[24]

To sum up: Hilbert's solvability thesis is not refuted by Gödel's incompleteness theorems, nor by the Continuum Hypothesis; however, they oblige us to rephrase it as follows: every problem is (potentially) solvable if it is endowed with well-defined truth-conditions, or, as Zermelo would put it, with a "reasonable" concept of truth.

Vojtěch Kolman
Department of Logic, Charles University
nám. Jana Palacha 2, 116 38 Praha 1, Czech Republic
vojtech.kolman@ff.cuni.cz

[24] One can possibly say that set theory has failed both of Frege's criteria for reference, as described so influentially by Quine, namely "to be is to be a value of a bound variable" and "no entity without identity" with "$|P(N)| =?$" taken as evidence.

References

Bridges, D., & Richman, F. (1987). *Varieties of Constructive Mathematics.* Cambridge: Cambridge University Press.

Ebbinghaus, H.-D. (2007). *Ernst Zermelo. An Approach to His Life and Work.* Berlin: Springer.

Ewald, W. (Ed.). (1996). *From Kant to Hilbert. A Source Book in the Foundations of Mathematics I–II.* Oxford: Clarendon Press.

Feferman, S. (1998). *In the Light of Logic.* Oxford: Oxford University Press.

Gödel, K. (1931). Über formal unentscheidbare Sätze der 'Principia Mathematica' und verwandter Systeme I. *Monatshefte für Mathematik und Physik*, *37*, 349–360.

Gödel, K. (1986). *Collected works I* (S. Feferman, J. Dawson, S. Kleene, G. Moore, R. Solovay, & J. van Heijenoort, Eds.). Oxford: Oxford University Press.

Gödel, K. (1995). Some basic theorems on the foundations of mathematics and their implications. In S. Feferman, J. Dawson, W. Goldfarb, C. Parsons, & R. Solovay (Eds.), *Collected Works III.* Oxford: Oxford University Press.

Gödel, K. (2003). *Collected Works V. Correspondence H–Z* (S. Feferman, J. Dawson, W. Goldfarb, C. Parsons, & W. Sieg, Eds.). Oxford: Oxford University Press.

Heath, L., Sir Thomas. (1931). *A Manual of Greek Mathematics.* Oxford: Clarendon Press.

Hilbert, D. (1900). Mathematische Probleme. In *Vortrag gehalten auf dem internationalen Mathematiker–Kongress zu Paris 1900* (pp. 253–297). Nachrichten von der Königlichen Gesellschaft der Wissenschaften zu Göttingen. (Page references are to the reprint in (Hilbert, 1935).)

Hilbert, D. (1918). Axiomatisches Denken. *Mathematische Annalen*, *78*, 405–415.

Hilbert, D. (1922). Neubegründung der Mathematik. Erste Mitteilung. *Abhandlungen aus dem mathematischen Seminar der Hamburgischen Universität*, *1*, 157–177. (Page references are to the reprint in (Hilbert, 1935).)

Hilbert, D. (1926). Über das Unendliche. *Mathematische Annalen*, *95*, 161–190.

Hilbert, D. (1930). Naturerkennen und Logik. *Die Naturwissenschaften*, *18*, 959–963. (Page references are to the reprint in (Hilbert, 1935).)

Hilbert, D. (1931). Die Grundlegung der elementaren Zahlentheorie. *Mathematische Annalen*, *104*, 485–495.

Hilbert, D. (1935). *Gesammelte Abhandlungen. Dritter Band: Analysis, Grundlagen der Mathematik, Physik, Verschiedenes.* Berlin: Springer.

Kolman, V. (n.d.). Is continuum denumerable? In M. Peliš (Ed.), *The Logica Yearbook 2007* (pp. 77–86). Praha: Filosofia.

Lorenzen, P. (1955). *Einführung in die operative Logik und Mathematik.* Berlin: Springer.

Lorenzen, P. (1962). *Metamathematik.* Mannheim: Bibliographisches Institut.

Lorenzen, P. (1968). *Methodisches Denken.* Frankfurt am Main: Suhrkamp.

Lorenzen, P. (1974). *Konstruktive Wissenschaftstheorie.* Frankfurt am Main: Suhrkamp.

Lorenzen, P. (1987). *Constructive Philosophy.* Amherst: The University of Massachusetts Press. (Edited and translated by K. R. Pavlovic.)

Lorenzen, P., & Lorenz, K. (1978). *Dialogische Logik.* Darmstadt: Wissenschaftliche Buchgesellschaft.

Schütte, K. (1960). *Beweistheorie.* Berlin: Springer.

Tarski, A. (1936). O pojęciu wynikania logicznego. *Przegląd Filozoficzny*, *39*, 56–68.

Weyl, H. (1921). Über die neue Grundlagenkrise der Mathematik. *Mathematische Zeitschrift*, *10*, 39–79. (Page references are to the reprint in (Weyl, 1968, vol. II.).)

Weyl, H. (1968). *Gesammelte Abhandlungen I-IV.* Berlin: Springer.

Zermelo, E. (1932). Über Stufen der Quantifikation und die Logik des Unendlichen. *Jahresbericht der Deutschen Mathematiker-Vereinigung*, *41*, 85–88.

Wittgenstein on Pseudo-Irrationals, Diagonal Numbers and Decidability

Timm Lampert*

In his early philosophy as well as in his middle period, Wittgenstein holds a purely syntactic view of logic and mathematics. However, his syntactic foundation of logic and mathematics is opposed to the axiomatic approach of modern mathematical logic. The object of Wittgenstein's approach is not the representation of mathematical properties within a logical axiomatic system, but their representation by a symbolism that identifies the properties in question by its syntactic features. It rests on his distinction of descriptions and operations; its aim is to reduce mathematics to operations. This paper illustrates Wittgenstein's approach by examining his discussion of irrational numbers.

1 Tractarian heritage

In the *Tractatus*, TLP for short, Wittgenstein distinguishes between operations and functions. As do Russell and Whitehead in the *Principia Mathematica*, PM for short, he uses "functions" in the sense of "propositional functions", which are representable by symbols of the form φx within a logical formalism. In contrast, the concept of operation is Wittgenstein's own creation. According to Wittgenstein, the "basic mistake" of the symbolism of PM is the failure to distinguish between propositional functions and operations (WVC p. 217, and TLP 4.126). In this respect, the syntax of PM suffers from the same deficiency as the syntax of ordinary language. Wittgenstein distinguishes between functions and operations by the criterion of the possibility of iterative application, TLP 5.25f.:

> (Operations and functions must not be confused with each other.)
>
> A function cannot be its own argument, whereas an operation can take one of its own results as its base.

*I am grateful to Victor Rodych for discussions and comments.

Due to its possible iterative application, an operation generates a series of internally related elements. This series is defined by an initial member, η, and an operation, $\Omega(\xi)$, that must be applied to generate a new member from a previous one ξ. The form of such a definition is $[\eta, \xi, \Omega(\xi)]$. This series is not defined as an "infinite extension" but by the iterative application of an operation that determines forms. The natural numbers, for example, are defined by the operation +1. Starting with 0 as initial member, this yields the series 0, 0 + 1, 0 + 1 + 1 etc., which is denoted by $[0, \xi, \xi + 1]$, cf. TLP 6.03. According to Wittgenstein's point of view numbers are forms defined by operations (cf. WVC, p. 223). They are neither objects denoted by names nor classes or classes of classes described by functions. While functions determine the extension of a property independent of its symbolic representation, operations determine the syntax of symbols. Operations do not refer to anything outside the symbols; they determine formal (internal) properties rather than material (external) properties. Operations do not state anything, but determine how to vary the form of their bases (inputs) without contributing any content. In contrast, functions, e.g., "x is human," state that their arguments have some property, which is not determined by the symbol of the arguments. A function determines an extension of objects, namely the "totality" or class of objects that satisfy the function.

Operations are internally related, they can "counteract the effect of another" and "cancel out another" (TLP 5.253); they form a system. In TLP Wittgenstein reconstructs so called "truth functions" such as negation, conjunction, disjunction and implication as "truth operations". They form the system of logical operations. Likewise, he understands addition, multiplication, subtraction and division as a system of "arithmetic operations". In both cases, this forces significant changes in the traditional symbolism of logic and arithmetic. In logic, he invents his ab-notation, in which the truth operators are not represented by $\neg, \wedge, \vee$ or $\rightarrow$ but by ab-operations, which assign a- and b-poles to a- and b-poles (cf., e.g., CL, letters 28, 32, NL, pp. 94–96, 102, MN, pp. 114–116, and TLP 6.1203). By this he intends to overcome within propositional logic the "basic mistake" of PM in failing to distinguish symbolically between operations and functions. In arithmetic he defines natural numbers by operations, cf. TLP 6.02–6.04, and indicates a symbolism of primitive arithmetic wholly resting on operations (cf. TLP 6.24f.). He explicitly opposes this to the Frege's and Russell's program to reduce mathematics to a "a theory of classes" (TLP 6.031), these classes being defined by propositional functions.

Wittgenstein called for a symbolism based on operations as a counter-program to Frege's and Russell's logicism. This still holds for his middle period. Instead of his peculiar term "operation," he frequently uses the common expression "law," and instead of the technical term "propositional function," he uses the less specific expression of "description". Yet, he still claims that mathematics is dealing with systems, operations or laws and not with totalities, functions or descriptions (cf., e.g., WVC, p. 216f., or MS 107, p. 116). Likewise, he claims that "the falsities in philosophy of mathematics" are based on a confusion of the "internal properties of a form", which are determined by operations, and "properties" in terms of material properties of daily life, which are identified by propositional functions, cf. PG, p. 476. He also calls the view that bases mathematics on functions the "extensional view" whereas he professes an "intensional view" that identifies mathematical properties by syntactic properties of an adequate symbolic representation (PG, pp. 471–474, RFM V, § 34–40).

In the following we go on to illustrate Wittgenstein's intensional view in his intermediate (1929–1934) discussion of irrational numbers. Finally, we will apply this discussion to diagonal numbers, as well as to the notions of enumerability, decidability and provability. We hereby want to address two challenges faced by Wittgenstein's program:

(i) How to apply it to other parts of mathematics besides primitive arithmetic?

(ii) How to relate it to the basic notions and impossibility results of modern mathematical logic?

2 Irrationals

Cauchy sequences

Irrationals are customarily defined as equivalence classes of identical Cauchy sequences. A Cauchy sequence is an *infinite sequence* of rational numbers $a_1, a_2, \dots$ such that the absolute difference $|a_m - a_n|$ can be made less than any given value $\epsilon > 0$ whenever the indices m, n are taken to be greater than some natural number k. Two Cauchy sequences $a_1, a_2, \dots$ and $a'_1, a'_2, \dots$ are identical if and only if for any given $\epsilon > 0$ there is some natural number k such that $|a_n - a'_n| < \epsilon$ for all n greater than k. The idea behind this definition is that all methods approximating the "true expansion" of an irrational number must once result in the same expansion up to a certain digit. For example, the methods illustrated in Tables 1 and 2 both approximate the true decimal expansion of $\sqrt{2}$ in a plain manner.

	a_1	a_2	a_3	a_4	a_5	a_6	a_7	a_8	a_9
$x^2 < 2$	1			1.25	1.375		1.40625		1.4140525
$x^2 > 2$		2	1.5			1.4375		1.421875	

Table 1. Method 1

	a_1	a_2	a_3	a_4	a_5	a_6	a_7	a_8	a_9	a_{10}
$x^2 < 2$	1		1.4		1.41		1.414		1.4141	
$x^2 > 2$		2		1.5		1.42		1.415		1.4142

Table 2. Method 2

At some point the methods come up with identical decimal expansions up to a certain digit. For example, from a_9 on both sequences begin with 1.41. Thus, going further and further one approximates more and more "the" expansion of the irrational number. However, no finite sequence will ever represent the "true expansion", as it is the limit of all sequences approximating it; the "true expansion" is beyond all finite sequences — it is infinite.

With respect to Wittgenstein's point of view, it is important to note that these methods of approximation do not generate the next digits by iteration. Instead, at any step it must be checked whether the square of the result is < 2 or > 2.

Wittgenstein's critique

Wittgenstein's main critique of the definition of irrational numbers in terms of Cauchy sequences is that this definition does not provide an identity criterion, which decides the identity of two real numbers (PR §§ 186, 187, 191, 195). The problem is that, on the standard conception of irrational numbers as infinite sequences of rational numbers, for any infinite sequence s there are infinite many sequences that are identical with s up to a certain digit k. However, the definition does not provide a method to specify some upper bound for k in comparing two arbitrary real numbers. Thus, no finite comparison is sufficient to decide whether two arbitrary sequences are identical. The definition has it that the "true expansion" lies beyond all finite sequences. Therefore, it provides only a sufficient criterion for a negative answer but no sufficient criterion for a positive answer to the question of identifying arbitrary real numbers. In this respect, we have the same situation as in the case of determining within a traditional logical calculus whether some formula of first order logic is not a theorem.

One might reply to this critique that one cannot claim the decidability of things that simply are not decidable; the nature of the real numbers as infinite sequences implies that one cannot decide upon the identity of two real numbers. However, in fact it is from the purported definition that the problem arises, and it is not carved in stone that this indeed captures the "nature" of real numbers. According to Wittgenstein's analysis the definition is nothing but a consequence of the extensional view of modern mathematics. This spuriously takes the designations of real numbers by ordinary language as *descriptions* of everyday properties, which determine a certain extension. For example, in the case of $\sqrt{2}$ one wrongly analyses the ordinary explanation in terms of "the number that when multiplied by itself is identical with 2" as a description of a material, non-symbolic property. This property is then conceived as being satisfied by the "true infinite expansion", which is approximated by multiplying finite sequences with themselves and comparing the result to 2. In order to come to understand Wittgenstein's point of view, it is crucial to recognize that there is an alternative to this conception that refers to known mathematics. According to this point of view, real numbers are not defined by extensions, but by laws in the sense of Wittgenstein's operations.

Wittgenstein's alternative

In order to come to understand Wittgenstein's position one must recognize that he rejects methods of approximation such as the above illustrated methods 1 and 2. Although these kinds of methods of approximation might be called "laws," they are not "laws" in terms of operations. They are not operations because they do not generate a sequence by iteration. How to go on does not simply depend on the previous members but on a comparison between the last member and some condition. For example, at each stage in the development of the decimal expansion of $\sqrt{2}$, one must consider whether squaring the last member is greater or smaller than 2. This method is incompatible with Wittgenstein's purely syntactic foundation of mathematical properties. In his program, any sequence must be definable by an operation that determines nothing but the syntax of the members of the sequence. Only in this way is the property constituting the sequence reduced to an internal property of forms that can be identified by the symbolic features of the members of the series.

Wittgenstein's well known rejection of "arithmetical experiments" is based on his requirement to define sequences by syntactic means alone, PR § 190:

> In this context we keep coming up against something that could be called an "arithmetical experiment". Admittedly the data determine the result, but I can't see *in what way* they determine it (cf., e.g., the occurrences of 7 in π.) The primes likewise come out from the method for looking for them, as the results of an experiment. To be sure, I can convince myself that 7 is a prime, but I can't see the connection between it and the condition it satisfies. — I have only found the number, not generated it.
>
> I look for it, but I don't generate it. I can certainly see a law in the rule which tells me how to find the primes, but not in the numbers

> that result. And so it is unlike the case $+\frac{1}{1!}$, $-\frac{1}{3!}$, $+\frac{1}{5!}$ etc., where I can see a law *in the numbers.*
>
> I must be able to write down a part of the series, in such a way that you can *recognize* the law.
>
> That is to say, no *description* is to occur in what is written down, everything must be represented.
>
> The approximations must themselves form what is *manifestly* a series.
>
> That is, the approximations themselves must obey a law.

The series of primes is Wittgenstein's paradigm of a series that cannot be generated by an operation. Although operations are available to generate an infinite series of primes, no operation is known to generate the primes in a certain order that ensures that *all* primes are enumerated. In his detailed discussions of primes in other places, Wittgenstein draws the consequence that we still lack of a clear concept of "the" primes. All we have is a concept of what "a" prime is, which allows us to decide whether a given number is prime or not (PR § 159, 161, cf. (Lampert, 2008)). For the same reason, he rejects the definition of a real number P as the dual fraction with $a_n = 1$ if n is prime and $a_n = 0$ otherwise (cf. PG, p. 475). This definition does satisfy the definition of real numbers by Cauchy sequences, but it does not satisfy Wittgenstein's criterion of being definable by an operation. In the quoted passage, Wittgenstein emphasizes that we do have a method to look for the next prime: we go through the series of natural numbers and decide one by one whether each member satisfies the condition to be divisible only by 1 and itself. However, this method does not satisfy his standard of a definition by operation. As long as we are not able to reduce the property of being a prime to some operation generating the series of primes by iteration, "we can't see the connection" between the members of the series and the condition they satisfy: we cannot "recognize the law" *in the series.* The problem is the same as with the above illustrated methods of approximating $\sqrt{2}$. Instead of generating the next member by iteration, we must decide whether some condition is satisfied or not in order to find the next member.

Wittgenstein's reference to the series of primes as an illustration of arithmetical experiments demonstrates that his concept of operation is not equivalent to that of primitive recursive function. Primes are definable by a primitive recursive function, but not by an operation.[1] Iteration in the case of operations means that the output of the nth application of an operation is

[1] The question in what sense Wittgenstein characterizes real numbers as "laws" is thoroughly discussed in the literature (cf. (Da Silva, 1993), (Frascolla, 1994, pp. 85–92), (Marion, 1998), (Rodych, 1999) and (Redecker, 2006, ch. 5.2)). However, the main reason why the identification of laws with Wittgenstein's notion of operations seemed to

itself the input of the $n+1^{\text{th}}$ application of the very same operation. In contrast, recursion in the case of primitive recursive functions means that the value of a primitive recursive function f for the successor of n, $S(n)$, is defined by referring to the value of the very same function f for n. This does not imply that the values of f are themselves their arguments. This is only true in case of the successor function, which itself is primitive recursive. However, the identity function, e.g., $I(x) = x$, and the zero function, $Z(x) = 0$, on which the definition of primitive recursive functions are based, are *functions* and do not define a series by iterative application. The same holds for primitive recursive characteristic functions. They have the form "$f(x) = 0$ if $\varphi(x)$ and $f(x) = 1$ otherwise." In Wittgenstein's terms, characteristic functions are a paradigm of "descriptions" and not of operations. In contrast, any iteration by applying operations has the form $a_n = \Omega' \vec{a_i}$ where $\vec{a_i}$ stands for members previous to a_n. For example, the series of Fibonacci numbers is defined by $a_n = a_{n-2} + a_{n-1}$. Recursion in the case of primitive recursive functions is part of a strategy of defining primitive recursive functions, whereas operations are not *defined* by iteration but *applied* iteratively. They are defined by some purely syntactic variation that generates a formal series of systematically varied members if iteratively applied. In the case of Fibonacci numbers, this operation consists of adding the last two members. Starting from 0 and 1, this generates the series 0, 1, $0+1$, $1+(0+1)$, $(0+1)+(1+(0+1))$, $(1+(0+1))+((0+1)+(1+(0+1)))$ etc.[2]

If not even primitive recursive functions satisfy Wittgenstein's standards of a purely syntactic foundation of mathematics, this causes doubts whether his programme is realizable at all. Likewise, his rejection of arithmetical experiments and his claim to "recognize the law" in the series has caused trouble. The decimal sequences of irrationals do not satisfy Wittgenstein's demand for sequences that manifestly obey a law. Do not irrationals contradict Wittgenstein's claim from their very nature? Thus, it seems unclear how Wittgenstein's point of view can even do justice to such basic irrational numbers as π and $\sqrt{2}$ (cf., e.g., (Redecker, 2006, p. 212)).

However, these problems only arise if one overlooks the fact that the possibility of definitions by operations depends on the mode of representation. In case of irrationals, the syntactic features of the decimal system are responsible for their "lawless" representation. However, this kind of representation is not essential; it obscures their lawful nature instead of revealing it. In MS 107 p. 91, Wittgenstein writes (translated by T. L.):

be insufficient to most commentators is that *operations* in Wittgenstein's sense were not distinguished sharply from the notion of primitive recursive *functions*.

[2] Brackets are merely introduced to identify a_{n-2} and a_{n-1}.

The procedure of extracting $\sqrt{2}$ *in the decimal system*, e.g., is an arithmetical experiment, too. However, this only means that this procedure is not completely essential to $\sqrt{2}$ and a representation must exist that makes the law recognizable.

To see the connection between the members of a sequence representing a real number and the condition or property that these members satisfy, one must refer to an equivalence transformation that reduces this property to an internal property of forms. There is no equivalence transformation between $\sqrt{2}$ and a decimal number. This already shows that it is impossible to represent $\sqrt{2}$ by the decimal system; whatever decimal number one generates, it cannot be identical with $\sqrt{2}$ — referring to "infinite extensions" is just another expression of this deficiency. However, using the representation by continued fractions, it is possible to represent $\sqrt{2}$ by an operation, cf. MS 107, p. 126 (translated by T. L., cf. MS 107, p. 99):

> [...] in $\frac{1}{2}$, $\frac{1}{2+\frac{1}{2}}$, $\frac{1}{2+\frac{1}{2+\frac{1}{2}}}$ etc. one can recognize the law one cannot recognize in the decimal development.

The connection between the property of $\sqrt{2}$ as "the number that multiplied with itself is identical with 2" and its definition by its continued fraction is due to equivalence transformation:

$$
\begin{array}{ll}
x^2 = 2 & \mid \sqrt{\ } \\
x = \sqrt{2} & \mid a = 1 + (a-1) \\
x = 1 + (\sqrt{2} - 1) & \mid a = \frac{1}{\frac{1}{a}} \\
x = 1 + \frac{1}{\frac{1}{\sqrt{2}-1}} & \mid \frac{1}{a-b} = \frac{a+b}{a^2-b^2} \\
x = 1 + \frac{1}{\frac{\sqrt{2}+1}{\sqrt{2}^2-1^2}} & \mid a = \frac{a}{\sqrt{2}^2-1^2} \\
x = 1 + \frac{1}{\sqrt{2}+1} & \mid x = \sqrt{2},\ a + b = b + a \\
x = 1 + \frac{1}{1+x} & \mid -1 \\
x - 1 = \frac{1}{1+x} & \mid 1 + x = 2 + (x-1) \\
x - 1 = \frac{1}{2+(x-1)} &
\end{array}
$$

Thus, $\sqrt{2}-1$ is representable by the operation $\frac{1}{2+(x-1)}$. Starting with $1-1$ for $x-1$, the iterative application of this operation yields the series $\frac{1}{2+(1-1)}$, $\frac{1}{2+\frac{1}{2+(1-1)}}$, $\frac{1}{2+\frac{1}{2+\frac{1}{2+(1-1)}}}$ etc. This is identical to the series Wittgenstein mentions if one eliminates $+(1-1)$ by an equivalence transformation. In the short notation of regular periodic continued fractions, $\sqrt{2}$ is definable by $[1;\overline{2}]$. A continued fraction of a real number is periodic if and only if the real number is a quadratic irrational (theorem of Lagrange). The notation of continued fraction identifies a common property of quadratic irrationals by a common syntactic feature, and thus shows that this property is an internal property. Other irrational numbers are representable by regular continued fractions that are not periodic but still definable by operations, such as the Euler number $e : [2; 1, 2, 1, 1, 4, 1, 1, 6, 1, 1, 8, \ldots]$. Another type of irrational numbers are not definable by operations within regular continued fractions but within irregular continued fractions such as $\frac{4}{\pi} = 1 + \cfrac{1^2}{2 + \cfrac{3^2}{2 + \cfrac{5^2}{\ddots}}}$. Furthermore, the continued fraction representation for a number is finite if and only if the number is rational. This shows that this mode of representation reveals by its syntactic properties internal properties of numbers that are not identified by the decimal number system. We learn more about "the laws of numbers", their internal structure, by representing them in the notation of continued fractions.

Mathematical proofs reveal this internal structure by equivalence transformations. Consider, for example, the golden ratio. Its representation as a decimal number does not show its exceptional nature. However, by an equivalence transformation resulting in an operation defining a continued fraction, internal properties of the golden ratio are identified by the syntactic features of this adequate representation. This procedure reduces the property that the ratio of two quantities a and b is identical to the ratio of the sum of them to the larger quantity a to an operation:

$$\phi = \frac{a}{b} = \frac{a+b}{a} = 1 + \frac{b}{a} = 1 + \frac{1}{\phi}. \tag{1}$$

By the operation $1 + \frac{1}{\phi}$, the periodic, regular continued fraction $[1; \overline{1}]$ is defined. By this representation it is proven that the golden ratio is "the most irrational and the most noble number," because these properties are identified by the lowest possible numbers in an infinite regular continued fraction. Furthermore, by this representation it is proven that the ratio of two neighboured Fibonacci numbers converges to the golden ratio. For the Fibonacci numbers are defined by $a_{n+1} = a_n + a_{n-1}$. Thus, with $a = a_n$ and $b = a_{n-1}$ we yield equation (1). The syntax of continued fractions provides symbolic connections that prove certain internal relations between numbers.

The continued fraction representation of any irrational number is unique. Thus, any definition of a real number by an operation (or "induction") defining a continued fraction satisfies Wittgenstein's criterion for representing a real number, MS 107 p. 89 (translation T. L.):

> I want a representation of the real number that reveals the number in an induction such that I have herewith the only proper, unique symbol.

It is by this property of uniqueness that the symbolic representation of irrationals by continued fractions serves as an identity criterion, which allows one to compare irrationals and rational numbers. The principle is the same as in the case of comparing fractions by converting them to fractions with identical denominators. The problem of deciding the identity of numbers results from a deficiency in their representation, allowing for ambiguity.

This does not mean that there must be one and only one proper notation for numbers. Nor does it mean that continued fractions are "the" proper notation of real numbers. Different internal properties of numbers, and herewith different types of numbers, may be identified by different systems of representation. And different types of numbers may be comparable within different modes of representation (cf. MS 107, p. 123). Natural numbers can be compared according to the conventions of the decimal system, fractions are comparable by converting them to fractions with identical denominator, rational numbers and quadratic irrationals are comparable by regular continued fractions etc. Furthermore, new proofs consist of making new symbolic connections. They invent new possibilities of comparing numbers and of revealing their internal relations. Not all internal relations of a number to other numbers must be revealed within only one notational system. For example, instead of representing π by an irregular continued fraction (as quoted above), $\frac{2}{\pi}$ can also be represented by $\frac{\sqrt{2}}{2} \cdot \frac{\sqrt{2+\sqrt{2}}}{2} \cdot \frac{\sqrt{2+\sqrt{2+\sqrt{2}}}}{2} \cdot \ldots$ or $\frac{\pi}{2}$ by $\frac{2}{1} \cdot \frac{2}{3} \cdot \frac{4}{3} \cdot \frac{4}{5} \cdot \frac{6}{5} \cdot \frac{6}{7} \cdot \frac{8}{7} \cdot \frac{8}{9} \cdot \ldots$. The internal properties of different numbers may call for operations referring to different modes of representation. There need not be a "system of irrational numbers" in the sense as there is a "system of natural numbers" or a "system of rational numbers" (cf. PG, p. 479, RFM, app. 3, § 33). As we have seen, only quadratic irrationals are definable by periodic, regular continued fractions, and another type of irrationals is not even definable by regular continued fractions. Different types of irrationals are definable by different kinds of operations within different modes of representation.

According to Wittgenstein's intensional point of view, our mathematical comprehension and knowledge depends on the syntax of mathematical representation. This is not due to psychological reasons. Instead, this is because mathematical proofs make symbolic connections between different modes of representation, and because the solvability of mathematical problems depends on imposing adequate notations. Instead of concluding from a specific, deficient mode of representation the lawless nature of irrational numbers, which makes it impossible to decide upon their identity and which invokes misconceptions such as "infinite extensions", one should look for adequate representations that reveal their lawful nature and make it possible to decide upon their identity. This is done by reducing their properties to operations instead of conceptualizing them in terms of functions. If such a reduction is not available, this means that one does not have a full understanding of the properties in question. We can then only refer to a vague understanding expressed within a deficient, descriptive symbolism. Only by imposing an adequate expression that depicts those properties by its syntactic features, can we be sure that those properties are properly defined.

This approach is in conflict with basic impossibility results of modern mathematical logic, such as the non-enumerability of the irrationals, the undecidability of first-order logic or the incompleteness of logical axiomatizations of arithmetics. This does not mean that Wittgenstein's point of view implies that these results are false in the sense that their negation is true. Instead, his intensional view implies that it does not make sense to speak of "the irrationals" unless an operation is known that allows us to generate them by iteration (and thus to enumerate "the irrationals"). This, of course, does not mean that he claims that such an operation is or must be available. Likewise, his intensional view implies that one cannot speak of decidability or provability in an absolute sense, such that one can say in advance that certain properties of formulae of a certain syntax are not decidable or provable, independent of the syntactic manipulations that might be invented to identify those properties. According to Wittgenstein "being a tautology" ("being true in all interpretations") or "being a theorem" of first order logic is not defined properly unless some sort of equivalence procedure is invented that converts first order formulae to an adequate representation that identifies their logical properties by its syntactic properties. From this point of view, it cannot be said that it is impossible to define such procedures, because the properties in question that are said to be undecidable or unprovable are not represented properly unless such procedures are available. Likewise, from Wittgenstein's point of view the incompleteness of axiomatic systems of arithmetic means in the first place that those systems do not properly represent the properties in question. It does not mean that we know that a certain property holds, but its formal representation is not derivable. Instead, it means that we have a deficient understanding of that property expressed by an inadequate representation. In the following, we will show that this conflict between Wittgenstein's point of view and the impossibility results can all be traced back to his rejection of "descriptions" in terms of characteristic functions as adequate forms to represent real numbers.

3 Pseudo-Irrationals

Wittgenstein illustrates his point of view by providing several definitions of pseudo-irrationals. These are definitions of irrationals in terms of Cauchy sequences. However, contrary to $\sqrt{2}$ or π no reductions to operations of these definitions are available. Thus, according to Wittgenstein there are no irrationals corresponding to those definitions. Besides the above mentioned definition of P as the dual fraction $0.a_1a_2\ldots$ with $a_n = 1$ if n is prime and $a_n = 0$ otherwise, Wittgenstein discusses the following definitions (cf. PG, p. 475):

π': The decimal number $a_1.a_2a_3\ldots$ with $a_na_{n+1}a_{n+2} = 000$ if $a_na_{n+1}a_{n+2} = 777$ in π; otherwise $a_n = a_n$ of π.

F: The dual fraction $0.a_1a_2a_3\ldots$ with $a_n = 1$ if $x^n + y^n = z^n$ is solvable for n $(1 \geq x, y, z \geq 100)$; otherwise $a_n = 0$.

All these definitions are intended to define an irrational number by a characteristic function. In this case, the dots "…" refer to an "infinite extension". Thus, they are ill-defined according to Wittgenstein's standards. They do not identify a number but describe an arithmetical experiment. Wittgenstein emphasizes that even if the characteristic functions become reducible to operations, this does not mean that this shows that the definitions in fact define irrational numbers. Instead, it means that vague definitions that do not identify numbers are replaced with exact definitions that are able to identify numbers. He, for example, considers the situation when Fermat's theorem is proven. Due to his rejection of descriptions, he does not analyse this situation in terms of coming to know the number F that before was only described. Instead, the proof allows one to replace the pseudo-definition of F, which does not identify a number (neither a rational nor an irrational one), with $F = 0.11$, which is a rational number (PG, p. 480). Before, it was not decidable whether "F" denotes a number such that $F = 0.11$ or not; the definition by description simply did not define rules to do this. This demonstrates the lack of meaning that is given to "F" by the previous definition. The proof, if it is valid, makes connections to other parts of mathematics that were not recognized before and thus gives "F" a clear meaning.

Cantor's proof of the non-enumerability of irrational numbers is based on defining a diagonal number by a characteristic function. Given some enumeration of dual fractions between 0 and 1, the proof of the non-enumerability of "all" of them is based upon the following diagonal number $\mathcal{D}$:

$\mathcal{D}$: The dual fraction $0.a_1a_2\ldots$ with $a_n = 0$ if the n'^{th} digit of the n'^{th} dual fraction is 1; otherwise $a_n = 1$.

To this definition, the same objections apply as to the definitions of P, π' or F: It is a definition by description in terms of a characteristic function. It describes an arithmetical experiment and does not identify a number, which can only be done by an operation. However, such an operation is not available. Thus, it is not meaningful to say that $\mathcal{D}$ is an "irrational number" not occurring in the assumed enumeration of irrationals. This, of course, does not mean that Wittgenstein claims that "the irrationals" are enumerable. Instead, he objects to identifying irrational numbers by non-periodic, infinite decimal or dual fractions. This criterion does not say anything about a certain type of numbers; it only says something about the deficiency of the decimal notation (PG, p. 474). This notation cannot serve as the unique notation for real numbers, as it does not make it possible to decide upon the identity of numbers. Likewise, Wittgenstein objects to the picture of a real number as a "point" on the "line" of real numbers. These items are elements of the extensional view. They arise from treating "is an irrational number" as well as "is a rational number" or "is a natural number" as concepts (propositional functions) identifying certain sets of numbers. This makes it possible to ask about the "cardinality" of those sets. This, in turn, allows one

(i) to use "infinite" as a number word and speak of "the infinite number" of objects satisfying some concept, and

(ii) to compare the cardinality of sets by coordinating their elements.

Finally, from this and the method of diagonalization one comes to speak of sets with a cardinality greater than that of the set of natural numbers. First and foremost, Wittgenstein's criticism is that this conceptual machinery is rather an expression of the extensional view than a description of the nature of numbers (RFM, app. 3, § 19). He cuts the roots of (transfinite) set theory by conceptualizing "types of numbers" in terms of "systems" instead of "sets". According to his intensional point of view, the criterion to identify a type of number is the possibility to generate them by an operation. As this implies their enumerability in terms of the iterative application of an operation, it does not make sense to speak of types of numbers that are not enumerable.

According to Church's thesis, the concept of decidability is representable by a primitive recursive characteristic function. Thus, on the basis of an enumeration of first-order logic formulae by their Gödel numbers, the property of being a theorem (or a tautology) is representable by the following number:

T: The dual fraction $0.a_1a_2\ldots$ with $a_n = 0$ if $\vdash \varphi_n$ (or $\models \varphi_n$); and $a_n = 1$ otherwise.

On the basis of diagonalization, undecidability proofs demonstrate that characteristic functions such as the one defining T cannot be primitive recursive. From Wittgenstein's point of view, these proofs are based upon a confusion of material and formal properties. As a formal property, theoremhood (or being a tautology) is not representable by a characteristic function. Instead, these properties are only represented adequately by a shared syntactic property in an ideal notation. This is illustrated by the representation of tautologies via truth tables or disjunctive normal forms of propositional logic as well as by means of Venn diagrams in monadic first order logic. Wittgenstein's conception calls for equivalence transformations to identify the truth conditions of logical formulae by means of syntactic properties of their proper representation. This conception differs from the traditional semantics of first-order logic. Presuming an endless enumeration of interpretations $\Im_1, \Im_2, \ldots$, each being either a model or a counter-model of a formula A, one might represent the truth conditions of A according to these interpretations by the following number:

$\theta(A)$: The dual fraction $0.a_1a_2\ldots$ with $a_n = 0$ if $\Im_n \models A$ and $a_n = 1$ otherwise.

On the contrary, Wittgenstein's approach calls for a representation of the truth conditions of a formula A that allows one to identify the truth conditions of A without deciding whether single interpretations are models or counter-models of A. Furthermore, the proper representation of first order formulae should reveal the internal relations of non-equivalent logical formulae by making it possible to generate the system of truth conditions by operations. To have an idea of what Wittgenstein envisages, one might think of a systematic generation of reduced disjunctive normal forms of the Quine–McCluskey algorithm,[3] that represent all possible truth functions of propositional logic. Likewise, the task of first order logic is to define analogous disjunctive normal forms and procedures for their unique reduction within first order logic. To claim that this is impossible presumes the extensional view that is rejected by Wittgenstein's endeavour.

Likewise, Gödel represents "x is a proof of y" by a primitive recursive function xBy in definition 45 of his incompleteness proof (cf. (Gödel, 1931, p. 358)). On this basis, he expresses "x is provable" by $\exists y yBx$ in definition 46. This is incompatible with Wittgenstein's claim that the internal relation of being provable (derivable) should be defined by operations instead of propositional functions. This, in turn, presumes a proof procedure in term

[3] Note that the reduced disjunctive normal forms of the Quine–McCluskey algorithm are unique; any equivalent propositional formula is represented by the same reduced disjunctive normal form. Ambiguity only comes into play in the second step of the Quine–McCluskey algorithm that intends to minimize reduced disjunctive normal forms.

of equivalence transformations to an adequate symbolism that makes such a definition possible, instead of a proof procedure in terms of logical derivations from axioms. The lack of such a definition means a deficiency in the syntactic representation of the formulae in question. According to Wittgenstein's point of view, the conclusion that must be drawn from Gödel's incompleteness proof is to look for a formal representation of arithmetic that is not based upon the concept of propositional function, which is at the heart of any logical formalization.

Wittgenstein's intensional reconstruction of mathematics is not meant to be a "refutation" of the extensional view of modern mathematical logic. Instead, first and foremost it intends to propose a decisive alternative conceptualization of mathematics that radically differs in its foundations. According to him, the fruit of this endeavour should be a clarification of the philosophical problems of modern mathematics that will have the same influence on the increase of mathematics as sunshine has on the growth of potato shoots (PG, p. 381).

Timm Lampert
University of Berne
`timm.lampert@philo.unibe.ch`

Abbreviations

CL Wittgenstein, L. (1997). *Cambridge Letters*, Oxford: Blackwell.

MN Wittgenstein, L. (1979). Notes dictated to G. E. Moore in Norway, in: *Notebooks 1914–1916*, pp. 108–119. Oxford: Blackwell.

MS 107 Wittgenstein, L. (2000) Manuscript 107 according to von Wright's catalogue. Published in *Wittgenstein's Nachlass: the Bergen Electronic Edition*. London: Oxford University Press.

NL Wittgenstein, L. (1979). Notes on Logic, in: *Notebooks 1914–1916*, pp. 93–107. Oxford: Blackwell.

PG Wittgenstein, L. (1974). *Philosophical Grammar*, Blackwell: London 1974.

PR Wittgenstein, L. (1975) *Philosophical Remarks*, Oxford: Blackwell.

RFM Wittgenstein, L. (1956) *Remarks on the Foundations of Mathematics*, Oxford: Blackwell.

TLP Wittgenstein, L. (1994). *Tractatus Logico-Philosophicus*. London: Routledge.

WVC Waismann, F. (1979) *Wittgenstein and the Vienna Circle*, Oxford: Blackwell.

References

Da Silva, J. J. (1993). Wittgenstein on irrational numbers. In K. Puhl (Ed.), *Wittgenstein's philosophy of mathematics* (pp. 93–99). Vienna: Hölder–Pichler–Tempsky.

Frascolla, P. (1994). *Wittgenstein's philosophy of mathematics.* Routledge: London.

Gödel, K. (1931). Über formal unentscheidbare Sätze der Principia Mathematica und verwandter Systeme. *Monatshefte für Mathematik und Physik*, *38*, 173–198.

Lampert, T. (2008). Wittgenstein on the infinity of primes. *Journal for the History and Philosophy of Logic*, *29*, 63–81.

Marion, M. (1998). Wittgenstein and finitism. *Synthese*, *105*, 141–176.

Redecker, C. (2006). *Wittgensteins Philosophie der Mathematik.* Frankfurt: Ontos.

Rodych, V. (1999). Wittgenstein on irrationals and algorithmic decidability. *Synthese*, *118*, 279–304.

What is the Definition of 'Logical Constant'?

Rosen Lutskanov*

The design of this paper is to motivate and introduce informally a definition of the notion of 'logical constant' which does not presuppose the analytic/synthetic distinction. To this end, I'm going to

1. explore the origin of this notion;
2. show why it is important to define it;
3. review some paradigmatic (but ostensibly unsatisfactory) alleged definitions;
4. hint at the true relation between the notions of 'logical constant' and 'analyticity';
5. make manifest the implicit rendering of analyticity which is nested in the classical definitions of logical constants;
6. discuss the strictly alternative construal of analyticity prominent in present-day philosophy of logic;
7. provide sketchy definition of logical constants that deviates from the first but remains true to the second.

1

It was Bolzano, who in the distant 1837 was probably the first to suggest that there are concepts belonging to logic alone: according to his own example, the fact that the question "whether coriander improves one's memory" obviously does not concern logic at all, suggests that it is not about coriander but studies something different (Hodges, 2006, p. 42). In his own view, the logic's subject matter is exhausted by the 'logical ideas' which affect the

*I would like to express my deep gratitute to the organizers of LOGICA 2008 for the grant they have awarded me for participation in the conference.

logical form of propositions (or 'sentences in themselves') and are to be singled out by the condition that their variance modifies the truth-value of any expression containing them (Siebel, 2002, p. 590). Then came Frege who in his "Begriffsschrift" (1879) had a clear-cut division of symbols, employed in formal languages, into two kinds: 'those that can be taken to mean various things' (variable arguments) and 'those that have a fully determinate sense' (constant functions). As far as logic is concerned, the second kind of symbols corresponds to Bolzano's 'logical ideas' because it represents those parts of the formal expression that have to remain invariant under replacement (Frege, 1960, p. 13). Finally our story reaches Russell, who in 1903 introduced the now familiar term 'logical constant' as substitute for Bolzano's 'logical ideas' and Frege's 'logical functions'. According to him these are the notions accountable for the truth of all propositions which we view as a priori justified. But he did not provide a formal characterization, only the following deliberately confusing explanation: "logical constants are all notions definable in terms of the following: Implication, the relation of a term to a class of which it is a member, the notion of such that, the notion of relation, and such further notions as may be involved in the general notion of propositions of the above form" (Russell, 1903, p. 3). The reason for such striking obscurity is the fact that he thought that "logical constants themselves are to be defined only by enumeration, for they are so fundamental that all the properties by which the class of them might be defined presuppose some terms of the class" (Russell, 1903, pp. 8–9).

2

Later Russell's invention survived the demise of his logicism, although its introduction was initially motivated as part of the attempt to show that all mathematical notions are reducible to the notions of logic (exemplified by the 'logical constants'). Today we are generally inclined to claim that "logical concept is what can be expressed by a logical constant in a language" hence the question "What is logic?" is to be answered by answering the question "What is a logical constant?" (Hodes, 2004, p. 134). On the other hand, we just cannot afford treating the notion of logical constant as indefinable as Russell did, since presently we have at our disposal alternative lists of logical constants imposing on us different conceptions about the subject matter of logic. Famously, Quine did his best to expel Russell's "relation of a term to a class of which it is a member" from the list of logical constants, claiming that the theory of the '$\in$'-relation is not logic but "set theory in sheep's clothing" (Quine, 2006, p. 66). This excommunication of set-membership from the province of logic is the sole difference between Quine's nominalistic preference for first-order logic and Russell's ontologically exuberant type theory. In the face of this manifest discrepancy, we have to admit that the only way to provide a motivated choice of logical framework is to exhibit justified definition of the term 'logical constant' and to show how our conception of logic stems out of it. The age-old 'laundry list' comprising the venerable members of the family of logical notions is not enough. So, which definitions of 'logical constant' are currently in circulation?

3

Luckily, we have plenty of answers of this toilsome question; regrettably, none of them fared very well. The first attempt to provide rigorous definition was provided by Carnap in his "Logical Syntax of Language" which was subsequently simplified by Tarski. His definition was founded on the concepts of 'premiss-class' and 'range' (Spielraum): two premiss-classes were said to be 'equipollent' if each of them is consequence of the other and the range was defined as class of premiss-classes M with the property that each premiss-class which is equipollent to a premiss-class belonging to M also belongs to M. Then Carnap explained that the range M of a proposition p represents "the class of all possible cases in which p is true" or "the domain of all possibilities left open by p" (Carnap, 1959, p. 199). In this setting, it seems natural to define the 'logical junctions' as simple set-theoretical operations on ranges: by 'supplementary' range of a given range M_1 we mean a range M_2 comprising those premiss-classes that don't belong to M_1; then for a proposition p_1 with range M_1 we can define its 'negation' p_2 as the proposition whose range coincides with the supplementary range of p_1. In the same vein, we can define the 'disjunction' of two propositions p_1 and p_2 as another proposition p_3 whose range is the union of the ranges of p_1 and p_2 (Carnap, 1959, p. 200). A year later Tarski showed that the definition can be simplified by substituting 'content' for 'range' (the content of p is the class of all non-analytic consequences of p): then p_2 is negation of p_1 iff they have exclusive contents and p_3 is disjunction of p_1 and p_2 iff its content is product of the contents of p_1 and p_2 (Carnap, 1959, p. 204). These attempted definitions of Carnap and Tarski were not conceived as satisfactory, probably because they founded the conceptual apparatus of logic on the conceptual apparatus of set theory. This is not an epistemologically flawless move: the operation of sentence negation seems more familiar than the intricate operation of class complementation; that is why the first is not to be defined by means of the second.

May be this is the reason why later Tarski took another course. In his famous lecture "What are logical notions" (1966) he proposed the now classical definition: logical are just these notions which are invariant under all permutations of the universe of individuals onto itself. This definition provoked severe criticisms because it treats as logical properties all cardinality features of the domain of discourse. Another painful defect was exposed by McGee who defined an operation of 'wombat disjunction' ($\cup_W$) such that '$p \cup_W q$' is true if '$p \vee q$' is true and there are wombats (there is an element of the domain of the model which satisfies the predicate 'is wombat') and false otherwise (Feferman, 1997, p. 9–10). Clearly, wombat disjunction is invariant under arbitrary permutations, but it is hard to admit that it is logical notion — in order to establish the truth or falsity of any proposition containing essential occurrences of wombat disjunction we need to corroborate a specific empirical assumption concerning the existence (or non-existence) of wombats. There are several well-known attempts to rectify Tarski's definition by replacing 'invariance under arbitrary permutations' with 'invariance under arbitrary bijections' (Mostowski, 1957), 'rigid invariance under arbitrary bijections' (McCarthy, 1981), and 'invariance under arbitrary homomorphisms' (Feferman, 1997). As far as we know, no one of these attempts is able to discriminate properly between the logical and the empirical (Mostowski's criterion qualifies 'unicorn' as logical notion) or the logical and the mathematical (Feferman's criterion renders 'there exist infinitely many' as belonging to logic). That is why, we can recapitulate this part of the discussion by noticing that "it seems inevitable to conclude that these proposals inspired by Tarski... do not even meet the minimal requirement of extensional adequacy" (Gomez-Torrente, 2002, p. 20).

A third variant for definition of the notion of logical constant stems from the works of Gentzen. His followers were inclined to claim that logical constants are to be identified solely by the introduction and elimination rules governing their inferential uses. Conjunction, for example, is nothing but this part of our lexicon that features in inferences like

$$\frac{A,\, B}{A \wedge B} \quad \text{and} \quad \frac{A \wedge B}{A,\, B}.$$

This bright idea was shattered by Prior, who provided his infamous tonk-counterexample dealing with a new particle 'tonk' governed by the following rules:

$$\frac{A}{A\,\text{tonk}\,B}\ (\text{tonk-Int}) \qquad \text{and} \qquad \frac{A\,\text{tonk}\,B}{B}\ (\text{tonk-Elim}).$$

The introduction of 'tonk' allows showing the formal language in question to be inconsistent: just substitute '$\neg A$' for 'B' and apply successively (tonk-Int) and (tonk-Elim). This was intended to mean that not any set of introduction and elimination rules defines a logical constant: something more had to be added. The mysterious additional ingredient was later identified as 'conservativity' (Belnap) or 'harmony' (Dummett). In Dummett's own explanation, "Let us call any part of a deductive inference where, for some logical constant c, a c-introduction rule is followed immediately by a c-elimination rule a 'local peak for c'. Then it is a requirement, for harmony to obtain between the introduction rules and elimination rules for c, that the local peak for c be capable of being leveled, that is, that there is a deductive path from the premises of the introduction rule to the conclusion of the elimination rule without invoking the rules governing the constant c" (Dummett, 1991, p. 248). As Dummett himself readily acknowledged, "The conservative extension criterion is not, however, to be applied to more than a single logical constant at a time. If we so apply it, we allow for the prior existence, in the practice of using the language, of deductive inference, since there are a number of logical constants" but "the addition of just one logical constant to a language devoid of them... cannot yield a conservative extension" since "if deductive inference is ever to be said to be able to increase our knowledge, then it must sometimes enable to recognize as true a statement that we should not, without its use, been able so to recognize" (Dummett, 1991, p. 220). This difficulty seems insurmountable: we can use the leveling of local peaks technique to identify a single particle as logical constant, but it is not possible to rely on the same strategy to delineate the realm of logical notions.

4

Up to this point we have reviewed three paradigmatic attempts to provide definition of the notion of logical constant. It appears that none of them is materially adequate:

(i) Carnap's set-theoretic approach construed logical notions using precise mathematical methods but did not even pose the question which operations on premiss-classes are to be viewed as belonging to logic;

(ii) Tarski's model-theoretic approach could not single out the class of logical constants and experienced serious difficulties with borderline cases such as non-existent objects and mathematical entities;

(iii) Dummett's proof-theoretic approach provided justified criteria for logicality of single connectives ('intrinsic harmony') but couldn't achieve generally applicable standard (for 'total harmony').

But the material inadequacy is not the sole or even the gravest shortcoming of these purported definitions. They all were devised with an eye on the notions currently recognized as 'logical' but were not couched in a broad theoretical framework, clarifying their interplay with some particular rendering of the notion of analyticity. If we turn back we shall see that the advent of logical constants was necessitated by the fact that the 'analytic program' (the attempt to identify 'logical truth' and 'analytical truth') was essential part of the 'logicist program': a truth is analytic if it can be reduced to general logical laws and definitions. In a nutshell this reduction establishes that only logical constants occur essentially in it and the logical constants were driven out on stage simply to provide a touchstone for termination of this reductive procedure. That is why, "The question 'What is a logical constant?' would be unimportant were it not for the analytic program" (Hacking, 1994, p. 3). Now we are able to perceive where the real problem lies: on the one hand, when we try to define logical constants and do logic, we silently presuppose that it is possible to discriminate rigorously between analytically true (true by virtue of linguistic conventions) and synthetically true (true by virtue of brute matters of fact); on the other hand, when we try to make sense of what we are doing and do philosophy of logic, we overtly blur the analytic/synthetic distinction. In the following two paragraphs I'll do my best to explain why this double-mindedness is so crucial in the present context.

5

When we do mathematical logic, we invariably and unwittingly stick to the 'Viennese' orthodoxy. The way formal languages are presented and logical symbols are employed was modeled upon the paradigm of Wittgenstein's Tractatus. Let us remember that his "fundamental idea" was that while all other words stand for objects, "the logical constants are not representatives" (Wittgenstein, 1963, prop. 4.0312). This conception was the sole basis of the idea that the 'real' propositions are empirically contentful 'pictures of reality', while the propositions of logic are representationally idle 'tautologies' (Wittgenstein, 1963, prop. 4.462). Carnap rehearsed the same line of thought in his works on formal semantics: he started with the suggestion that "we must distinguish between descriptive signs and logical signs which do not themselves refer to anything in the world of objects, but serve in sentences about empirical objects" (Carnap, 1958, p. 6) and concluded that it is possible to classify any sentence as 'L-sentence' (that is, 'logical' = 'analytic' = 'true or false on logical grounds') or 'F-sentence' ('factual' = 'synthetic' = 'true or false by virtue of facts of the world'). Although developed in different setting, Tarski's model-theoretic approach to formal semantics reiterates the same steps which are mirrored in the two types of clauses in his recursive truth definition: on the one side, we have a base clause introducing a valuation function that assigns truth-values to atomic sentences in the model (here sentences receive truth-values on extra-logical reasons; if we have in mind some particular interpretation of the language we can say that they are 'true or false by virtue of facts of the world'); on the other side, we have recursive clause which determines in what way the truth-values of complex sentences built from atomic ones and logical constants depend on the truth-values already assigned to atomic sentences (here sentences receive truth-values on intra-logical reasons, the definitional sub-clauses for the particular logical connectives are analytically true linguistic conventions fixing the meaning of logical vocabulary). Finally, if we take a look at the rival proof-theoretic approach championed by Dummett, we would see the same pattern. The insistence on 'conservativity' in dealing with introduction and elimination rules for logical constants could be motivated only by the idea of the purely tautologous character of logically valid inferences. The local peaks should be in principle 'levelable', precisely because the manipulation with logical vocabulary adds no substantive new information about the world — in short, because logic is analytic and has nothing to do with sentences, true by virtue of facts of the world.

6

When we do philosophy of logic, we are often said completely different things, incompatible with the idea that logical truth (conceived as a paradigmatic case of analyticity) is to be demarcated from factual truth. Starting with Wittgenstein again, we see that all his later development — from "Some Remarks on Logical Form" where he admits that "we can only arrive at a correct analysis by what might be called, the logical investigation of the phenomena themselves" (Wittgenstein, 1993b, p. 30) to "On Certainty" where he denied the possibility to distinguish from the outset logical from empirical propositions because "the river-bed of thoughts may shift" (Wittgenstein, 1993a, p. 15) — can be seen as rejection of the previous sharp division of all locutions into vacuously true 'tautologies' and meaningful 'pictures of reality'. Tarski himself, as early as 1930, was committed to the same line of thought: in a note of Carnap's diary, dated February 22, 1930 we read: "8–11 with Tarski at a Cafe. About monomorphism, tautology, he will not grant that it says nothing about the world; he claims that between tautological and empirical statements there is only a mere gradual and subjective distinction" (Mancosu, 2005, pp. 328–329). Several years later, in "On the concept of following logically", we read: "At the foundation of our whole construction lies the division of all terms of a language into logical and extra-logical. I know no objective reasons which would allow one to draw a precise dividing line between the two categories of terms... the division of terms into logical and extra-logical exerts an essential influence on the definition also of such terms as 'analytic' and 'contradictory'; yet the concept of an analytic sentence... to me personally seems rather murky

(Tarski, 2002, pp. 188–189). Still later, in 1944 Tarski confessed in a letter to Morton White that he is inclined to think that "logical and mathematical truths don't differ in their origin from empirical truths — both are results of accumulated experience... [and we have to be prepared to] reject certain logical premises (axioms) of our science in exactly the same circumstances in which I am ready to reject empirical premises (e.g., physical hypotheses)" (White, 1987, p. 31). It would not be strange, if these words sound familiar: the same critiques were formulated by Quine, who met Carnap in Prague in 1933 and forced him to admit the untenability of the analytic/synthetic distinction: "Is there a difference in principle between logical axioms and empirical sentences? He [Quine] thinks not. Perhaps I [Carnap] seek a distinction just for its utility, but it seems he is right: gradual difference: they are sentences we want to hold fast" (Quine, 2004, p. 55). In "Truth by Convention" (Quine, 1936) stressed that some analytically true statements — definitional conventions — can be overthrown for empirical reasons, and in "Two dogmas of empiricism" (1951) introduced the field metaphor that obliterates completely the analytic/synthetic distinction, making evident that it is "folly to seek a boundary between synthetic statements, which hold contingently on experience, and analytic statements, which hold come what may" (Quine, 1961, p. 50). Generally, the destruction of this distinction was effected in Harvard: from the 1940 disputes of Carnap, Tarski and Quine, to White's early "The analytic and the synthetic: an untenable dualism" (1950), Quine's ground-braking "Two dogmas of empiricism" and Goodman's reflective equilibrium theory developed in "The new riddle of induction" (1954).

7

The definitions of logical constants we have discussed were shown to be motivated by the untenable assumption that we are capable of discriminating rigorously between analytic (true by virtue of linguistic conventions) and synthetic (true by virtue of matters of fact) propositions. It seems to me that it is justified to search for a definition of logical constants that conforms to the mainstream philosophy of logic, a fortiori a definition which does not presuppose the analytic/synthetic distinction. Needless to say, everything I can suggest on this topic up to the present moment is sketchy and inconclusive. First of all, I admit that logic is concerned with the codification of inferential practices which are generally 'out there' before we try to impose normative restrictions on them. These practices produce what Brandom calls 'material inferences' — inferences that are not justified with recourse to the features of logical vocabulary but seem as immediately acceptable. Any chain of material inferences can be called an 'argument' — this suggests that in general the material inferences are serially ordered and aim at something — the claim that needs to be established as true or false. Any argument can be modeled naturally in a slight modification of the framework developed in Gupta and Belnap's "Revision Theory of Truth" (Gupta & Belnap, 1993). Let us consider a formal language L and a model M_0 that assigns to some sentences in L the value 'true': these are the 'axioms' (in their ancient interpretation as 'sentences proposed for consideration') that we temporarily accept as true. Then the set of possible material inferences with premises true in M_0 defines a jump-operator correlating with it another model of L (let us designate it as 'M_1') containing all those sentences that have to be accepted as true on the basis of the bootstrap model M_0 (in general, we do not suppose that the jump operator is monotone: some previously accepted sentences can be refuted at later stages). The same procedure can be applied again and again which gives rise to indefinitely extendible series of models M_0, M_1, M_2, etc. which we shall call 'an argument' (whose premises are the axioms, defined by M_0). In the course of any typical argument A there shall be sentences that at some stage of its development (say M_n) receive constant interpretation (these are the fix-points of the jump operator); for those that are evaluated as true in all successive stages (M_{n+1}, M_{n+2}, etc.) we shall say that they are 'rendered stably true' (by the argument A). Now, instead of a single argument, let us consider a bunch of arguments A_1, A_2, etc. and suppose that there is a class of sentences, classified as stably true by any one of them; I propose these sentences to be called 'rendered valid' (by the set of arguments A_1, A_2, etc.). My suggestion is to equate logicality with the just defined concept of 'validity'; in this way we remain fair to some of traditionally recognized distinctive features of logical truth:

(i) it is topic-neutral (because it is not relative to a particular argumentative setting);

(ii) it is necessary (because it inevitably shows up in any train of reasoning belonging to the general argumentative setting);

(iii) it is analytic (because given a valid sentence and a set of argumentative premises we can demonstrate by means of analysis of the accepted material inferences that it is genuine proposition of logic).

Moreover, this division of sentences into analytic (rendered valid) and synthetic (not rendered valid) cannot be drawn from the outset because in general the question "is the sentence s rendered valid by the set of arguments A_1, A_2, etc.?" is not decidable.

After we have secured a workable notion of logicality, we can ask ourselves again: what is a logical constant? The answer is that a lexical unit is to be treated as piece of logical vocabulary when it is invariably interpretable component of some set of valid sentences. Then we can hunt down the logical constants using the 'inverse' logical approach developed by van Benthem who suggested that instead of choosing some predefined set of logical constants and asking what types of inferences are validated by them, we can take some intuitively convincing set of (material) inferences that validate particular propositions and search for the specific constants that are accountable for them. This methodological shift from predefined normative accounts of logicality to purely descriptive explorations of inferential practices was named "Copernican revolution in logic" (Benthem, 1984, p. 451). What I've tried to do here, was to show that the revolution must go on...

Rosen Lutskanov
Institute for Philosophical Research, Bulgarian Academy of Sciences
6 Patriarch Evtimii Blvd., Sofia 1000, Bulgaria
rosen.lutskanov@gmail.com
http://www.philosophybulgaria.org/en/Sekcii/Logika/Sastav.php

References

Benthem, J. v. (1984). Questions about quantifiers. *Journal of Symbolic Logic*, *49*(2), 443–466.

Carnap, R. (1958). *Introduction to symbolic logic and its applications.* New York: Dover Publications.

Carnap, R. (1959). *The logical syntax of language.* Paterson, NJ: Littlefield, Adams and Co.

Dummett, M. (1991). *The logical basis of metaphysics.* Cambridge, MA: Harvard University Press.

Feferman, S. (1997). *Logic, logics, and logicism.* (Retrieved from: http://math.stanford.edu/~feferman/papers.)

Frege, G. (1960). Begriffsschrift. In P. Geach & M. Black (Eds.), *Translations from the philosophical writings of Gottlob Frege* (pp. 1–20). Oxford: Basil Blackwell.

Gomez-Torrente, M. (2002). The problem of logical constants. *Bulletin of Symbolic Logic*, *8*(1), 1–37.

Gupta, A., & Belnap, N. (1993). *The revision theory of truth.* Cambridge, MA: MIT Press.

Hacking, I. (1994). What is logic? In D. Gabbay (Ed.), *What is a logical system?* (pp. 1–34). Oxford: Clarendon Press.

Hodes, H. (2004). On the sense and reference of a logical constant. *The Philosophical Quarterly*, *54*(214), 134–165.

Hodges, W. (2006). The scope and limits of logic. In D. Jacquette (Ed.), *Handbook of the philosophy of science. Philosophy of logic* (pp. 41–64). Dordrecht: North–Holland.

Mancosu, P. (2005). Harvard 1940–1941: Tarski, Carnap and Quine on a finitistic language of mathematics for science. *History and Philosophy of Logic*, *26*, 327–357.

McCarthy, T. (1981). The idea of a logical constant. *Journal of Philosophy*, *78*, 499–523.

Mostowski, A. (1957). On a generalization of quantifiers. *Fundamenta Mathematicae*, *44*, 12–36.

Quine, W. (1936). Truth by convention. In O. Lee (Ed.), *Philosophical essays for A. N. Whitehead* (pp. 90–124). New York: Longmans.

Quine, W. (1961). Two dogmas of empiricism. In *From a logical point of view* (pp. 39–52). Cambridge, MA: Harvard University Press.

Quine, W. (2004). Two dogmas in retrospect. In R. Gibson (Ed.), *Quintessence: Readings from the philosophy of W. V. Quine* (p. 54-63). Cambridge, MA: Harvard University Press.

Quine, W. (2006). *Philosophy of logic* (second ed.). Cambridge, MA: Harvard University Press.

Russell, B. (1903). *The principles of mathematics* (Vol. I). Cambridge: Cambridge University Press.

Siebel, M. (2002). Bolzano's concept of consequence. *The Monist*, *85*(4), 580–599.

Tarski, A. (2002). On the concept of following logically. *History and Philosophy of Logic*, *23*, 155–196.

White, M. (1987). A philosophical letter of Alfred Tarski. *The Journal of Philosophy*, *84*(1), 28–32.

Wittgenstein, L. (1963). *Tractatus logico–philosophicus.* London: Routledge and Kegan Paul.

Wittgenstein, L. (1993a). *On certainty* (G. Anscombe & G. Wright, Eds.). Oxford: Basil Blackwell.

Wittgenstein, L. (1993b). Some remarks on logical form. In J. Klagge & A. Nordmann (Eds.), *Philosophical occasions, 1912–1951* (pp. 29–36). Indianapolis: Hackett Publishing Company.

Epistemic Logic with Relevant Agents

Ondrej Majer Michal Peliš*

1 Introduction

The aim of epistemic logics is to formalize epistemic states and actions of (possibly human) rational agents. A traditional means for representing these states and actions employs the framework of modal logics, where knowledge corresponds to some necessity operator. Modal axioms (K, T, 4, 5,...) then correspond to structural properties of the agent's knowledge. Employing strong modal systems such as S5 leads to representations of agents who are too ideal in many respects — they are logically omniscient, they have a perfect reflection of their both positive and negative knowledge (positive and negative introspection) etc. Sometimes these representations are called epistemic logics of *potential* rather than actual knowledge.

Frameworks representing only perfect agents have been frequently criticized, see (Fagin, Halpern, Moses, & Vardi, 2003) and (Duc, 2001), and some steps towards more realistic representations have been made (e.g., (Duc, 2001)). We also attempt to represent agents in an environment more realistically. Our motivation is epistemic, we shall concentrate on an agent working with experimental scientific data.

A realistic agent

Our agent is a scientist undertaking experiments or observations. Her typical environment is an experimental setup and her knowledge is usually experimental data (inputs and outputs of an experiment/observation) and some generalizations extracted from the experimental data.

* Work on this text was supported in part by grant no. 401/07/0904 of the Grant Agency of the Czech Republic and in part by grant no. IAA900090703 (Dynamic formal systems) of the Grant Agency of the Academy of Sciences of the Czech Republic. We wish to thank to Timothy Childers for valuable comments.

We assume the observations ('facts') are typically represented by atoms and their conjunctions and disjunctions, while generalizations ('regularities') are represented by conditionals (and their combinations). A conditional is supposed to record a regularly observed connection between the facts represented by the antecedent and the facts represented by the consequent.

It seems to be clear that for many reasons the material implication is not an appropriate representation of such a conditional. One of the main reasons is that the material implication may connect any two arbitrary formulas α, β. For example,

1. $\alpha \to (\beta \to \alpha)$,
2. $(\alpha \wedge \neg\alpha) \to \beta$,
3. $\alpha \to (\beta \vee \neg\beta)$,

are tautologies of classical logic. In our epistemic interpretation the material implication would make a 'law' from every two 'facts', which would obviously make the representation useless. It has other undesirable properties. It cannot deal with errors in the data, which result to contradictory facts (a situation which may very well happen in the scientific practice due to equipment errors). One such error corrupts all the remaining data (from a contradiction everything follows — see 2). It also admits 'laws' which are of no use as their consequent is a tautology (as in 3 — a tautology follows from anything)

The tautologies 1–3 are just examples of the paradoxes of material implication. As these 'paradoxes' were completely solved only in the systems of relevant logics, the obvious choice for a conditional for our scientific agent is relevant implication.

2 Relational semantics for relevant logics

Our point of departure will be the distributive relevant logic R of Anderson and Belnap (1975). The most natural way to introduce relevant logics is certainly proof theoretical (see, e.g., (Paoli, 2002)). However we would like to follow the modal tradition in representing an agent's epistemic states as a set of formulas and make the agent's knowledge dependent not only on the current epistemic state, but also on the states epistemic alternatives. Technically speaking we want to use a relational semantics. This cannot be a standard Kripke semantics with possible worlds and a binary accessibility relation, but a more general relational structure.

Formally our framework will be based on the Routley–Meyer semantics, as developed by Mares (Mares, 2004), Restall (Restall, 1999), Paoli (Paoli,

2002), and others, to which we shall add epistemic modalities. This semantics has been under constant attack for its seeming unintuitivness, but we believe it fits very well our motivations.

We give an informal exposition of structures in the relevant frame and definition of connectives (for formal definitions see the appendix A).

Relevant frame

A relevant frame is a structure $\mathbf{F} = \langle S, L, C, \trianglelefteq, R \rangle$, where S is a non-empty set of situations (states), $L \subseteq S$ is a non-empty set of designated *logical situations*, $C \subseteq S^2$ is a *compatibility* relation, $\trianglelefteq \subseteq S^2$ is a relation of *involvement*, $R \subseteq S^3$ is a *relevance* relation.

A model $\mathbf{M}$ is a relevant frame with the relation $\Vdash$, where $s \Vdash \varphi$ has the same meaning as in Kripke frames — that s carries the information that the formula φ is true ($\varphi \in s$ if we consider states to be sets of formulas).

Situations Situations or information states play the same role as possible worlds in Kripke frames. We assume, they consist of data immediately available to the agent. Like possible worlds, we can see situations as sets of formulas, but, unlike possible worlds, situations might be incomplete (neither φ nor $\neg\varphi$ is true in s) or inconsistent (both φ and $\neg\varphi$ are true in s).

Conjunction and disjunction Classical (weak) conjunction and disjunction correspond to the situation when the agent combines data immediately available to her, i. e. data from her current situation. They behave in the same way as in the case of classical Kripke frames — their validity is given locally:

$$s \Vdash \psi \wedge \varphi \text{ iff } s \Vdash \psi \text{ and } s \Vdash \varphi$$

$$s \Vdash \psi \vee \varphi \text{ iff } s \Vdash \psi \text{ or } s \Vdash \varphi$$

Weak connectives are the only ones which are defined locally. The truth of negation and implication depends also on the data in situations, related to the actual ones, so they are modal by nature. It is possible to define strong conjunction and disjunction as well (see appendix A).

Implication Implication is a modal connective in the sense that its truth depends not only on the current situation, but also on its neighborhood. It can be again understood in analogy with the standard modal reading. We say that an implication $(\varphi \rightarrow \psi)$ holds necessarily in a Kripke frame iff in all worlds where the antecedent holds, the consequent holds as well. In other words, the implication $(\varphi \rightarrow \psi)$ holds through all the neighborhood of the actual world. In the relevant case the neighborhood of a situation s is given by pairs of situations y, z such that s, y, z are related by the ternary relation R. We shall call y, z antecedent and consequent situations, respectively. We say that the implication $(\varphi \to \psi)$ holds at the situation s iff it is the case that for every antecedent situation y where φ (the antecedent of the implication) holds, ψ (the consequent of the implication) holds at the corresponding consequent situation z.

$$s \Vdash (\varphi \to \psi) \quad \text{iff} \quad (\forall y, z)(Rsyz \text{ implies } (y \Vdash \varphi \text{ implies } z \Vdash \psi))$$

The relation R reflects in our interpretation actual experimental setups. Antecedent situations correspond to some initial data (outcome of measurements or observations) of some experiment, while the related consequent situations correspond to the corresponding resulting data of the experiment. Implication then corresponds to some (simple) kind of a rule: if I observe in my current situation, that at every experiment (represented by a couple antecedent–consequent situation) each observation of φ is followed by an observation of ψ, then I accept 'ψ follows φ' as a rule.

Logical situations The framework we presented so far is very weak: there are just few tautologies valid in all situations and some of the important ones — those being usually considered as basic logical laws — are missing. For example the widely accepted identity axiom $(\alpha \to \alpha)$ and the Modus Ponens rule fail to hold in every situation.

This is connected to the question of truth in a relevance frame (model). If we take a hint from Kripke frames, we should equate truth in a frame with truth in every situation. But this would gives us an extremely weak system with some very unpleasant properties (cf. (Restall, 1999)). Designers of relevant logics took a different route — instead of requiring truth in all situations, they identify the truth in a frame just with the truth in all logically well behaved situations. These situations are called *logical*. In order to satisfy the 'good behavior' of a situation l it is enough to require that all the information in any antecedent situation related to l is contained in the corresponding consequent situation as well: for each $x, y \in S$, $Rlxy$ implies $|x| \subseteq |y|$, where $|s|$ is the set of all formulas, which are true in the situation s.

It is easy to see that situations constrained in this way validate both the identity axiom and (implicative) Modus Ponens.

Involvement Involvement is a relation resembling the persistence relation in intuitionistic logic — we can see it as a relation of information growth. However not every two situations which are in inclusion with respect to the validated formulas are in the involvement relation. We require that such an inclusion is observed or witnessed. Not every situation can play the role of the witness — only the logical situations can.

$$x \trianglelefteq y \quad \text{iff} \quad (\exists l \in L)(Rlxy)$$

Negation In Kripke models the *negation* of a formula φ is true at a world iff φ is not true there. As situations can be incomplete and/or inconsistent, this is not an option any more. Negation becomes a modal connective and its meaning depends on the worlds related to the given world by a binary modal relation C known as *compatibility*. Informally we can see the compatible situations as information sources our scientist wants to be consistent with. (Imagine the data of research groups working on related subjects.)

The formula $\neg\varphi$ holds at $s \in S$ iff it is not 'possible' (in the standard modal sense with respect to the relation C) that φ: at no situation s', compatible with ('accessible from') the situation s, it is the case that φ (either s' is incomplete with respect to φ or $\neg\varphi$ holds there).

$$s \Vdash \neg\varphi \quad \text{iff} \quad (\forall s' \in S)(sCs' \text{ implies } s' \nVdash \varphi)$$

Informally speaking, the agent can explicitly deny some hypothesis (a piece of data) only if no research group in her neighborhood claims it is true. This condition also has a normative side: she has to be skeptical in the sense that she denies everything not positively supported by any of her colleagues (in the situations related to her actual situation).

If we want to grant negative facts the same basic level as positive facts, we can read the clause for the definition of compatibility in the other direction: the agent can relate her actual situation just to the situations which do not contradict her negative facts.

Depending on the properties of the compatibility relation we obtain different kinds of negations. We shall shortly comment on them.

The compatibility relation is in general not reflexive: inconsistent situations are not self-compatible and so reflexivity holds only for consistent situations. It is clear that for an inconsistent self-compatible situation the clause for negation would not work. On the other hand, inconsistent situations can be compatible with some incomplete situations.

Nor is C transitive. Let us have situations x, y, z such that $x \Vdash \varphi$, $z \Vdash \neg\varphi$, and y does not include either φ or $\neg\varphi$. Assume that xCy and yCz. Then according to the definition of negation it cannot be that xCz.

It is quite reasonable to assume that C is *symmetric*. This condition implies that we get only one negation (otherwise we would get left and right negation) and we get the 'unproblematic' half of the law of double negation (if $x \Vdash \varphi$, then $x \Vdash \neg\neg\varphi$).

We also assume C is *directed* and *convergent.* Directedness means that there is at least one compatible situation for each $x \in S$. Convergence says that there is a maximal compatible situation $x^\star$. (See appendix A.)

Maximal compatible situations (with respect to x) can be inconsistent about everything not considered in x. From the symmetry of C we obtain $x \trianglelefteq x^{\star\star}$. If we assume, moreover, $x \trianglerighteq x^{\star\star}$, then we get the operation $\star$ with the property $x = x^{\star\star}$, i.e. the *Routley star.* The definition of negation-validity is then written in the form:

$$x \Vdash \neg\varphi \quad \text{iff} \quad x^\star \nVdash \varphi$$

The Routley star has been one of the controversial points of the Routley–Meyer semantics, but in our motivation it has a quite natural explanation: if compatible situations represent colleagues from different research groups our agent collaborates with, then the maximal compatible situation correspond to a colleague ('boss') who has all the information the other colleagues from the group have. Then if the agent wants to accept some negative clause she does not have to speak to each of the colleagues and ask his/her opinion, she just asks the 'boss' directly and knows that bosses opinion represents the opinions of the entire compatible research group.

This completes our exposition of relational semantics for relevant logics. We now move to epistemic modalities.

3 Knowledge in relevant framework

There have been some attempts to combine an epistemic and relevant framework (see (Cheng, 2000) and (Wansing, 2002)), but they have a different aim then our approach.

From a purely technical point of view there are a number of ways to introduce modalities in the relevant framework — Greg Restall in (Restall, 2000) provides a nice general overview. As we mentioned, the relevant framework already contains modal notions. We therefore decided to use these notions to introduce epistemic modalities rather than to introduce new ones.

In the classical epistemic frame what an agent knows in a world w is defined as what is true in all epistemic alternatives of w, which are given by the corresponding accessibility relation. Our idea of the agent as a scientist processing some kind of data requires a different approach.

We assume our agent in her current situation s observes (has a direct approach to) some data, represented by formulas which are true at s. She is aware of the fact that these data might be unreliable (or even inconsistent). In order to accept some of the current data as knowledge the agent requires a confirmation from some 'independent' resources.

In our approach resources are situations dealing with the same kind of data available in the current situation. A resource shall be more elementary than the current situation, i.e., it should not contain more data (a resource is below s in the $\trianglelefteq$-relation). Also the data from the resource should not contradict the data in the current situation (a resource is compatible with s).

Definition 1 (Knowledge).

$$s \Vdash K\varphi \quad \text{iff} \quad (\exists x)(sC^{\triangleleft}x \text{ and } x \Vdash \varphi),$$

where $sC^{\triangleleft}x$ iff sCx and $x \trianglelefteq s$ and $x \neq s$.

In short, φ is known iff there is a resource ('lower' compatible situation different from the actual one) validating φ.

We allowed our agent to deal with inconsistent data in order to get a more realistic picture. However, the agent should be able to separate inconsistent data. The modality we introduced provides us with such an appropriate filter. Let us assume both φ and $\neg\varphi$ are in s (e.g., our agent might receive such inconsistent information from two different sources). The agent considers both φ and $\neg\varphi$ to be possible, but neither of them is confirmed information as according to the definition, no situation compatible to s can contain either φ or $\neg\varphi$.

Basic properties

It is to be expected that our system blocks all the undesirable properties of both material and strict implication. Moreover, we ruled out the validity of some of the properties of 'classical' epistemic logics that we have criticized, in particular, both positive and negative introspection, as well as some closure properties.

Let us have a relevant frame $\mathbf{F} = \langle S, L, C, \trianglelefteq, R\rangle$. Recall that the truth in the frame $\mathbf{F}$ corresponds to the truth in the logical situations of $\mathbf{F}$ (under any valuation). We will also use the stronger notion of truth in all situations of $\mathbf{F}$ (under any valuation). From the viewpoint of our motivation the latter notion is more interesting as our agent might happen to be in other situations than the logical ones.

Our approach makes the 'truth axiom' **T** valid. For any situation $s \in S$, if φ is known at s ($s \Vdash K\varphi$), then there is a $\trianglelefteq$-lower compatible witness with φ true, which makes φ to be true at s as well. Thus, formula

$$K\alpha \rightarrow \alpha$$

is valid.

The axiom **K** and the necessity rule, common to all normal epistemic logics, fail. First, let us assume that φ is valid formula. The necessity rule

$(\frac{\varphi}{K\varphi})$ would imply the validity of $K\varphi$. $\models \varphi$ means that φ is true in every logical situation l. However, for $l \Vdash K\varphi$ a confirmation from a different resource is required, there must be a situation x such that $x \Vdash \varphi$ and $lC^{\triangleleft}x$, which, in general, does not need to be the case.

Second, in our interpretation the validity of axiom **K** is not well motivated and does not hold. **K** is in fact a 'distribution of confirmation': If an implication is confirmed then the confirmation of the antecedent implies the confirmation of the consequent.

$$\not\models K(\alpha \rightarrow \beta) \rightarrow (K\alpha \rightarrow K\beta)$$

Introspection As we defined knowledge as independently confirmed data, the epistemic axioms **4** and **5** correspond in our framework to a 'second order confirmation' rather than to introspection. It is easy to see that both axioms fail.

$$\not\models K\alpha \rightarrow KK\alpha,$$
$$\not\models \neg K\alpha \rightarrow K\neg K\alpha$$

Necessity and possibility We do not introduce possibility using the standard definition $M\varphi \stackrel{\text{def}}{\equiv} \neg K\neg\varphi$. Our idea of epistemic possibility is that our agent considers all the data available at the current situation as possible. If we introduce formally $s \Vdash M\varphi$ as $s \Vdash \varphi$, then it follows from the **T** axiom that in all situations necessity implies possibility:

$$(\forall s \in S)(s \Vdash K\varphi \rightarrow M\varphi)$$

However for the standard dual possibility this is not true.

$$\not\models K\varphi \rightarrow \neg K\neg\varphi$$

Let us comment on the relation of negation and necessity in our framework. There is a difference between $s \not\Vdash K\varphi$ and $s \Vdash \neg K\varphi$. The former simply says that φ is not confirmed at the current situation s, while the latter says that φ is not confirmed in the situations compatible with s. From this point of view it is uncontroversial that both $K\varphi$ (confirmation in the current situation) and $\neg K\varphi$ (the lack of confirmation in the compatible situations) might be true in some situation s (the necessary condition is that s is not compatible with itself).

Closure properties It is easy to see that the modal Modus Ponens

$$\frac{K\alpha \quad K(\alpha \rightarrow \beta)}{K\beta}$$

does not hold (for the reasons given in the section on **K** axiom). However, its weaker version

$$\frac{K\alpha \quad K(\alpha \to \beta)}{\beta}$$

holds not only in logical situations, but in all situations. If $K\alpha$ and $K(\alpha \to \beta)$ are true in any $s \in S$, then $s \Vdash \beta$. Axiom **T** and the assumption $Rsss$ are crucial here.

Contradiction in our system is non-explosive: φ and $\neg\varphi$ might hold in a contradictory situation, which need not be connected to any situation where ψ holds.

$$\not\models (\varphi \wedge \neg\varphi) \to \psi$$

On the other hand, the knowledge of contradiction implies anything (as a contradiction is never confirmed):

$$\models K(\varphi \wedge \neg\varphi) \to \psi$$

Modal adjunction also does not hold — if $K\alpha$ and $K\beta$ are true in s, then obviously $(\alpha \wedge \beta)$ is true there because of the truth axiom but $K(\alpha \wedge \beta)$ does not need to be true in s. (If each of α and β is confirmed by some resource, there still might be no resource confirming their conjunction.)

4 Conclusion

We introduced a system of epistemic logic based on the framework of relevant logic. We gave an epistemic interpretation of the relational semantics for relevant logics and defined epistemic modalities motivated by this interpretation. Instead of introducing additional relations into the framework, we argued in favor of using modalities based on the relations already contained in the frame.

The whole project is at an initial stage: there is much to be done both technically and in the area of interpretation. In particular we shall develop in a more detail the epistemic interpretation of our framework, give an axiomatization of our system, and characterize its formal properties.

A Relevant logic R

There are more formal systems that can be called relevant logic. From the proof-theoretical viewpoint, all of them are considered to be substructural logics (see (Restall, 2000) and (Paoli, 2002)). Here we present the axiom system and (Routley–Meyer) semantics from (Mares, 2004) with some elements from (Restall, 1999).

Syntax

We use the language of classical propositional logic with signs for atomic formulas $\mathcal{P} = \{p, q, \dots\}$, formulas being defined in the usual way:

$$\varphi ::= p \mid \neg\psi \mid \psi_1 \vee \psi_2 \mid \psi_1 \wedge \psi_2 \mid \psi_1 \rightarrow \psi_2$$

Axiom schemes

1. $A \rightarrow A$
2. $(A \rightarrow B) \rightarrow ((B \rightarrow C) \rightarrow (A \rightarrow C))$
3. $A \rightarrow ((A \rightarrow B) \rightarrow B)$
4. $(A \rightarrow (A \rightarrow B)) \rightarrow (A \rightarrow B)$
5. $(A \wedge B) \rightarrow A$
6. $(A \wedge B) \rightarrow B$
7. $A \rightarrow (A \vee B)$
8. $B \rightarrow (A \vee B)$
9. $((A \rightarrow B) \wedge (A \rightarrow C)) \rightarrow (A \rightarrow (B \wedge C))$
10. $(A \wedge (B \vee C)) \rightarrow ((A \wedge B) \vee (A \wedge C))$
11. $\neg\neg A \rightarrow A$
12. $(A \rightarrow \neg B) \rightarrow (B \rightarrow \neg A)$

Strong logical constants $\otimes$ (group conjunction, fusion) and $\oplus$ (group disjunction) are definable by implication and negation:

- $(A \oplus B) \stackrel{\text{def}}{\equiv} \neg(\neg A \rightarrow B)$
- $(A \otimes B) \stackrel{\text{def}}{\equiv} \neg(\neg A \oplus \neg B)$

Rules

Adjunction From A and B infer $A \wedge B$.

Modus Ponens From A and $A \rightarrow B$ infer B.

Routley–Meyer semantics

An R-frame is a quintuple $\mathbf{F} = \langle S, L, C, \trianglelefteq, R \rangle$, where S is a non-empty set of situations and $L \subseteq S$ is a non-empty set of logical situations. The relations $C \subseteq S^2$, $\trianglelefteq \subseteq S^2$, and $R \subseteq S^3$ were introduced in section 2, here we sum up their properties.

Properties of the relation R The basic property of R:

$$\text{if } Rxyz,\ x' \trianglelefteq x,\ y' \trianglelefteq y, \text{ and } z \trianglelefteq z', \quad \text{then } Rx'y'z'.$$

This means that the relation R is monotonic with respect to the involvement relation.

Moreover it is required that:

(r1) $Rxyz$ implies $Ryxz$;

(r2) $R^2(xy)zw$ implies $R^2(xz)yw$, where R^2xyzw iff $(\exists s)(Rxys$ and $Rszw)$;

(r3) $Rxxx$;

(r4) $Rxyz$ implies $Rxz^\star y^\star$.

Properties of the relation C Compatibility between two states is inherited by the states involved in them ('less informative states'):

$$\text{If } xCy,\ x_1 \trianglelefteq x, \text{ and } y_1 \trianglelefteq y, \text{ then } x_1Cy_1.$$

Moreover, we require the following properties:

(c1) (symmetricity) xCy implies yCx;

(c2) (directedness) $(\forall x)(\exists y)(xCy)$;

(c3) (convergence) $(\forall x)(\exists y(xCy)$ implies $(\exists x^\star)(xCx^\star$ and $\forall z(xCz$ implies $z \trianglelefteq x^\star)))$;

(c4) $x \trianglelefteq y$ implies $y^\star \trianglelefteq x^\star$;

(c5) $x^{\star\star} \trianglelefteq x$.

Model R-model $\mathbf{M}$ is a R-frame $\mathbf{F}$ with a valuation function $v\colon \mathcal{P} \to 2^S$. The truth of a formula at a situation is defined in the following way:

- $s \Vdash p$ iff $s \in v(p)$,
- $s \Vdash \neg\varphi$ iff $s^\star \nVdash \varphi$,

- $s \Vdash \psi \wedge \varphi$ iff $s \Vdash \psi$ and $s \Vdash \varphi$,
- $s \Vdash \psi \vee \varphi$ iff $s \Vdash \psi$ or $s \Vdash \varphi$,
- $s \Vdash (\varphi \rightarrow \psi)$ iff $(\forall y, z)(Rsyz$ implies $(y \Vdash \varphi$ implies $z \Vdash \psi))$.

As we already said, the truth of a formula in a model and in a frame, respectively, is defined as truth in all logical situations of this model/frame. As usual, R-tautologies are formulas true in all relevant frames. Whenever φ is a R-tautology, we write $\models \varphi$ and say that φ is a valid formula.

The condition (r1) validates the implicative version of *Modus Ponens* (axiom schema 3). It does not validate the conjunctive version $(A \wedge (A \rightarrow B)) \rightarrow B$, which requires (r3).

(r2) corresponds to the '*exchange rule*' $(A \rightarrow (B \rightarrow C)) \rightarrow (B \rightarrow (A \rightarrow C))$, which is derivable from the axioms given above.

(r4) validates *contraposition* (axiom schema 12). If we work without the Routley star, this can be rewritten as:

$$Rxyz \text{ implies } (\forall z' Cz)(\exists y' Cy)(Rxy'z').$$

Directedness and convergence conditions are necessary for the definition of the Routley star. From (c1) we obtain the validity of $(A \rightarrow \neg\neg A)$ and from the last condition (c5) we get the axiom schema 11.

Ondrej Majer
Institute of Philosophy, Academy of Sciences of the Czech Republic
Jilská 1, 110 00 Praha 1
majer@site.cas.cz
http://logika.flu.cas.cz

Michal Peliš
Institute of Philosophy, Academy of Sciences of the Czech Republic
Jilská 1, 110 00 Praha 1
pelis@ff.cuni.cz
http://logika.flu.cas.cz

References

Cheng, J. (2000). A strong relevant logic model of epistemic processes in scientific discovery. In E. Kawaguchi, H. Kangassalo, H. Jaakkola, & I. Hamid (Eds.), *Information modelling and knowledge bases XI* (pp. 136–159). Amsterdam: IOS Press.

Duc, H. N. (2001). *Resource-bounded reasoning about knowledge.* Unpublished doctoral dissertation, Faculty of Mathematics and Informatics, University of Leipzig.

Fagin, R., Halpern, J., Moses, Y., & Vardi, M. (2003). *Reasoning about knowledge.* Cambridge, MA: MIT Press.

Mares, E. (2004). *Relevant logic.* Cambridge: Cambridge University Press.

Mares, E., & Meyer, R. (1993). The semantics of **R4**. *Journal of Philosophical Logic*, *22*, 95–110.

Paoli, F. (2002). *Substructural logics: A primer.* Dordrecht: Kluwer.

Restall, G. (1993). Simplifeid semantics for relevant logics (and some of their rivals). *Journal of Philosophical Logic*, *22*, 481–511.

Restall, G. (1995). Four-valued semantics for relevant logics (and some of their rivals). *Journal of Philosophical Logic*, *24*, 139–160.

Restall, G. (1996). Information flow and relevant logics. In *Logic, language and computation: The 1994 Moraga proceedings. CSLI Lecture Notes* (Vol. 58, pp. 463–477). Stanford, CA: CSLI.

Restall, G. (1999). Negation in relevant logics: How I stopped worrying and learned to love the Routley star. In D. Gabbay & H. Wansing (Eds.), *What is negation?* (Vol. 13, pp. 53–76). Dordrecht: Kluwer.

Restall, G. (2000). *An introduction to substructural logics.* London–New York: Routledge.

Wansing, H. (2002). Diamonds are a philosopher's best friends. *Journal of Philosophical Logic*, *31*, 591–612.

Betting on Fuzzy and Many-valued Propositions

Peter Milne

1 Introduction

In a 1968 article, 'Probability Measures of Fuzzy Events', Lotfi Zadeh proposed accounts of absolute and conditional probability for fuzzy sets (Zadeh, 1968). Where P is an ordinary ("classical") probability measure defined on a σ-field of Borel subsets of a space X, and μ_A is a fuzzy membership function defined on X, i.e. a function taking values in the interval $[0, 1]$, the probability of the fuzzy set A is given by

$$P(A) = \int_X \mu_A(x)\,\mathrm{d}P.$$

The thing to notice about this expression is that, in a way, there's nothing "fuzzy" about it. To be well defined, we must assume that the "level sets"

$$\{x \in X : \mu_A(x) \leq \alpha\}, \quad \alpha \in [0, 1],$$

are P-measurable. These are ordinary, "crisp", subsets of X. And then $P(A)$ is just the expectation of the random variable μ_A. — This is entirely classical. Of course, you may *interpret* μ_A as a fuzzy membership function but really we have, if you'll pardon the pun, in large measure lost sight of the fuzziness.

So you might ask:

- is this the only way to define fuzzy probabilities?

The answer, I shall argue, is yes.

Defining conditional probability Zadeh offered

$$P(A|B) = \frac{P(AB)}{P(B)}, \qquad \text{when} \quad P(B) > 0,$$

where

$$\forall x \in X \quad \mu_{AB}(x) = \mu_A(x) \times \mu_B(x).$$

One might wonder:

- is this the only way to define conditional probabilities?

The answer, I shall suggest, is no, it is not the *only* way but it is the only *sensible* way.

Zadeh assigns probabilities to sets. What I offer here, using Dutch Book Arguments, is a vindication of Zadeh's specifications when probability is assigned to propositions rather than sets. (But translation between proposition talk and set and event talk is straightforward. It's just that proposition talk fits better with betting talk.)

2 Bets and many-valued logics

I apply "the Dutch Book method", as Jeff Paris calls it (Paris, 2001), to fuzzy and many-valued logics that meet a simple linearity condition. I shall call such logics additive.

Additivity

For any valuation v and for any sentences A and B

$$v(A \wedge B) + v(A \vee B) = v(A) + v(B)$$

where '$\wedge$' and '$\vee$' the conjunction and disjunction of the logic in question.

Additivity is common: the Gödel, Łukasiewicz, and product fuzzy logics are all additive, as are Gödel and Łukasiewicz n-valued logics.

In order to employ Dutch Book arguments, we need a betting scheme suitably sensitive to truth-values intermediate between the extreme values 0 and 1. Setting out the classical case the right way makes one generalization obvious.

Rather than betting odds, which are algebraically less tractable, we use, as is standard, a "normalized" betting scheme with fair betting quotients. Classically, with a bet on A at betting quotient p and stake S:

- the bettor gains $(1-p)S$ if A;
- the bettor loses pS if not-A.

Taking 1 for truth, 0 for falsity, and $v(A)$ to be the truth-value of A, we can summarise this scheme like this:

the pay-off to the bettor is $(v(A) - p)S$.

And now we see how to extend bets to the many valued case: we adopt the same scheme but allow $v(A)$ to have more than two values. The slogan is: the pay-off is the larger the more true A is.[1]

Using this betting scheme, we obtain Dutch Book arguments for certain seemingly familiar principles of probability, seemingly familiar in that formally they recapitulate classical principles.

- $0 \leq \Pr(A) \leq 1$;
- $\Pr(A) = 1$ when $\models A$;
- $\Pr(A) = 0$ when $A \models$;
- $\Pr(A \wedge B) + \Pr(A \vee B) = \Pr(A) + \Pr(B)$.

Here $\wedge$ and $\vee$ are the conjunction and disjunction, respectively, of an additive fuzzy or many-valued logic.

Other principles that may or may not be independent, depending on the logic:

- $\Pr(A) + \Pr(\neg A) = 1$ when $v(\neg A) = 1 - v(A)$;
- $\Pr(A) \geq x$ when, under all valuations, $v(A) \geq x$;
- $\Pr(A) \leq x$ when, under all valuations, $v(A) \leq x$;
- $\Pr(A) \leq \Pr(B)$ when $A \models B$.

I'll show how two of the arguments go as there's an interesting connection with the standard Dutch Book arguments used in the classical, two-valued case.

We let x range over the possible truth-values (which all lie in the interval $[0, 1]$). Clearly, for given p, we can choose a value for the stake S that makes

$$G_x = (x - p)S$$

negative, for *all* values of x in the interval $[0, 1]$, if, and only if, p is less than 0 or greater than 1. Hence

$$0 \leq \Pr(A) \leq 1.$$

[1] The suggested pay-off scheme is, of course, only the most straightforward way to implement the slogan. One could distort truth values: take a strictly increasing function $f\colon [0,1]^2 \rightarrow [0,1]$ with $f(0) = 0$, $f(1) = 1$, and take pay-offs to be given by $(f(v(A)) - p)S$. Analogously, Zadeh could have taken $\int_X f(\mu_A(x))\,\mathrm{d}P$ to define distorted probabilities. — And the point is that such "probabilities" *are* distorted for when f is not the identity function it may be that $P(A) < c$ even though $\mu_A(x) > c$, for all $x \in X$.

So far so good, but here's the cute bit:

$$G_x = xG_1 + (1-x)G_0,$$

so G_x is negative for all values of $x \in [0,1]$ *if*, and only if, G_1 and G_0 are both negative. From the classical case, we know that the necessary and sufficient condition for the latter is that p lie outside the interval $[0,1]$. It suffices to look at the classical extremes to fix what holds good for all truth-values in the interval $[0,1]$.

Next we consider four bets:

1. a bet on A, at betting quotient p with stake S_1;
2. a bet on B, at betting quotient q with stake S_2;
3. a bet on $A \wedge B$, at betting quotient r with stake S_3;
4. a bet on $A \vee B$, at betting quotient s with stake S_4.

We assume that for all allowed values of $v(A)$ and $v(B)$,

$$v(A \wedge B) + v(A \vee B) = v(A) + v(B) \quad \text{and} \quad v(A \wedge B) \leq \min\{v(A), v(B)\}.$$

Then, where x, y, and z are the truth-values of A, B and $A \wedge B$ respectively, the pay-off is

$$G_{x,y} = (x-p)S_1 + (y-q)S_2 + (z-r)S_3 + ((x+y-z)-s)S_4.$$

This can be rewritten as

$$G_{x,y} = zG_{1,1} + (x-z)G_{1,0} + (y-z)G_{0,1} + (1-x-y+z)G_{0,0}.$$

The co-efficients are all non-negative and cannot all be zero. Thus $G_{x,y}$ is negative, for all allowable x, y, and z, *just in case* $G_{1,1}$, $G_{1,0}$, $G_{0,1}$, and $G_{0,0}$ are all negative. From the standard Dutch Book argument for the two-valued, classical case, we know this to be possible if, and only if, $p+q \neq r+s$. Hence

$$\Pr(A \wedge B) + \Pr(A \vee B) = \Pr(A) + \Pr(B).$$

3 The classical expectation thesis for finitely-many-valued Łukasiewicz logics

As an initial vindication of Zadeh's account, we find that in the context of a finitely-many-valued Łukasiewicz logic, all probabilities are *classical expectations*. That is, the probability of a many-valued proposition is the expectation of its truth-value *and* that a proposition has a particular truth-value is expressible using a *two-valued* proposition. So in this setting, in analogy with Zadeh's assignment of absolute probabilities to fuzzy sets, *all* probabilities are expectations defined over a classical domain.

In all Łukasiewicz logics, conjunction and disjuction are evaluated by the functions $\max\{0, x+y-1\}$ and $\min\{1, x+y\}$, respectively.

Employing Łukasiewicz negation and one or more of Łukasiewicz conjunction, disjunction, and implication, one can define a sequence of $n+1$ formulas of a single variable, $J_{n,0}(p), J_{n,1}(p), \ldots, J_{n,n}(p)$, which have this property (Rosser & Turquette, 1945): in the semantic framework of $(n+1)$-valued Łukasiewicz logic it is the case that for every formula A, for all k, $0 \leq k \leq n$, and for every valuation v,

$$v(J_{n,k}(A)) = 1, \text{ if } v(A) = \frac{k}{n};$$

$$v(J_{n,k}(A)) = 0, \text{ if } v(A) \neq \frac{k}{n}.$$

In the semantic framework of $(n+1)$-valued Łukasiewicz logic, for all sentences A,

$$\models J_{n,0}(A) \vee_{Ł} J_{n,1}(A) \vee_{Ł} \cdots \vee_{Ł} J_{n,n}(A) \quad \text{and}$$

$$J_{n,i}(A) \wedge_{Ł} J_{n,j}(A) \models, \quad 0 \leq i < j \leq n. \quad (*)$$

From the probability axioms, we have, for all sentences A, that

$$\sum_{0 \leq i \leq n} \Pr(J_{n,i}(A)) = 1.$$

The propositions of the form $J_{n,i}(A)$ are two-valued, so, $(n+1)$-valued Łukasiewicz logic reducing to classical logic on the values 0 and 1, the logic of these propositions is classical. Thus, when restricted to these propositions and their logical compounds, the probability axioms give us a classical, finitely additive, probability distribution. What we show next is that this classical probability distribution determines the probabilities of all propositions in the language.

Theorem 1 (Classical Expectation Thesis). *In the framework of* $(n+1)$-*valued Łukasiewicz logic,*

$$\Pr(A) = \frac{1}{n} \sum_{0 \leq i \leq n} i \Pr(J_{n,i}(A)).$$

Proof. From (*) and the two-valuedness of the $J_{n,i}(A)$'s we have

$$A \dashv\models (A \wedge_{Ł} J_{n,0}(A)) \vee_{Ł} (A \wedge_{Ł} J_{n,1}(A)) \vee_{Ł} \cdots \vee_{Ł} (A \wedge_{Ł} J_{n,n}(A)).$$

From our probability axioms it follows that logically equivalent propositions must receive the same probability, so

$$\Pr(A) = \sum_{0 \leq i \leq n} \Pr(A \wedge_{Ł} J_{n,i}(A)). \tag{†}$$

We consider two bets, one on $A \wedge_{Ł} J_{n,k}(A)$ at betting quotient p and stake S_1, the other on $J_{n,k}(A)$ at betting quotient q with stake S_2. The pay-offs are:

$$G_{=\frac{k}{n}} = \left(\frac{k}{n} - p\right) S_1 + ((1-q)S_2) \quad \text{when } A \text{ has truth-value } \frac{k}{n},$$

$$G_{\neq\frac{k}{n}} = -pS_1 - qS_2 \quad \text{when } A \text{ has truth-value other than } \frac{k}{n}.$$

Setting $S_2 = -\frac{k}{n}S_1$ gives a pay-off, independent of the truth-value of A, of $\left(\frac{qk}{n} - p\right) S_1$, which can be made negative by choice of S_1 provided $p \neq \frac{qk}{n}$. On the other hand, for arbitrary S_1 and S_2, when $p = \frac{qk}{n}$ the two pay-offs are

$$G_{=\frac{k}{n}} = (1-q)\left[\frac{k}{n}S_1 + S_2\right] \quad \text{when } A \text{ has truth-value } \frac{k}{n}, \quad \text{and}$$

$$G_{\neq\frac{k}{n}} = -q\left[\frac{k}{n}S_1 + S_2\right] \quad \text{when } A \text{ has truth-value other than } \frac{k}{n}.$$

These cannot both be negative. Hence

$$\Pr(A \wedge_{Ł} J_{n,k}(A)) = \frac{k}{n} \Pr(J_{n,k}(A)).$$

Substituting in (†), we obtain:

$$\Pr(A) = \frac{1}{n} \sum_{0 \leq i \leq n} i \Pr(J_{n,i}(A)).$$

□

Two comments

Firstly, having been obtained by an independent Dutch Book argument, the Classical Expectation Thesis may seem to be an additional principle. In fact it is not; it is derivable from our axioms for probability. To show this we have to introduce a propositional constant, introduced into Łukasiewicz logic by Słupecki in order to obtain expressive completeness (Słupecki, 1936).

In the semantics of $(n+1)$-valued Łukasiewicz logic, in which all formulas are assigned values in the set $\{0, \frac{1}{n}, \frac{2}{n}, \ldots, \frac{n-1}{n}, 1\}$, the propositional constant t has this interpretation:

$$\text{under all valuations } v, \quad v(t) = \frac{n-1}{n}.$$

Let t_1 be the $(n-2)$-fold $\wedge_Ł$-conjunction of t with itself. For $1 < k \leq n$, let t_k be the $(k-1)$-fold $\vee_Ł$-disjunction of t_1 with itself. $v(t_1) = \frac{1}{n}$ and $v(t_k) = \frac{k}{n}$. Since we have

$$\begin{aligned} t_k \wedge_Ł t_1 &\models, \quad 1 \leq k < n, \qquad \text{and} \\ &\models t_n, \end{aligned}$$

from our probability axioms we obtain:

$$\begin{aligned} &\Pr(t_k) = k \Pr(t_1), \quad 1 \leq k \leq n, \qquad \text{and} \\ &\Pr(t_n) = 1, \end{aligned}$$

hence

$$\Pr(t_k) = \frac{k}{n}, \qquad 1 \leq k \leq n.$$

Using the t_i's we can derive the Classical Expectation Thesis. (I'll skip the details here.)

Secondly, the Dutch Book argument for the Classical Expectation Thesis goes through with *any* notion of conjunction for which $v(A\&B) = v(A)$ when $v(B) = 1$ and $v(A\&B) = 0$ when $v(B) = 0$. Also, the $J_{n,i}(A)$'s being truth-functional, the Classical Expectation Thesis holds good of every proposition in the semantic framework, not just those expressible using the Łukasiewicz connectives.

4 The extension to infinitely many truth-values (a sketch)

For any rational number x in the interval $[0, 1]$, there is a formula $\phi(p)$ of a single propositional-variable p, constructed using Łukasiewicz negation and any one or more of Łukasiewicz conjunction, disjunction, or implication, such that, under *any* valuation taking values in $[0, 1]$, $v(\phi(A/p)) = 0$ if $v(A) \leq x$ and $v(\phi(A/p)) > 0$ otherwise (McNaughton, 1951).

Employing the Gödel negation,[2] then, we have,

[2] The Gödel negation is, to be sure, not usually taken to be part of the vocabulary of the Łukasiewicz logics. Semantically, however, it can be defined in the Łukasiewicz fuzzy/many-valued frameworks as the *external* negation that maps 0 to 1 and all other values to 0.

- for each interval $[0, x]$ with x rational, a formula $J_{[0,x]}(A)$ that takes the value 1 under any valuation v for which $v(A) \leq x$ and otherwise takes the value 0;

- for each half-open interval $(x, y]$ with rational endpoints x and y, $x < y$, a formula $J_{(x,y]}(A)$ that takes the value 1 under a valuation v when $v(A) \in (x, y]$ and otherwise takes the value 0.

Given a strictly increasing, finite sequence $x_0, x_1, \ldots, x_{n-1}$ of rational numbers in the open interval $(0, 1)$, consider the family of $n + 1$ bets:

- a bet on A at betting quotient q with stake S;

- a bet on $J_{[0,x_1]}(A)$ at betting quotient p_1 with stake S_1;

- a bet on $J_{(x_{i-1},x_i]}(A)$ at betting quotient p_i with stake S_i, $1 < i < n$;

- a bet on $J_{(x_i,1]}(A)$ at betting quotient p_n with stake S_n.

$$\sum_{2\leq i\leq n} x_{i-1} \Pr(J_{(x_{i-1},x_i]}(A)) \leq \Pr(A) \leq$$

$$\leq x_1 \Pr(J_{[0,x_1]}(A)) + \sum_{2\leq i\leq n} x_i \Pr(J_{(x_{i-1},x_i]}(A)),$$

where $x_n = 1$. So by taking finer and finer partitions we can more closely approximate the probability of A from above and below. This may not quite do to fix $\Pr(A)$ exactly. For that we *may* also need the probabilities of at most a countable infinity of (two-valued) statements of the form

$$v(A) \leq x$$

where x is an irrational number.[3]

With these in hand, we then find that

$$\Pr(A) = \int_0^1 x \, \mathrm{d}F_A(x),$$

where F_A is the ordinary, "classical" distribution function determined by the probabilities of the $J_{[0,x]}(A)$'s, $J_{(x,y]}(A)$'s and however many $v(A) \leq x$'s with x irrational we have used.

By introducing a countably infinite family of logical constants, we can *derive* this classical representation from the previously given principles of probability together with the principle

[3] Recall Zadeh's assumption regarding the P-measureability of "level sets".

- for any proposition A logically constrained to take only the values 0 and 1 and for rational values of x in the interval $[0,1]$, $\Pr(t_x \wedge A) = x \Pr(A)$,

where t_x takes the value x under all valuations v.

The really neat feature of infinitely many-valued Łukasiewicz logics is that this principle is derivable from the basic principles

- $0 \leq \Pr(A) \leq 1$;
- $\Pr(A) = 1$ when $\models A$;
- $\Pr(A) = 0$ when $A \models$;
- $\Pr(A \wedge_Ł B) + \Pr(A \vee_Ł B) = \Pr(A) + \Pr(B)$.

5 Conditional probabilities

In the classical setting, a bet on A conditional on B is a bet that goes ahead if, and only if, B is true and is then won or lost according as to whether A is true or not. The pay-offs for such a conditional bet with stake S at betting quotient p are:

- the bettor gains $(1-p)S$ if A and B;
- the bettor loses pS if not-A and B;
- the bettor neither gains nor loses if not-B.

We can summarise this betting scheme like this:

$$v(B)(v(A) - p)S.$$

And so, as with ordinary bets, we now know one way to extend the scheme for conditional bets on classical, two-valued propositions to many-valued propositions.

A straightforward Dutch Book argument, which again piggy-backs on the proof in the two-valued case, then tells us that

$$\Pr(A \wedge_\times B) = \Pr(A|B) \times \Pr(B)$$

where

$$v(A \wedge_\times B) = v(A) \times v(B).$$

— Allowing for the change of setting, just what Zadeh said.

You can, if you are so minded, generalize the classical scheme using *any* many-valued or fuzzy conjunction that is "classical at the extremes":

$$(v(A \wedge B) - v(B)p)S.$$

A Dutch Book argument — in all essentials, the *same* Dutch Book argument — will then deliver:

$$\Pr(A \wedge B) = \Pr(A|B) \times \Pr(B).$$

However, $\Pr(\cdot|B)$ satisfies the axioms for an absolute probability measure *only* when the product conjunction, $\wedge_{\times}$ is used.[4]

Peter Milne
Department of Philosophy, University of Stirling
Stirling FK4 9LA, United Kingdom
peter.milne@stir.ac.uk

References

McNaughton, R. (1951). A theorem about infinite–valued sentential logic. *Journal of Symbolic Logic*, *16*, 1–13.

Paris, J. (2001). A note on the dutch book method. In G. De Cooman, T. Fine, & T. Seidenfeld (Eds.), *ISIPTA '01, Proceedings of the second international symposium on imprecise probabilities and their applications, Ithaca, NY, USA* (pp. 301–306). Maastricht: Shaker Publishing. (A slightly revised version is available on-line at http://www.maths.manchester.ac.uk/~ jeff/papers/15.ps)

Rosser, J., & Turquette, A. (1945). Axiom schemes for m–valued propositional calculi. *Journal of Symbolic Logic*, *10*, 61–82.

Słupecki, J. (1936). Der volle dreiwertige Aussagenkalkül. *Comptes rendues des séances de la Société des Sciences et Lettres de Varsovie*, *29*, 9–11.

Zadeh, L. A. (1968). Probability measures of fuzzy events. *Journal of Mathematical Analysis and Applications*, *23*, 421–427.

[4] Beyond the classical, two-valued case, product conjunction requires that there be an infinity of truth-values.

Inferentializing Consequence

Jaroslav Peregrin*

The proof of correctness and completeness of a logical calculus w.r.t. a given semantics can be read as telling us that the tautologies (or, more generally, the relation of consequence) specified in a model-theoretic way can be equally well specified in a proof-theoretic way, by means of the calculus (as the theorems, resp. the relation of inferability of the calculus). Thus we know that both for the classical propositional calculus and for the classical predicate calculus theorems and tautologies represent two sides of the same coin. We also know that the relation of inference as instituted by any of the common axiom systems of the classical propositional calculus coincides with the relation of consequence defined in terms of the truth tables; whereas the situation is a little bit more complicated w.r.t. the classical predicate calculus (the coincidence occurs if we restrict ourselves to closed formulas; otherwise $\forall xFx$ is inferable from Fx without being its consequence). And of course we also know cases where a class of tautologies of a semantic system does not coincide with the class of theorems of any calculus. (The paradigmatic case is the second-order predicate calculus with standard semantics.)

This may make us consider the problem of "inferentializability". Which semantic systems are "inferentializable" in the sense that their tautologies (their relation of consequence, respectively) coincide with the class of theorems (the relation of inferability, respectively) of a calculus? One answer is ready: it is if and only if the set of tautologies is recursively enumarable. But this answer is not very informative, indeed saying that the set is recursively enumerable is only reiterating that it conicides with the class of theorems of a calculus. Moreover, paying due attention to the terms such as "calculus" and "inference" shows us that it is possible to relate them to various "levels", whereby the problem of inferentializability becomes quite nontrivial.

* Work on this paper was supported by the grant No. 401/07/0904 of the Czech Science Foundation.

1 Consequence

Consequence, as the concept is usually understood, amounts to truth-preservation, i.e., A is a consequence of $A_1, \ldots, A_n$ iff the truth of all of $A_1, \ldots, A_n$ brings about the truth of A, i.e., iff any truth valuation mapping all of $A_1, \ldots, A_n$ on 1 maps also A on 1.[1] It is obvious that the "any" from the previous sentence cannot mean "any whatsoever" (of course there *does* exist a function mapping all of $A_1, \ldots, A_n$ on 1 and A on 0!), it must mean something like "any admissible". Hence there must be some concept of admissibility in play: some mappings of sentences of $\{0, 1\}$ will be admissible, others not. But, of course, that if we take the sentences to be sentences of a meaningful language, such a division of valuations is forthcoming: if $A_1, \ldots, A_n$ are *Fido is a dog* and *Every dog is a mammal* (hence $n = 2$), A is *Fido is a mammal*, then the valuation mapping the former two sentences on 1 and the latter one on 0 is not admissible — it is not compatible with the semantics of English.

Hence we assume that any semantics of any language provides for the division of the sentences of the language into true and false, thereby dividing the space of the mappings of the sentences on $\{0, 1\}$ into admissible and inadmissible. (In fact I maintain a much stronger thesis, namely that any semantics can be *reduced* to such a division, but I am not going to argue for this thesis here — I have done so elsewhere, see (Peregrin, 1997).) Thereby it also establishes the relation of consequence, as the relation of truth-preservation for all admissible valuations. If we use the sentences $S_1, S_2, \ldots$ of the language in question to mark columns of the following table using all possible truth-valuations as its rows, we can look at the delimitation of the admissible valuations as striking out rows of the table.

	S_1	S_2	S_3	S_4	$\cdots$
v_1	0	0	0	0	$\cdots$
~~v_2~~	~~1~~	~~0~~	~~0~~	~~0~~	~~$\cdots$~~
v_3	0	1	0	0	$\cdots$
v_4	1	1	0	0	$\cdots$
v_5	0	0	1	0	$\cdots$
~~v_6~~	~~1~~	~~0~~	~~1~~	~~0~~	~~$\cdots$~~
$\vdots$	$\vdots$	$\vdots$	$\vdots$	$\vdots$	$\ddots$

[1] See (Peregrin, 2006).

A more exact articulation of these notions yields the following definition:

Definition 1. A *semantic system* is an ordered pair $\langle S, V \rangle$, where S is a set (the elements of which are called *sentences*) and $V \subseteq \{0, 1\}^S$. The elements of $\{0, 1\}^S$ are called *valuations* (of S). (A valuation will be sometimes identified with the set of all those elements of S that are mapped on 1 by it.) The elements of V are called *admissible valuations of* $\langle S, V \rangle$, the other valuations (i.e. the elements of $\{0, 1\}^S \setminus V$) are called *inadmissible.* The relation of *consequence* induced by this system is the relation $\models$ defined as follows

$X \models A$ iff $v(A) = 1$ for every $v \in V$ such that $v(B) = 1$ for every $B \in X$.

2 Varieties of inference

Now consider the stipulation of an inference, $A_1, \ldots, A_n \vdash A$ (for some elements $A_1, \ldots, A_n$, A of S). Such a stipulation can be seen as excluding certain valuations: namely all those that map $A_1, \ldots, A_n$ on 1 and A on 0. (Thus, for example, the exclusions in the above table might be the result of stipulating $S_1 \vdash S_2$.) Hence if we call the pair constituted by a finite set of elements of S and an element of S an *inferon*, we can say that inferons exclude valuations and ask which sets of valuations can be demarcated by means of inferons.

Definition 2. An *inferon* (*over* S) is an ordered pair $\langle X, A \rangle$ where X is a finite subset of S and A is an element of S. An inferon is said to *exclude* an element v of $\{0, 1\}^S$ iff $v(B) = 1$ for every $B \in X$ and $v(A) = 0$. An ordered pair $\langle S, \vdash \rangle$ such that S is a set and $\vdash$ is a finite set of inferons (i.e. a binary relation between finite subsets of S and elements of S) will be called an *inferential structure.* An inferential structure is said to *determine* a semantic system $\langle S, V \rangle$ iff V is the set of all and only elements of $\{0, 1\}^S$ not excluded by any element of $\vdash$. A semantic system is called an *inferential* system iff it is determined by an inferential structure.

Now an obvious question is which semantic systems are inferential. But before we turn our attention to it, we will consider various possible generalizations of the concept of inference. First, let a *quasiinferon* differ from an inferon in that its second component is not a single statement, but a finite set of statements. A *quasiinferon* will exclude every valuation that maps every element of its first component on 1 and every element of its second component on 0. (Of course the concept of quasiinferon defined in this way is closely connected with the concept of *sequent* as introduced by (Gentzen, 1934) and (Gentzen, 1936).[2]) Second, let a *semiinferon* differ from an inferon in that its first component is not necessarily finite. A *semiquasiinferon* will be a quasiinferon with both its first and its second component not necessarily finite. Third, let a *protoinferential* structure be an inferential structure with its second component not necessarily finite (and think of the concepts of *protosemiinferential*, *protoquasiinferential* and *protosemiquasiinferential* structure analogously).

In the following definition, we abbreviate the prefixes, which have already become somewhat monstrous:

Definition 3. An element of $\mathrm{Pow}(S) \times \mathrm{Pow}(S)$ is called an *SQI-on over S*. It is called a *QI-on* if it is an element of $\mathrm{FPow}(S) \times \mathrm{FPow}(S)$ (where $\mathrm{FPow}(S)$ is the set of all finite subsets of S), it is called an *SI-on* if it is an element of $\mathrm{Pow}(S) \times S$ and it is called an *I-on* if it is an element of $\mathrm{FPow}(S) \times S$.[3] The ordered pair $\langle S, \vdash \rangle$ where $\vdash$ is a set of SQI-ons (QI-ons, SI-ons, I-ons) will be called a *PSQI-structure* (*PQI-structure*, *PSI-structure*, *PI-structure*). It is called an *SQI-structure* (*QI-structure*, *SI-structure*, *I-structure*) iff $\vdash$ is finite. An SQI-on $\langle X, Y \rangle$ is said to *exclude* an element v of $\{0,1\}^S$ iff $v(B) = 1$ for every $B \in X$ and $v(A) = 0$ for every $A \in Y$. A (P)(S)(Q)I-structure $\langle S, \vdash \rangle$ is said to *determine* a semantic system $\langle S, V \rangle$ iff V is the set of all and only elements of $\{0,1\}^S$ not excluded by any element of $\vdash$. A semantic system is called a *(P)(S)(Q)I-system* iff it is determined by a (P)(S)(Q)I-structure.

Summarizing the concepts introduced in this definition, we have the following table:

$\langle S, \vdash \rangle$ is a...	iff $\vdash$ is a...	$\vdash$ thus being a subset of
I-structure	a finite set of I-ons	$\mathrm{FPow}(S) \times S$
QI-structure	a finite set of QI-ons	$\mathrm{FPow}(S) \times \mathrm{FPow}(S)$
SI-structure	a finite set of SI-ons	$\mathrm{Seq}(S) \times S$
PI-structure	a set of I-ons	$\mathrm{FPow}(S) \times S$
SQI-structure	a finite set of SQI-ons	$\mathrm{Pow}(S) \times \mathrm{Pow}(S)$
PQI-structure	a set of QI-ons	$\mathrm{FPow}(S) \times \mathrm{FPow}(S)$
PSI-structure	a set of SI-ons	$\mathrm{Seq}(S) \times S$
PSQI-structure	a set of SQI-ons	$\mathrm{Pow}(S) \times \mathrm{Pow}(S)$

[2] For an exposition of sequent calculus and its relationship to the more straightforwardly inferential approach as embodied in natural deduction see, e.g., (Negri & Plato, 2001).

[3] Throughout the whole paper we identify singletons with their respective single elements; hence we often write simply v instead of $\{v\}$.

Our aim now is to find criteria of the various levels of inferentializability. Before we state and prove theorems crucial in this respect, we introduce some more definitions.

3 Criteria of inferentializability

Definition 4. Let U be a set of valuations of a semantic system $\langle S, V\rangle$ (i.e. a subset of $\{0,1\}^S$). $T(U)$ (the set of *U-tautologies*) will be the set of all those elements of S which are mapped on 1 by all elements of U; and analogously $C(U)$ (the set of *U-contradictions*) will be the set of all those elements of S which are mapped on 0 by all elements of U. Let X and Y be subsets of S. The *cluster generated by X and Y*, $\mathrm{Cl}[X,Y]$, will be the set of all the valuations that map all elements of X on 1 and all elements of Y on 0. Generally, U is a *cluster* iff it contains (and hence is identical with) $\mathrm{Cl}[T(U), C(U)]$. A cluster U is called *finitary* iff both $T(U)$ and $C(U)$ are finite, it is called *inferential* iff $C(U)$ is a singleton.

Now it is clear that a semantic system $\langle S, V\rangle$ is a PSQI-system iff $\{0,1\}^S \setminus V$ is a union of clusters. (Hence every semantic system is a PSQI-system, for every single valuation constitutes a cluster.) The reason is that a system is a PSQI-system if its inadmissible valuations are determined by a set of SQI-ons and what an SQI-on excludes is a cluster of valuations. If we use specific kinds of SQI-ons, such as SI-ons, we will have a specific kind of clusters, like inferential clusters; and if we allow for only a finite number of SQI-ons, we will have to count with only finite unions. This yields us the facts summarized in the following table:

$\langle S, V\rangle$ is a ...	iff $\{0,1\}^S \setminus V$ is a union of ...
PSQI-system	clusters
PSI-system	inferential clusters
PQI-system	finitary clusters
SQI-system	a finite number of clusters
PI-system	finitary inferential clusters
SI-system	a finite number of inferential clusters
QI-system	a finite number of finitary clusters
I-system	a finite number of finitary inferential clusters

Theorem 1. *A semantic system $\langle S, V\rangle$ is a PSI-system iff V contains every $v \in \{0,1\}^S$ such that for every $A \in C(v)$ there is a $v' \in V$ such that $T(v) \subseteq T(v')$ and $A \in C(v')$.*

Proof. A semantic system $\langle S, V\rangle$ is a PSI-system system iff $\{0,1\}^S \setminus V$ is a union of inferential clusters. This is to say that it is a PSI-system iff for every $v \in \{0,1\}^S \setminus V$ there is a set $X \subseteq T(v)$ and a sentence $A \in C(v)$ such that no valuation v' such that $X \subseteq T(v')$ and $A \in C(v')$ is admissible. In other words, $\langle S, V\rangle$ is a PSI-system iff for every $v \notin V$ there is a set $X \subseteq T(v)$ and a sentence $A \in C(v)$ such that V does not contain any v' such that $X \subseteq T(v')$ and $A \in C(v')$. By contraposition, $\langle S, V\rangle$ is a PSI-system iff the following holds: given a valuation v, if for every set $X \subseteq T(v)$ and every sentence $A \in C(v)$ there is a $v' \in V$ such that $X \subseteq T(v')$ and $A \in C(v')$, then $v \in V$. This condition can obviously be simplified to: given a valuation v, if for every sentence $A \in C(v)$ there is a $v' \in V$ such that $T(v) \subseteq T(v')$ and $A \in C(v')$, then $v \in V$. □

Theorem 2. *A semantic system $\langle S, V\rangle$ is a PQI-system iff V contains every v such that for every finite $X \subseteq T(v)$ and finite $Y \subseteq C(v)$ there is a $v' \in V$ such that $X \subseteq T(v')$ and $Y \subseteq C(v')$.*

Proof. A semantic system $\langle S, V\rangle$ is a PQI-system system iff $\{0,1\}^S \setminus V$ is a union of finite clusters. This is to say that it is a PQI-system iff for every $v \in \{0,1\}^S \setminus V$ there are finite sets $X \subseteq T(v)$ and $Y \subseteq C(v)$ such that no valuation v' such that $X \subseteq T(v')$ and $Y \subseteq C(v')$ is admissible. In other words, $\langle S, V\rangle$ is a PQI-system iff for every $v \notin V$ there are sets $X \subseteq T(v)$ and $Y \subseteq C(v)$ such that V does not contain any v' such that $X \subseteq T(v')$ and $Y \subseteq C(v')$. By contraposition, $\langle S, V\rangle$ is a PQI-system iff the following holds: given a valuation v, if for every sets $X \subseteq T(v)$ and $Y \subseteq C(v)$ there is a $v' \in V$ such that $X \subseteq T(v')$ and $Y \subseteq C(v')$, then $v \in V$. This condition can obviously be simplified to: given a valuation v, if for every finite $X \subseteq T(v)$ and finite $Y \subseteq C(v)$ there is a $v' \in V$ such that $X \subseteq T(v')$ and $Y \subseteq C(v')$, then $v \in V$. □

We leave out the proof of the following theorem, as it is straightforwardly analogous to the proofs of the previous two.

Theorem 3. *A semantic system $\langle S, V\rangle$ is a PI-system iff V contains every v such that for every finite $X \subseteq T(v)$ and every $A \in C(v)$ there is a $v' \in V$ such that $X \subseteq T(v')$ and $A \in C(v')$.*

Hence we have necessary and sufficient conditions for a semantic system being a PSI-, a PQI-, or a PI-system. Unfortunately, we do not have such conditions for its being an SQI-, an SI-, a QI-, or an I-system. However, we are able to formulate at least a useful *necessary* condition for its being an SQI-system.

Theorem 4. *A semantic system $\langle S, V\rangle$ is an SQI-system only if V contains no v such that for every finite $X \subseteq T(v)$ and finite $Y \subseteq C(v)$ there is a $v' \notin V$ such that $X \subseteq T(v')$ and $Y \subseteq C(v')$.*

Proof. A semantic system $\langle S, V\rangle$ is a PQI-system iff $\{0,1\}^S \setminus V$ is a finite union of clusters. Hence if it is a PQI-system, there must exist a finite set I and two collections $\langle X^i\rangle_{i\in I}$, $\langle Y^i\rangle_{i\in I}$ of subsets of S so that

$$\{0,1\}^S \setminus V = \bigcup_{i\in I} \mathrm{Cl}[X^i, Y^i].$$

This is the case iff V equals the complement of $\bigcup\limits_{i\in I} \mathrm{Cl}[X^i, Y^i]$, hence iff

$$V = \bigcap_{i\in I} \overline{\mathrm{Cl}[X^i, Y^i]}.$$

But as $\mathrm{Cl}[X^i, Y^i] = \{v : X^i \subseteq T(v) \text{ and } Y^i \subseteq C(v)\}$,

$$\begin{aligned}\overline{\mathrm{Cl}[X^i, Y^i]} &= \{v : X^i \nsubseteq T(v) \text{ or } Y^i \nsubseteq C(v)\} = \\ &= \{v : X^i \cap C(v) \neq \varnothing \text{ or } Y^i \cap T(v) \neq \varnothing\} = \\ &= \{v : X^i \cap C(v) \neq \varnothing\} \cup \{v : Y^i \cap T(v) \neq \varnothing\} = \\ &= \bigcup_{x\in X^i} \{v : x \in C(v)\} \cup \bigcup_{y\in Y^i} \{v : y \in T(v)\} = \\ &= \bigcup_{x\in X^i} \mathrm{Cl}[\varnothing, \{x\}] \cup \bigcup_{y\in Y^i} \mathrm{Cl}[\{y\}, \varnothing].\end{aligned}$$

Now using the generalized de Morgan's law saying that

$$\bigcap_{j\in I}\bigcup_{j\in J} Z_i^j = \bigcup_{f\in F}\bigcap_{j\in I} Z_{f(j)}^j$$

where $F = I^J$, we can see that

$$V = \bigcup_{f\in F}\bigcap_{j\in f^+} \mathrm{Cl}[f(j), \varnothing] \cap \bigcap_{j\in f^-} \mathrm{Cl}[\varnothing, f(j)]$$

where F is the set of all functions mapping every $i \in I$ on an element of $f(i)$ of $X_i \cup Y_i$, and f^+, and f^-, respectively, are the sets of all those elements of I that are mapped by f on elements of X_i, and Y_i, respectively. It further follows that

$$V = \bigcup_{f\in F} \mathrm{Cl}[X^f, Y^f]$$

where $X^f = \{f(j) : j \in f^+\}$ and $Y^f = \{f(j) : j \in f^-\}$. As both f^+ and f^- are finite, this means that V is a union of finite clusters. It follows that for every $v \in V$ there are finite sets $X \subseteq T(v)$ and $Y \subseteq C(v)$ such that every valuation v' such that $X \subseteq T(v')$ and $Y \subseteq C(v')$ is admissible. In other words, for every $v \in V$ there are sets $X \subseteq T(v)$ and $Y \subseteq C(v)$ such that V contains every v' such that $X \subseteq T(v')$ and $Y \subseteq C(v')$. By contraposition: given a valuation v, if for every set $X \subseteq T(v)$ and $Y \subseteq C(v)$ there is a $v' \notin V$ such that $X \subseteq T(v')$ and $Y \subseteq C(v')$, then $v \notin V$. This condition can obviously be simplified to: given a valuation v, if for every finite $X \subseteq T(v)$ and finite $Y \subseteq \mathrm{C}(v)$ there is a $v' \notin V$ such that $X \subseteq T(v')$ and $Y \subseteq C(v')$, then $v \notin V$. □

4 A hierarchy of semantic systems

Let us introduce some more definitions.

Definition 5. A semantic system $\langle S, V\rangle$ is called

- *saturated* iff V contains every v such that for every $A \in C(v)$ there is a $v' \in V$ such that $T(v) \subseteq T(v')$ and $A \in C(v')$;
- *compact* iff V contains every v such that for every finite $X \subseteq T(v)$ and finite $Y \subseteq C(v)$ there is a $v' \in V$ such that $X \subseteq T(v')$ and $Y \subseteq C(v')$;
- *co-compact* iff V contains no v such that for every finite $X \subseteq T(v)$ and finite $Y \subseteq C(v)$ there is a $v' \notin V$ such that $X \subseteq T(v)$ and $Y \subseteq C(v')$.
- *compactly saturated* iff V contains every v such that for every finite $X \subseteq T(v)$ and every $A \in C(v)$, there is a $v' \in V$ such that $X \subseteq T(v')$ and $A \in C(v')$.

Given these, we can rephrase the theorems we have proved in the following way:

Theorem 5. *A semantic system $\langle S, V\rangle$ is*

- *always a PSQI-system;*
- *a PSI-system iff it is saturated;*
- *a PQI-system iff it is compact;*
- *an SQI-system only if it is co-compact;*
- *a PI-system iff it is compactly saturated;*

Moreover, easy corollaries of the theorems are the following necessary conditions for a system being an SI-, a QI- and an I-system:

Corollary 2. *A semantic system $\langle S, V\rangle$ is*

- *an SI-system only if it is saturated and co-compact;*

- *a QI-system only if it is compact and co-compact;*
- *an I-system only if it is compactly saturated and co-compact.*

The kinds of semantic systems we have introduced can be arranged into the following diagram, where the arrows indicate containment in the sense that an arrow leads from a concept to a different one if the extension of the former includes that of the latter.

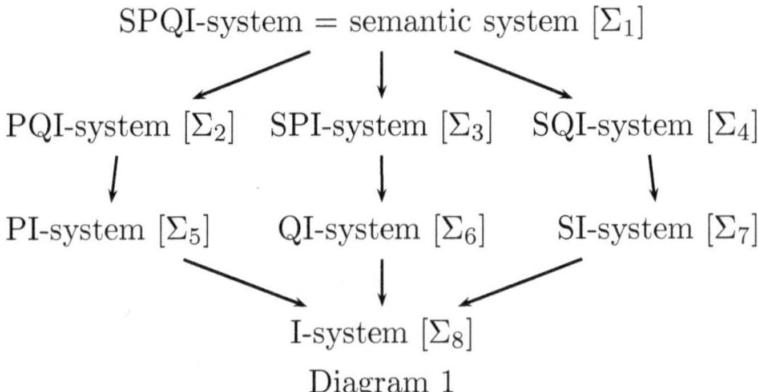

Diagram 1

What we are going to show now is that all the inclusions are *proper*. The symbols in brackets following each kind term is the name of a semantic system which will witness the properness. The systems are the following (S is supposed to be an infinite set):

- $\Sigma_1 = \langle S, \{v \in \mathrm{Pow}(S) : T(v) \text{ is finite}\}\rangle$;
- $\Sigma_2 = \langle S, \{\varnothing\}\rangle$;
- $\Sigma_3 = \langle S, \{v \in \mathrm{Pow}(S) : C(v) \text{ is finite}\}\rangle$;
- $\Sigma_4 = \langle S, \mathrm{Pow}(S) \setminus \{S\}\rangle$;
- $\Sigma_5 = \langle S, \{S\}\rangle$;
- $\Sigma_6 = \langle \{A, B\}, \{\{A\}, \{B\}\}\rangle$;
- $\Sigma_7 = \langle S, \{v \in \mathrm{Pow}(S) : C(v) = A\}\rangle$ for a fixed $A \in S$;
- $\Sigma_8 = \langle \{A, B\}, \{\{A, B\}, \{B\}\}\rangle$.

To show that they do fit into the very slots of Diagram 1 where we have put them, let us first give one more definition:

Definition 6. A valuation is called *full* if it maps every sentence on 1. (In other words, the full valuation is S.) A valuation is called *empty* if it maps every sentence on 0. (In other words, the empty valuation is $\varnothing$.)

Σ_1 is not saturated, for V does not contain the full valuation, f, though for every $A \in C(f)$ there is a $v \in V$ such that $T(f) \subseteq T(v)$ and $A \in C(v)$. (As there is no $A \in C(v)$, this holds trivially. It follows that no system not admitting the full valuation is saturated.) Hence it is not a PSI-system. It is not compact, because V does not contain the full valuation, but for every finite subset X of $T(f)$ it contains a v' such that $X \subseteq T(v')$ (whereas $Y \subseteq C(v')$ for every finite subset Y of $C(f)$ holds trivially); hence it is not a PQI-system. Moreover, it is not co-compact, for V contains the empty valuation, whereas as V cannot contain any valuation mapping only a finite number of sentences on 0, there is, for every finite subset Y of S, a $v' \notin V$ such that $Y = C(v')$. Hence it is not an SQI-system.

Σ_2 is a PQI-system, for it is determined by the infinite set of QI-ons $\{\langle\{A\}, \varnothing\rangle : A \in S\}$. However, it is not saturated, for V does not contain the full valuation, hence it is not a P(S)I-system. Also it is not co-compact, for Vcontains the empty valuation, whereas for every finite subset Y of S there is a $v' \notin V$ such that $X \subseteq C(v')$ (whereas that $Y \subseteq T(v')$ for every finite subset Y of $T(f)$ holds trivially); hence it is not a (S)QI-system.

Σ_3 is a PSI-system, for it is determined by the infinite set of SI-ons $\{\langle X, A\rangle : X \subseteq S$ and X is infinite$\}$. However, it is not compact, because V does not contain the empty valuation, but for every finite subset $Y$of S it contains a v' such that $Y = C(v)$; hence it is not a P(Q)I-system. Moreover, it is not co-compact, for V contains the full valuation, whereas for every finite subset $X$of S there is a $v' \notin V$ such that $X = T(v')$, hence it is not an S(Q)I-system.

Σ_4 is an SQI-system, for it is determined by the SQI-on $\langle S, \varnothing\rangle$. However, it is not saturated, for V does not contain the full valuation, hence it is not a (P)SI-system. It is not compact, because V does not contain the full valuation, but for every finite subset $X$of S it contains a v' such that $X = T(v')$; hence it is not a (P)QI-system.

Σ_5 is a PI-system for it is determined by the infinite set of I-ons $\{\langle\varnothing, \{A\}\rangle$: $A \in S\}$. But it is not co-compact, for V contains the full valuation, whereas for every finite subset X of S there is a $v' \notin V$ such that $X = T(v')$, hence it is not a (S)(Q)I-system.

Σ_6 is a QI-system for it is determined by the finite set of QI-ons $\{\langle\varnothing, \{A, B\}\rangle, \langle\{A, B\}, \varnothing\rangle\}$. But it is not saturated, for the supervaluation of V is the empty valuation, hence it is not a (P)(S)I-system.

Σ_7 is an SI-system for it is determined by the single SI-on $\langle S \setminus \{A\}, A\rangle$. But it is not compact, for V contains, for every finite subset Y of $S \setminus \{A\}$, a v' such that $T(v') = Y$ and $C(v') = A$. Hence it is not a (P)(Q)I-system.

Σ_8 is an I-system, for it is determined by the I-on $\langle\varnothing, \{B\}\rangle$.

5 Consequence revisited

If what we are interested in is the relation of consequence, then our classificatory hierarchy becomes excessively fine-grained. In particular, we are going to show that for every (P)(S)QI-system there exists a (P)(S)I-system with the same relation of consequence. To do this let us define a concept introduced by (Hardegree, 2006):

Definition 7. Let U be a set of valuations of the class S of sentences. The *supervaluation* of U is the valuation such that $T(v) = T(U)$.

The next lemma shows that our Theorem 3 is equivalent to one of Hardegree's results:

Lemma 1. *A semantic system $\langle S, V\rangle$ is a (P)(S)QI-system iff V contains supervaluations of all its subsets.*

Proof. This follows directly from the fact that $\langle S, V\rangle$ is a (P)(S)QI-system iff it is saturated, for it can be easily seen that it is saturated iff V contains supervaluations of all its subsets. □

Lemma 2. *Extending admissible valuations of a semantic system by supervaluations does not change the relation of consequence.*

Proof. Let $\langle S, V\rangle$ be a semantic system and $\models$ the relation of consequence induced by it. Let v be a supervaluation of a subset of V and let $\models^*$ be the relation of consequence induced by $\langle S, V \cup \{v\}\rangle$. Suppose the two relations do not coincide; then there is a subset X of S and an element A of S so that $X \models A$, but not $X \models^* A$. This means that it must be the case that $v(B) = 1$ for every $B \in X$ and $v(A) = 0$, but that every $v' \in V$ such that $v'(B) = 1$ for every $B \in X$ is bound to be such that $v'(A) = 1$. But as v' is the supervaluation of an $U \subseteq V$, elements of U map all elements of X on 1, whereas at least one of them maps A on 0; which is a contradiction. □

This gives us the following reduced version of Diagram 1:

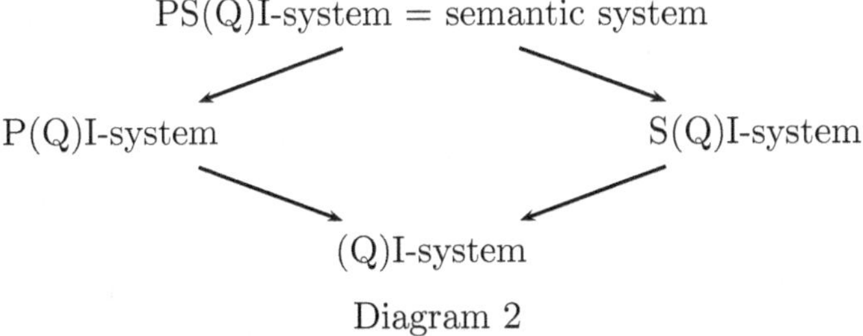

Diagram 2

Hence from the viewpoint of consequence, we have four types of semantic systems:

- Systems that are neither P(Q)I, nor S(Q)I. These are systems of the kind of Σ_1 and Σ_3.
- P(Q)I-systems that are not (Q)I-systems. Examples are Σ_2 and Σ_5.
- S(Q)I-systems that are not (Q)I-systems. Examples are Σ_4 and Σ_7.
- (Q)I-systems. Systems of the kind of Σ_6 and Σ_8.

Consequence as induced by the truth tables of classical propositional logic or by the model theory of the classical first-order predicate logic, of course, fall into the last category. Indeed any logic that has a strongly sound and complete axiomatization must trivially belong here. But even among the semantic systems studied by logicians there are some that fall outside this range ((Tarski, 1936) made this into a deep point — consequence, according to him, cannot be in general captured in terms of inferential rules).

From Diagram 2 we can see that there are two ways to go beyond the boundaries of I-systems: we may either alleviate the requirement of finiteness of antecedents of inferences, or alleviate the requirement of finiteness of the whole relation of inference. The ω-rule, which is often discussed in connection with the formalization of arithmetic, is an example of the former way; the axiom scheme of induction, that comprises an infinity of concrete axioms, is the example of the latter.

For a more specific example, consider the language of Peano arithmetic with the single admissible valuation determined by the intended interpretation within the standard model (let me call this system *true arithmetic*, TA). As it turns out, this system is a PQI-system. Indeed, it can be determined by the PQI-structure whose relation of inference consists of the I-ons of the form $\langle \varnothing, A \rangle$ for every true sentence A plus the QI-ons of the form $\langle \{B\}, \varnothing \rangle$ for every false sentence B. (We know that it is not an I-system, as we know that the truths of TA are not recursively enumerable.) Call the single admissible valuation of the system t.

If we extend the (single-element) set of admissible valuations of TA by the full valuation, it becomes saturated (indeed the supervaluation of every subset of the set of its admissible truth valuations will be admissible: the supervaluation of the empty set as well as the singleton of the full valuation is the full valuation, whereas the supervaluation of the two remaining sets is the valuation t). Hence this system is a PI-system (indeed, it is determined by the PI-structure the relation of inference of which consists of the I-ons of the form $\langle \varnothing, A \rangle$ for every sentence A true according to t plus the I-ons of the form $\langle \{B\}, C \rangle$ for every sentence B false according to t, and every sentence C) but has the same relation of consequence as the previous system.

6 Further steps

I hope to have shown how we can set up a useful framework for a systematic confrontation of proof theory and semantics, especially of inference and consequence; and that I have also indicated that this framework lets us prove some nontrivial and interesting results. However, it should be added that to bring results immediately concerning the usual systems of formal logic, our classificatory hierarchy will have to be made still more fine-grained.

The point is that while we only distinguished between systems that are determined by structures with a finite number of (S)(Q)I-ons (i.e. (S)(Q) I-systems) and those where the finiteness requirement is alleviated (the P(S)(Q)I-systems), we would need to consider systems in between these two extremes. The usual systems of formal logic can be considered as generalizing over inferential (as opposed to *pseudo*inferential) structures in two steps. First, they allow for an infinite number of (S)(Q)I-ons, which are, however, instances of a finite number of schemata. (This is, of course, possible only when we, unlike in the present paper, take into account some structuring of the set of sentences and consequently of the sentences themselves — if we consider the sentences as generated from a vocabulary by a set of rules.) This can be accounted for in terms of *parametric* SQI-ons, or p(S)(Q)I-ons. p(S)(Q)I-systems, then, fall in between (S)(Q)I-systems and P(S)(Q)I-system. Thus for example the semantic system of PA is a p(Q) I-system, for the infinity of its axioms is the union of instances of a finite number of axiom schemas. The semantic system of TA is a pSI-system, for we know that we can have its sound and complete axiomatization if we extend the axiomatic system with the omega-rule, which is, in our terminology, a pSI-on. Second they allow for infinite sets of (S)(Q)I-ons that are generated by a finite number of metainferential rules from sets of instances of finite number of schemata.

Jaroslav Peregrin
Department of Logic, Institute of Philosophy, Academy of Sciences
Jilská 1, 110 00 Praha 1, Czech Republic
peregrin@ff.cuni.cz
http://jarda.peregrin.cz

References

Gentzen, G. (1934). Untersuchungen über das logische Schliessen 1. *Mathematische Zeitschrift*, *39*, 176-210.

Gentzen, G. (1936). Untersuchungen über das logische Schliessen 2. *Mathematische Zeitschrift*, *41*, 405-431.

Hardegree, D. M. (2006). Completeness and super-valuations. *Journal of Philosophical Logic*, *34*, 81-95.

Negri, S., & Plato, J. von. (2001). *Structural proof theory.* Cambridge: Cambridge University Press.

Peregrin, J. (1997). Language and its models. *Nordic Journal of Philosophical Logic*, *2*, 1-23.

Peregrin, J. (2006). Meaning as an inferential role. *Erkenntnis*, *64*, 1-36.

Tarski, A. (1936). Über den Begriff der logischen Folgerung. *Actes du Congrés International de Philosophique Scientifique*, *7*, 1-11.

Meaning and Compatibility: Brandom and Carnap on Propositions

Martin Pleitz

1 Brandom's and Carnap's semantics

Robert Brandom in his Locke Lectures[1] has developed a new formal semantics that is based entirely on the one primitive notion of the incoherence of sets of sentences (Brandom, 2008, pp. 117–175). A language, i.e. a set of atomic sentences, is formally interpreted by an incoherence partition of its power-set. The incoherence partition must satisfy only one condition that Brandom calls "persistence", i.e., all sets containing an incoherent subset must themselves be incoherent. The incompatibility of two sentences is then defined as the incoherence of their union. Two sentences are called *incompatibility-equivalent* (or *I-equivalent* for short) if and only if they are incompatible with the same sets of sentences. In that case the two sentences are said to be *synonymous*. Therefore, the proposition expressed by a sentence can be represented by the *incompatibility-set* of the sentence, i.e., by the set of sets of sentences that are incompatible with it (Brandom, 2008, pp. 123ff.). This formal framework is a *semantics* because the basic notion of incoherence suffices to give an account of the meaning of sentences.

More than sixty years earlier, Rudolf Carnap had already proposed to represent propositions as sets of sets of sentences, although in a different way. Like Brandom, Carnap builds a language from a set of atomic sentences. To this language are added the connectives of propositional logic.[2] Carnap then defines a *state-description* as a set such that for each atomic sentence, either the sentence or its negation is an element of it (Carnap, 1947/1956,

[1] Brandom describes incompatibility semantics at length in the fifth of his John Locke Lectures, which he held 2006 in Oxford and 2007 in Prague and that have been published as (Brandom, 2008). The basic idea is sketched first in (Brandom, 1985), and mentioned repeatedly in (Brandom, 1994).

[2] Their meaning is given by the usual truth-tables.

p. 9). Two sentences are said to be *L-equivalent* if and only if they are members of the same state-descriptions. In that case the two sentences are said to be *synonymous*. Thus, the proposition expressed by a sentence can be represented by its *range*, i.e., by the set of state-descriptions that it is an element of (Carnap, 1947/1956, pp. 7–32, p. 181).[3] Carnap intends his state-descriptions to "represent Leibniz' possible worlds or Wittgenstein's possible states of affairs" (Carnap, 1947/1956, p. 9). But state-description semantics differs crucially from the mainstream of possible worlds semantics. In general, possible worlds are seen as basic objects.[4] State-descriptions, by contrast, are set-theoretical constructions from linguistic entities. Therefore state-description semantics is a precursor of theories that reduce possible worlds to maximally compatible sets of sentences.[5]

The fact that state-descriptions are *reductionist* possible worlds brings out a first similarity between Brandom's and Carnap's formal semantics: Both are *semantics without the world.* Meaning is *not* modeled as a relation between language and essentially non-linguistic objects. Rather, both incompatibility semantics and state-description semantics represent meaning by set-theoretical constructions built from linguistic objects alone.[6] It is the aim of my talk to show that the similarities between Brandom's and Carnap's semantics are more than superficial. I hope this comparison will shed some light on both theories.

2 Modifying state-description semantics

Before starting the comparison, a final adjustment must be made to Carnap's theory. Carnap cannot achieve his own aims, because his theory does

[3] Strictly speaking, this is true only for atomic sentences. For non-atomic sentences, we have to lay down the recursive definition of what it means for a sentence to hold in (i.e. to be true at) a state-description.

[4] On this point, radical modal realists like David Lewis, who holds that possible worlds are concrete objects (Lewis, 1986, pp. 81ff.), and moderate modal realists like Saul Kripke, who takes possible worlds to be stipulated, have to agree (Kripke, 1972/1980, pp. 15ff.).

[5] Reductionism is endorsed by Robert Adams, Alvin Plantinga, Robert Stalnaker, Andrew Roper, Phillip Bricker and Maxwell Cresswell ((Adams, 1979), (Plantinga, 1974), (Stalnaker, 1979), (Roper, 1982), (Bricker, 1987), (Cresswell, 2006)). The clearest statement of the theory was perhaps given by one of its opponents: David Lewis describes reductionism — that he calls "linguistic ersatzism" — from a critical perspective, but in great detail (Lewis, 1986, pp. 142–165).

[6] A similarity that is more than superficial concerns modality. Brandom and Carnap share a global understanding of modality, where there is no equivalent of a Kripkean relation of accessibility, and necessity conforms to the axioms of S5 (Brandom, 2008, pp. 129ff., 141ff.), (Carnap, 1947/1956, pp. 10, 174f., 186). For a demonstration that incompatibility semantics can be modified to accommodate other concepts of necessity (as it turns out, of B), see (Peregrin, 2007), as well as (Göcke, Pleitz, & von Wulfen, 2008) and (Pleitz & von Wulfen, 2008).

not rule out state-descriptions that contain sentences that intuitively are incompatible. This can be shown to an example taken from *Meaning and Necessity*.

Consider the atomic sentences "Scott is human" and "Scott is a rational animal". According to Carnap's definition, there will be a state-description containing the first and the negation of the second. Therefore the two sentences will not be L-equivalent and hence not synonymous. But they should! Carnap stipulates the English words "human" and "rational animal" to mean the same (Carnap, 1947/1956, p. 4f.). And, only a few pages later, he explicitly states that the predicates "human" and "rational animal" are coextensional in every state-description (Carnap, 1947/1956, p. 15). This is not a small technical point. As David Lewis has pointed out, any theory that reduces possible worlds to sets of sentences must rely on a primitive modal notion, because the sentences that represent possible worlds must be *coherent* in a sense that exceeds mere logical consistency. Lewis gives the example of "the positive and negative charge of point particles" (Lewis, 1986, p. 154).

As a preliminary solution, let us modify state-description semantics by ruling out those state-descriptions that contain atomic sentences that intuitively are incompatible. This modification will turn out to play a crucial role in the comparison of state-description semantics to incompatibility semantics (Sections 5–7). Therefore it is important to note that the *justification* of this modification of state-description semantics is independent of the enterprise of comparing it to incompatibility semantics: The criticism of Carnap's text is immanent, and Lewis's argument is perfectly general.

3 Mutual simulation

With this modification in place (Section 2), the comparison of Brandom's and Carnap's semantics can start. We have seen that, in incompatibility-semantics, incompatibility-sets explicate[7] propositions, and I-equivalence explicates synonymy. In state-description semantics, ranges explicate propositions, and L-equivalence explicates synonymy (Section 1). So there are two questions: What is the relation of the *objects representing propositions*, i.e. of incompatibility-sets and ranges? What is the relation of the *criteria of synonymy*, i.e. of I-equivalence and L-equivalence? I will answer the first question with an informal recipe for transforming incompatibility-sets into ranges and *vice versa* and the second with a theorem which says that I-equivalence and L-equivalence are coextensional.

Both Brandom and Carnap represent propositions by sets of sets of sentences. But, at least superficially, ranges and incompatibility-sets differ.

[7] For this use of "explicate", cf. (Peregrin, 2007, p. 13).

Intuitively, the range of a sentence will contain only sets of sentences *compatible* with it, while the incompatibility-set of a sentence will contain only *incompatible* sets of sentences. But nevertheless, we can move back and forth freely between incompatibility-sets and ranges. The incompatibility-set of a sentence can be represented by its complement, i.e. the *compatibility*-set of the sentence. The compatibility-set in turn can be stratified into a bundle of maximally coherent sets of sentences containing the sentence, which is the same as the range of the sentence. (Stratification is the procedure of taking out of the compatibility-set all sets of sentences contained in other sets of sentences of the compatibility-set.) To transform a range into an incompatibility-set, we just have to retrace those steps. This recipe for the transformation of incompatibility-sets and ranges is only informal, because the compared concepts stem from different theories. As yet, incompatibility-sets are defined only in incompatibility semantics and ranges are defined only in state-description semantics.

For the same reason, the theorem of coextensionality can not yet be stated or proved in either semantic theory.[8] We first have to define surrogates of the Carnapian concepts in Brandom's semantics and surrogates of the Brandomian concepts in Carnap's semantics (Appendices A & B).

In Brandom's incompatibility semantics, state-descriptions* can be defined as maximally coherent sets of sentences.[9] This gives us a definition of L-equivalence* as membership in the same maximally coherent sets. The theorem of coextensionality can now be stated and proved in incompatibility semantics. It says that two sentences are L-equivalent* just in case they are I-equivalent, i.e., that two sentences are included in the same maximally coherent sets just in case they are incompatible with the same sets of sentences (Appendix A). This has already been shown in a similar way by Jaroslav Peregrin, in his Comments on Brandom's Fifth Locke Lecture (Peregrin, 2007, p. 17f.).

In Carnap's state-description semantics, we can define incoherence* and incompatibility* on the basis of membership in state-descriptions. A pair of sentences is defined as *incompatible** if and only if there is no state-description that contains both.[10] (This concept of incompatibility* can already be found under the name of "L-exclusiveness" in Carnap's 1942

[8] As yet, the relation of I-equivalence is defined only in incompatibility semantics and the relation of L-equivalence only in state-description semantics.

[9] An asterisk (*) indicates a notion that is defined in a foreign setting. — State-descriptions can safely be treated as maximally coherent sets because of the exclusion of state-descriptions that contain incompatible sentences (Section 2).

[10] This formal notion of incompatibility* in modified state-description semantics is linked to our intuitive notion of compatibility, because we have used the intuitive notion of compatibility to rule out some of the original state-descriptions. In *unmodified* state-description semantics, the definition of compatibility* entails that *any two atomic sentences* are compatible*.

Introduction to Semantics (Carnap, 1942/1961, pp. 70, 94).) On the basis of the defined Brandomian concepts, it is possible to state and prove the theorem of coextensionality in state-description semantics (Appendix B).

In sum, it can be proved both in incompatibility semantics and in state-description semantics that two sentences are L-equivalent just in case they are I-equivalent.

4 Towards a genuine comparison

It is important to see that this does not show that Brandom's and Carnap's semantics are equivalent *simpliciter*. The results so far are that in incompatibility semantics, I-equivalence and a *defined notion* of L-equivalence* are coextensional and that in state-description semantics, L-equivalence and a *defined notion* of I-equivalence* are coextensional. Why is this not enough? Speaking metaphorically, for a comparison to be genuine it must be conducted *from the outside* of both semantic theories. Internal comparison is always in danger of working with poor substitutes.

As both theories try to give formal models of meaning, I suggest that the common ground for a genuine comparison is *a language that is intuitively interpreted*. The intuitive interpretation of a language is given by a translation of its terms into (a fragment of) natural language.[11] We will therefore have to work with particular examples of languages. Only if incompatibility semantics and state-description semantics assign the same relations of synonymy *for every intuitively interpreted language*, will it be appropriate to say that the two semantic theories are equivalent *simpliciter*.

There are three preconditions for a *fair* comparison of what incompatibility semantics and state-description semantics make of a particular intuitively interpreted language. Firstly, both sides should have shared intuitions about the relations between basic modal concepts.[12] Secondly, they should have

[11] "Interpretation" is an umbrella term for any procedure that associates meaning with expressions of a language. Here, in the context of formal languages, we can distinguish the following senses of "interpretation":

1. Interpretation (of the expressions of any formal language) by translation into the natural language we use,
2. interpretation (of the sentence letters of propositional logic) by truth tables,
3. interpretation (of the expressions of a predicate logical language) by ranges of state-descriptions, i.e. interpretation in state-description semantics, and
4. interpretation (of sentence letters) by an incoherence partition, i.e. interpretation in incompatibility semantics.

While 2, 3 and 4 are themselves part of formal theories, (i) makes use of our language, and hence can be called "intuitive" or "pre-theoretical" interpretation.

[12] For the pre-theoretical concepts of necessity, compatibility and a possible world, there should be agreement about principles like the following: A sentence is necessary if and

shared intuitions about modality in particular cases, as well.[13] Thirdly, we will need a shared formal framework, in the following sense: The intuitive interpretation of a particular language must, in each theory, uniquely determine a formal interpretation of the language. For Brandom, the formal interpretation of a language is given by its incoherence partition, and for Carnap, by the set of state-descriptions.[14]

5 Filters on state-description tables

To see how an incoherence partition relates to a set of state-descriptions, let us take another look at state-description semantics in its original form. This will provide a helpful contrast to understand formal interpretation in modified state-description semantics.

According to Carnap's original definition, a state-description is a set of sentences that, for every atomic sentence, contains either the sentence or its negation. The idea behind this definition goes back to what Ludwig Wittgenstein wrote about truth-tables in the *Tractatus.* He states that each row of a truth-table represents a possible state of affairs, and together they represent *all* possibilities (e.g., Tractatus, 4.2 & 4.3). The analogy with truth-tables provides a systematic technique for giving a complete list of state-descriptions of a given language, which we may call a "state-description table". How must a state-description table be changed to accommodate *modified* state-description semantics? The modification introduced into Carnap's original framework rules out those state-descriptions containing intuitively incompatible sentences. This amounts to *crossing out rows* in the truth-table-like list of all potential state-descriptions. Let us call a list of all rows that are crossed out a "*filter* on a state-description table".[15]

only if it is true in every possible world. Two sentences are compatible if and only if there is a possible world where both are true. A set of sentences represents a possible world if and only if it is maximally compatible. — In the context of the comparison of Brandom's and Carnap's semantics, it is important to realize that the *justification* of these principles rests in neither theory, but in our shared intuitions about modality in general.

[13] E.g., the historical Carnap might well disagree with Brandom whether it really is impossible that a blackberry be red and ripe (cf. (Brandom, 2008, p. 123)). As we want to compare semantic theories, not particular opinions about modality, we will have to abstract away from such disagreements.

[14] We should understand our intuitions about a particular language in terms of an intuitive notion of compatibility. Intuitive compatibility naturally leads to an incoherence partition and, at least according to the modification of Carnap's framework (Section 2), it determines what state-descriptions there are.

[15] This notion seems to be exactly the same as what Peregrin calls "inadmissible valuations", cf. Jaroslav Peregrin, "Inferentializing semantics and consequence", talk given on June 17, 2008, at *Logica 2008* in Hejnice.

The method of filters on state-description tables captures the formal interpretation of a language as given by modified state-description semantics. So, how does this relate to formal interpretation in incompatibility semantics, i.e. to an incoherence partition? As an incoherence partition obviously puts specific restrictions on the admissible distributions of truth-values over the atomic sentences, it uniquely determines a filter on the set of potential state-descriptions. But does a filter uniquely determine an incoherence partition? The answer to this question is not obvious, because an incoherence partition determines the coherence of *every* set of sentences, while the filter concerns only sets of maximal length.[16] But if according to the filter a (potential) state-description is coherent, then by persistence all its subsets are coherent, too. In particular, all subsets consisting of *unnegated* atomic sentences will be coherent.[17] Therefore a filter gives us all coherent sets of atomic sentences and thus uniquely determines an incoherence partition.

So we have found a genuine similarity of incompatibility semantics and modified state-description semantics: From an intuitive interpretation of a particular language, we reach an incoherence partition in Brandom's semantics and a filter on (potential) state-descriptions in Carnap's modified semantics. As there is a bijection between incoherence partitions and filters, there is a genuine equivalence between both kinds of formal interpretation.

6 Atomic sentences

Nonetheless, the equivalence of incoherence partitions and filters on state-descriptions hinges on shared intuitions about the concept of compatibility. Here there are contrary tendencies in Brandomian and Carnapian semantics.

This is obvious in the example of a simple language where all sets of sentences are said to be intuitively compatible. At least *prima facie*, Brandom and Carnap will make entirely different sense of this language. Carnap will understand compatibility as logical independence and accordingly assign the maximal number of different state-descriptions. Consequently, there will be different ranges for all atomic sentences; no two atomic sentences are synonymous. Brandom, by contrast, can make not much sense of this language, because in incompatibility semantics, the coherence of all sentences *trivializes* a language: All atomic sentences will have the same incompatibility-set, namely the empty set. So all atomic sentences will be

[16] What is more, unlike the sets dealt with by an incoherence partition, most state-descriptions contain negated sentences.

[17] Brandom has shown that the addition of the propositional connectives to incompatibility semantics is *conservative* (Brandom, 2008, p. 127). We therefore do not need to compute the incompatibility-sets of conjunctions and negations to answer the question whether a filter uniquely determines an incoherence partition.

I-equivalent and hence synonymous.[18] Why are Brandom and Carnap led to so different results?

To clarify this issue, let us take another historical step backwards, to what Wittgenstein says about elementary sentences (*Elementarsätze*) in the *Tractatus*. Apart from their syntactical atomicity, Wittgenstein's elementary sentences have the following properties:

> Property of Base: Any distribution of truth-values over all elementary sentences fixes the truth values of *all* sentences of the language. (Tractatus, 4.51, 5.3)

Wittgenstein's elementary sentences are *logically independent* in the sense that they have the following two properties:

> Property of No Entailment: No elementary sentence is logically entailed by any other elementary sentence. (Tractatus, 5.134)
>
> Property of Global Compatibility: No two elementary sentences are logically incompatible. (Tractatus, 4.211)

We can ask of every formal language whether its atomic sentences have the properties of Base, of No Entailment and of Global Compatibility. In the case of propositional logic, all three questions must of course be answered positively (that's why the method of truth-tables works). The differences of the three semantic theories in discussion can now be expressed in the following way:

The property of Base does not help to distinguish between the theories. In all three systems, sentences built only with the help of propositional logic are based on the atomic sentences, while sentences containing quantifiers or modal operators are not.[19] But the properties of No Entailment and Global Compatibility provide distinctive criteria: According to *original state-description semantics*, atomic sentences have both the property

[18] The example of the simple language $L = \{p, q, r\}$ where the set $\{p, q, r\}$ is coherent, illustrates that the notions of range* and L-equivalence* as defined in incompatibility semantics in some cases are *poor substitutes* for the notions of range and L-equivalence of state-description semantics (Section 4). In L, there will be only one state-description*, i.e. only one maximally coherent set, namely $\{p, q, r\}$, and accordingly only one range*, namely $\{\{p, q, r\}\}$, Consequently, p, q and r are L-equivalent*. So I-equivalence and L-equivalence* are indeed coextensional (Section 3). But all the same, the example of the language of three compatible sentences intuitively contradicts the equivalence of incompatibility semantics and (even modified) state-description semantics, because to the same intuitively interpreted language Carnap would assign eight different state-descriptions and three different ranges.

[19] The property of Base is lost in predicate logic (Peregrin, 1995, ch. 5). In modal propositional logic, the property of Base is lost, as well. The truth-values of all atomic sentences do not in general determine the truth-value of sentences of the form "necessarily α", because the modal operators are not truth-functional.

of No Entailment and of Global Compatibility.[20] According to *modified state-description semantics*, atomic sentences need not have the property of Global Compatibility (Section 2). And according to *incompatibility semantics*, atomic sentences need neither have the property of Global Compatibility nor the property of No Entailment. Thus the difference between modified state-description semantics and incompatibility semantics that emerged in the example of the simple language hinges on the property of No Entailment.[21]

But even this last difference could be smoothed out if state-description semantics was *fully* modified by giving up the property of No Entailment, as well. I see the following reason to do this. We allowed atomic sentences to be incompatible because we wanted to exclude state-descriptions containing particles bearing negative and positive charge (Section 2). But for a similar reason we may want to exclude state-descriptions containing whales that are not mammals (and the like). This amounts to restricting the class of state-descriptions further, to respect not only intuitive incompatibilities, but intuitive entailments, as well.

7 Holistic negation and the simulation of logical independence

Nonetheless it may seem a high price to give up all kinds of logical independence between atomic sentences. So let us turn again to incompatibility semantics. Why does Brandom accept the very low degree of logical independence? I see a reason that concerns the *holistic* character of Brandom's semantics and can be explained in the special case of negation.

To illustrate the holistic character of negation in incompatibility semantics let us return to the simple language of three compatible sentences (Section 6), that may be translated as "Carnap is a philosopher", "Scott is a philosopher" and "Brandom is a philosopher". In the simple language, the three sentences are valid and hence, their negations are incoherent. In other words, we cannot coherently say that Carnap is *not* a philosopher. Let us now *enlarge* the simple language by adding the sentences "Carnap is a florist", "Scott is a florist" and "Brandom is a florist". Let us furthermore stipulate (somewhat counterfactually) that it is incompatible to be a philosopher and a florist. Now we have reached a *different* language

[20] Carnap's original definition is thus in full agreement with Wittgenstein's ideas about elementary sentences. This is not surprising, as his state-description semantics was inspired by Wittgenstein's ideas about logical possibilities.

[21] In incompatibility semantics there are not only incompatibilities between atomic sentences, but, according to the definition of incompatibility-entailment, there may be relations of entailment between atomic sentences. In modified state-description semantics there may be incompatible atomic sentences, but at least *prima facie* this is no reason to assume that there are entailments between atomic sentences.

where there are six different incompatibility-sets for the six atomic sentences and their negations are coherent. In other words, only the addition to the language of an incompatible sentence like "Carnap is a florist" makes the negation of "Carnap is a philosopher" coherently expressible. This is an example of the holistic character of negation as defined by Brandom: The meaning of non-p depends on what may be said that is incompatible with p.

Another thing we now have achieved is that, in the six-sentence language, the original three sentences are *logically independent* of each other. From our example we can thus read off a general recipe for simulating logically independent atomic sentences in incompatibility semantics. We just have to start with an even number of atomic sentences, and lay down that each consecutive pair of sentences is incompatible, but the set of all even-numbered sentences is coherent.[22] Then all even-numbered sentences will be logically independent. So, though in incompatibility semantics there is no language where all atomic sentences are logically independent, there *are* languages where *half* of them are. In this case, we can also simulate state-descriptions that satisfy Carnap's *original* definition. The maximally coherent sets will be I-equivalent to those sets of sentences that, for each one of the logically independent sentences, contain either the sentence or its negation.

The possibility of simulating logically independent sentences and original state-descriptions helps to bring out the important difference between incompatibility semantics and *original* state-description semantics: Brandom can make sense of a lot *more* languages than Carnap originally could. So, to give up the logical independence of atomic sentences broadens the scope of formal semantics.

8 The similarity of Brandom's and Carnap's semantics

In sum, there are deep similarities between Brandom's incompatibility semantics and Carnap's modified state-description semantics. Not only does the theorem of coextensionality hold in both theories (Section 3), but for every intuitively interpreted language, we can reach equivalent formal interpretations: an incoherence partition or a filter on state-descriptions (Section 5). This equivalence holds if state-description semantics is fully modified, i.e., if it abandons the requirement that there are no entailments between atomic sentences (Section 6). And there are good reasons to do this (Sections 6 & 7).

In order to be able to spell out the result of my comparison in the form of a slogan, let me introduce the concept of a *minimally incoherent set.* Remember that a maximally coherent set is a coherent set such that, for every sentence that is not a member of it, the set plus that sentence is incoherent. A minimally incoherent set is the mirror-image of this, because it is defined

[22] The set of all queer-numbered sentences must of course be coherent, as well.

as an *in*coherent set such that, for every sentence that *is* a member of it, the set *minus* that sentence is *coherent.* Because of persistence, a complete list of minimally incoherent sets is a non-redundant way to specify *all* incoherent sets — that is: an incoherence partition.

With the help of this concept I can sum up my result in the following way: When we formally interpret a language, it is one and the same thing whether we give a complete list of *all maximally coherent sets* or a complete list of *all minimally incoherent sets.* Thus, Brandom's incompatibility semantics amounts to the same as the reductionist version of possible worlds semantics turned inside-out.

Martin Pleitz
Department of Philosophy, University of Münster
Domplatz 23, D–48143 Münster, Germany
martinpleitz@web.de

A The theorem of coextensionality in incompatibility semantics

Definition 1. A set S of sentences of L is called a *state-description** iff S is maximally coherent, i.e., iff for every sentence α, either α is a member of S or α is incompatible with S.

Two sentences α and β called *L-equivalent** iff they are included in the same maximally coherent sets.

First Theorem of Coextensionality. *The sentences α and β are L-equivalent* iff α and β are I-equivalent.*

Proof. "$\Leftarrow$": Let α and β be I-equivalent. Then, by the definition of I-equivalence, they are incompatible with the same sets of sentences (i.e. $\text{Inc}(\alpha) = \text{Inc}(\beta)$). But then they are *compatible* with the same sentences. They therefore are contained in the same maximally compatible sets of sentences, i.e., in the same state-descriptions* and consequently are L-equivalent*.

"$\Rightarrow$": Let α and β be *not* I-equivalent. Then there is a set X such that $X \cup \{\alpha\}$ is coherent while $X \cup \{\beta\}$ is incoherent. As for every compatible set of sentences there is a maximally compatible set of sentences that contains it, there is a state-description* S such that $X \cup \{\alpha\} \subseteq S$, and therefore $\alpha \in S$. As S contains X, it cannot include β, because the incoherence of $X \cup \{\beta\}$ would by persistence transfer to S. Therefore $\beta \notin S$. As $\alpha \in S$ and $\beta \notin S$, α and β are not L-equivalent*.[23] □

[23] For a similar proof, cf. (Peregrin, 2007, p. 17f.).

B The theorem of coextensionality in state-description semantics

Definition 2. A set of sentences is *incoherent** iff there is no state-description that it is a subset of (note that incoherence* is persistent). A pair of sentences is *incompatible** iff there is no state-description that contains both.

Two sentences α and β are *I-equivalent** iff α and β are incompatible* with the same sets of sentences.

Second Theorem of Coextensionality. *The sentences α and β are I-equivalent* iff they are L-equivalent.*

Proof. "$\Leftarrow$": Let α and β be L-equivalent. Then they are elements of the same state-descriptions. As compatibility* is defined by recourse to membership in state-descriptions, α and β therefore are compatible* with the same sets of sentences. Hence they are incompatible* with the same sets of sentences. So α and β are I-equivalent*.

"$\Rightarrow$": Let α and β be *not* L-equivalent. Then there is a state-description D such that α belongs to D while β does not belong to D. (Or *vice versa*. For reasons of symmetry it suffices to deal with one case.) Now let δ be the conjunction that completely describes D. Then, according to the definition of compatibility*, α and δ are compatible* while β and δ are incompatible*. Consequently, α and β are not I-equivalent*. □

References

Adams, R. M. (1979). Theories of actuality. Ithaca & London: Cornell University Press.

Brandom, R. (1985). Varieties of understanding. In N. Rescher (Ed.), *Reason and rationality in natural science.* Lanham: University Press of America.

Brandom, R. (1994). *Making it explicit.* Cambridge, MA–London: Harvard University Press.

Brandom, R. (2008). *Between saying and doing: Towards an analytic pragmatism.* Oxford–New York: Oxford University Press.

Bricker, P. (1987). Reducing possible worlds to language. *Philosophical Studies, 52*, 331–355.

Carnap, R. (1942/1961). *Introduction to semantics.* Cambridge, MA: Harvard University Press.

Carnap, R. (1947/1956). *Meaning and necessity.* Chicago: The University of Chicago Press.

Cresswell, M. J. (2006). From modal discourse to possible worlds. *Studia Logica, 82*, 307–327.

Göcke, B., Pleitz, M., & von Wulfen, H. (2008). How to Kripke Brandom's notion of necessity. In B. Prien & D. Schweikard (Eds.), *Robert Brandom: Analytic pragmatist* (pp. 135–161). Frankfurt: Ontos.

Kripke, S. (1972/1980). *Naming and necessity.* Cambridge, MA: Harvard University Press.

Lewis, D. (1986). *On the plurality of worlds.* Oxford–New York: Blackwell.

Loux, M. (1979). *The possible and the actual.* Ithaca & London: Cornell University Press.

Peregrin, J. (1995). *Doing worlds with words. formal semantics without formal metaphysics.* Dordrecht–Boston–London: Kluwer.

Peregrin, J. (2007). *Brandom's incompatibility semantics. comments on Brandom's Locke lecture V.* (Retrieved 22. 12. 2007 from http://jarda.peregrin.cz/mybibl/mybibl.php)

Plantinga, A. (1974). *The nature of necessity.* Oxford: Clarendon.

Pleitz, M., & von Wulfen, H. (2008). Possible worlds in terms of incompatibility: An alternative to Robert Brandom's analysis of necessity. In M. Peliš (Ed.), *The Logica Yearbook 2007* (pp. 119–131). Prague: Filosofia.

Roper, A. (1982). Towards an eliminative reduction of possible worlds. *The Philosophical Quarterly, 32*, 45.

Stalnaker, R. C. (1979). Possible worlds. Ithaca & London: Cornell University Press.

Wittgenstein, L. (1989). Tractatus logico–philosophicus. In *Werkausgabe* (Vol. 1). Frankfurt am Main: Suhrkamp.

Inference and Knowledge

Dag Prawitz*

We sometimes acquire new knowledge by making inferences. This fact may be seen as so obvious that it sounds strange to put as a problem how and why we get knowledge in that way. Nevertheless, there is no standard account of the epistemic significance of valid inferences.

For this discussion, I shall assume that a person's knowledge takes the form of a judgement that she has grounds for – an assumption related to the idea that a person knows that p, only if she has good grounds for holding the proposition p to be true. The question how inferences give knowledge may then be put: how may one get in possession of grounds for judgements by making inferences? One would expect that there is an easy answer to this question by just referring to how the concepts involved are understood. But, as I shall argue, there is little hope of answering this question when the notion of valid inference is understood in the traditional way. To account for the epistemic significance of valid inferences, we seem to need another approach to what it is for an inference to be valid and what it is to make an inference. I shall describe one such approach.

Given a valid argument or a valid inference from a judgement A to a judgement B, it may be possible for an agent who is already in possession of a ground for A to use this inference to get a ground for B, too. But the agent is not ensured a ground for B, just because of the inference from A to B being valid and the agent being in possession of a ground for A. The agent may simply be ignorant of the existence of this valid inference, in which case its mere existence does not make her justified in making the judgement B. A question that has to be answered is therefore what relation the agent has to have to the inference to make her justified in making the judgement B.

* Many of the ideas presented here were worked out while I was a fellow at the Institute of Advanced Studies at Università di Bologna in the spring of 2007 and were presented in lectures given at the Philosophy Department of Università di Bologna.

I shall restrict myself here to deductive inferences and conclusive grounds, and when speaking of "inference" and "ground", I shall always mean deductive inference and conclusive ground.

1 The problem

The question under what condition an agent, call her P, gets a ground for the conclusion of a valid inference can be formulated more explicitly as follows, where, for brevity, I restrict myself to the case when there is only one premiss:
Given that

there is a valid inference J from a judgement A to a judgement B, (a)

and that

the agent P has a ground for A, (b)

what further condition has to be satisfied in order for it to be the case that

P has a ground for B? (d)

The problem is thus to state a further condition (c) such that (a)–(c) imply (d). Obviously, as already remarked, (a) and (b) alone do not imply (d). Hence, we need to specify a condition (c), in other words a relation between an agent P and an inference J, which describes how the agent arrives at a ground for the conclusion of J.

2 A first attempt to find a condition (c)

Since the mere existence of a valid inference J to B from a judgement A, for which the agent has a ground, is not sufficient to give her a ground for B, one may think[1] that the extra condition that has to be satisfied is that

the agent P knows that the inference J from A to B is valid (c_k)

(the subscript k for 'knowing').

But clearly we do not normally establish the validity of an inference before we use it. If it were a necessary condition always to do so, a regress would result. The argument used to establish the validity would need some inferences, and if the validity of them had again to be established to give the argument any force, there would be a need of yet a further argument and so on. Unless there were some inferences whose validity could be known without any argument, we would be involved in an endless regress of trying

[1] This seems to be taken for granted by, e.g., (Etchemendy, 1990, p. 93).

to establish the validity of inferences before any could be used to get a ground for its conclusion, and hence we would never be able to acquire knowledge by inferences. At least in case the validity of an inference is defined in a model-theoretical way, analogously to how logical consequence is defined, it cannot be maintained that knowledge of validity is immediate and does not require an argument to be known.

I want to remark in passing that the regress noted above is different from the well-known Bolzano–Lewis regress.[2] This latter regress cast doubts not only on the necessity of the condition (c_k) but even on whether the condition is sufficient to guarantee the agent a ground for the conclusion. Why should we think (c_k) to be sufficient? Presumably because having a ground for the judgement A and hence knowing that the proposition occurring in A (i.e., the one affirmed by A) is true *and* knowing that the inference from A to B is valid and hence that it is truth preserving (in the sense that if the proposition occurring in A is true then so is the proposition occurring in B), the agent can infer B by simply applying modus ponens. This gives her a ground for B, one may think. If so, the reason for saying that the agent has a ground for B, when she knows the inference to be valid, seems to be that there is another inference than the original one from A to B, namely, an inference from two premisses, one of which is the agent's knowledge of the validity of the original inference. It is because of this new inference that the agent is claimed to get a ground for B. But by the same reasoning, what really guarantees the agent to have a ground for B is her knowledge of the validity of this new inference. In other words, there is a third inference with *three* premisses, one of which is the agent's knowledge of the validity of this second inference, and so on.

Already from the first regress discussed above we must conclude that (c_k) is not the right condition that we are seeking to describe how we generally acquire knowledge or grounds by inferences. At least, we must conclude this if "knows" in (c_k) means something like having established by argument. One may suggest that there is another concept of knowledge that is relevant here, for instance knowledge based on immediate evidence or implicit knowledge manifested in behaviour, like the implicit knowledge of meaning that Michael Dummett has called attention to. I do not want to deny that there may be a notion of validity of inference for which one can clarify such a concept of knowledge so that (c_k) becomes an appropriate condition. As already said, it would certainly require a departure from what now seems to be the dominant understanding of the validity of an inference in terms of truth preservation for all variations of the meaning of the non-logical expressions involved in the inference. Anyway, lacking a concept of knowledge

[2] (Bolzano, 1837) and (Carroll, 1895). I have discussed this regress more thoroughly in (Prawitz, 2009).

(and of validity) that makes (c_k) appropriate — to develop on would be a major task — I shall now leave this first attempt to find a condition (c).

3 A second attempt to find a condition (c)

Realizing that (c_k) is not the right condition, one may see the proposal of it as an overreaction to the simple observation first made, viz. that an agent need to stand in some relation to the inference, if it is to provide her with a ground for its conclusion. Of course, the mere existence of a valid inference cannot automatically provide the agent with a ground for the conclusion, one may say. She has to *do* something. But she does not need to establish the validity of the inference. All that is needed is that she actually uses the inference. Then the validity of the inference does provide the agent with a ground for the conclusion, given that she already has one for the premiss. One may thus suggest that the condition sought for should simply be

$$P \text{ makes the inference } J, \text{ that is, infers } B \text{ from } A. \tag{c}$$

There is clearly something right in this suggestion. One should distinguish between *an inference act*, and an *inference* in the sense of an argument determined by a number of premisses and a conclusion. It is first when an agent makes an inference, i.e., carries out an inference act, that the question arises whether she is justified in making the assertion that occurs as conclusion of the inference.

But it must then be asked what is meant by making an inference or by inferring a conclusion B from a premiss A. We usually announce the result of such an act verbally by simply first making the assertion A, then saying "hence" or "therefore" B, or, in the reverse order, we first make the assertion B, and then say "since" or "because" A. An inference act, looked upon as verbal behaviour, can be seen as a kind of complex speech act in which we do not only make an assertion but also give a reason for the assertion in the form of another assertion or some other assertions from which it is (implicitly or explicitly) claimed to follow.

However, if this is all that is meant by inferring a conclusion from a premiss, then one cannot expect that conditions (a), (b), and (c) together with reasonable explications of the notions involved are sufficient to imply that the person in question has a ground for the conclusion B. To see this, one may consider a scenario where a person announces an inference in the way described, say as a step in a proof, but is not able to defend the inference when it is challenged. Such cases occur actually, and the person may then have to withdraw the inference, although no counter example may have been given. If it later turns out that the inference is in fact valid, perhaps by a long and complicated argument, the person will still not be considered to have had a ground for the conclusion at the time when she asserted it, and the proof that she offered will still be considered to have had a gap at that time. This would be a situation in which conditions (a), (b) and (c) were all satisfied, but (d) would not be said to hold.

If the inference from A to B is generally recognized as valid, then, sociologically speaking so to say, an agent who has a ground for A and makes the assertion B, giving A as her reason, will certainly be considered to have a ground for her assertion. If instead the inference is not obviously valid even to experts, the agent is not considered to have obtained a ground for B because of making the assertion B and giving A as her reason. But we are of course not satisfied with a sociological description of when an agent is considered to have a ground (obtained, e.g., by adding as a condition that the validity of the inference should be generally recognized by experts in the field).

It thus still remains to state the appropriate condition under which a valid inference gives an agent a ground for the conclusion of an inference. One may think that it must be basically right that we get a ground for a judgement by inferring it form other judgements for which we already have grounds, and that hence condition (c) is rightly stated as above. But then "to infer" or "to make an inference" must mean something more than just stating a conclusion and giving premisses as reasons. The basic intuition is, I think, that to infer is to "see" that the proposition occurring in the conclusion must be true given that the propositions occurring in the premisses are true, and the problem is how to get a grip of this metaphoric use of "see".

4 The nature of the problem

At this point it may be good to pause and consider in more detail the nature of the problem that I have posed. I have used the term ground in connection with judgements to have a name on what a person needs to be in possession of in order that her judgement is to be justified or count as knowledge, following the Platonic idea that true opinions do not count as knowledge unless one has grounds for them. The general problem that I have posed is how inferences may give us such grounds.

As I use the term ground, a person's judgement is justified or counts as knowledge when she in possession of a ground for the judgement. Consequently, one does not need to show that one is in possession of a ground for a judgment in order to be justified in making the judgement, it is enough that in fact one is in possession of such a ground. Justifications must end somewhere, as Wittgenstein puts it. And the point where they must end is exactly when one has got in possession of what counts as a justifications or a ground; something would be wrongly called ground, if it was not enough to be in possession of it — in other words, if one had yet to show something concerning the ground — in order that one's judgement was to be considered justified.

Hence, it is not the agent P who has to state an adequate condition (c) and show that (d) holds, i.e., that she has a ground for B, when the conditions (a)–(c) are satisfied; as just said, she is justified when she is in possession of a ground, regardless of what she can show about it. It is we as philosophers who have to state an adequate condition (c) and then derive (d) from (a)–(c) to give an account of the epistemic significance of valid inferences. The point of making inferences is to acquire knowledge, and philosophy of logic would not be up to its task, if it could not explain how this comes about. To explain this is to say under what conditions a valid inference can supply us with grounds.

Since the fact that we acquire knowledge by making inference is such a basic feature of logic, one should expect the account of this fact to be quite simple, once we have understood rightly the key concepts involved here, in particular the notions *valid inference, inferring* or *making an inference*, and *ground*. When these notions have been explicated appropriately, one should expect it to be a simple conceptual truth that (a)–(c) imply (d).

What is surprising is that there is no generally accepted account of the epistemic significance of inferences and that puzzling problems seem to arise when such an account is attempted. This is a sign that our usual understanding of the key concepts involved is faulty.

5 Logical consequence and logically valid inference

Since it is *valid* inferences that allow epistemic progress, a crucial ingredient in the account must be to give that notion an adequate meaning. The concept of valid inference is traditionally connected with that of logical consequence and with necessary truth-preservation. Often one simply says that an inference is valid if and only if the conclusion is a logical consequence of the premisses, which in turn is equated with it being necessarily the case that the conclusion is true if all premisses are. However, I have been following Frege in taking the premisses and the conclusion of an act of inference to be speech acts in which a proposition is judged to be true, hence taking the premisses and conclusion of an inference to be judgements.[3] The traditional idea of inference as necessarily truth preserving is then better formulated by saying that an inference is valid if and only it necessarily holds

[3] In more recent time, the point that premisses and conclusions are not propositions but judgements or assertions has especially been emphasized by Per Martin-Löf (Martin-Löf, 1985); see also (Sundholm, 1998). In contrast to Frege and Martin-Löf, however, I shall also consider the case when the premisses and conclusion are judgements made under assumptions.

that when all the propositions affirmed in the premisses are true, then so is the proposition affirmed in the conclusion.

Since Alfred Tarski's (1936) revival of Bernard Bolzano's (1837) definition of logical consequence, it has been common to interpret the modal notion of necessity in this context extensionally, saying in effect, as we all know, that a proposition (or sentence) B is a logical consequence of set a Γ of propositions (or sentences) if and only if for all variations of the content of the non-logical notions occurring in B and in the elements of Γ, it is in fact the case that B is true when all the elements of Γ are.

How does the validity of an inference contribute, together with the other two conditions (b) and (c), to the agent being in possession of a ground for the conclusion? This is the crucial question that any proposed notion of validity has to face. In particular, why should the fact that an inference is truth-preserving contribute to our getting a ground for the conclusion? As already noted, an agent's *knowledge* that an inference is truth preserving would contribute to her getting a ground for the conclusion of the inference, but such knowledge should not presumed, and the *fact* that it is truth preserving is irrelevant.

Now, nobody suggests that the validity of an inference is to be defined in terms of just truth preservation. Following Bolzano and Tarski, the model-theoretical definition says that an inference is valid if it is truth preserving regardless of how the contents of non-logical expressions are varied. But this does not essentially change the situation. Why should the additional fact that the inference is truth-preserving also when the content of the non-logical expressions is varied be relevant to question whether the agent can see that the proposition affirmed in the *actual* situation (where the content is not varied) is true? The same can be said of validity defined in terms of necessary truth preservation, if the necessity is understood ontologically. Why should the fact that an inference is truth preserving in *other* possible worlds help the agent to see that the proposition affirmed in the conclusion is true in the *actual* world? It is difficult to see how anything but *knowledge* of this fact could be relevant here (and if knowledge is assumed, it is sufficient to know that the truth of the propositions in the premisses materially implies the truth of the proposition in the conclusion — i.e., no variation of content is needed). Therefore, there seems to be little hope that one can find an appropriate condition (c) when validity of inference is defined in the traditional way.

It is different if the necessity is understood epistemically and this is taken to mean that the truth of the propositions asserted by the premisses, which the agent is assumed to know, somehow guarantees that the agent can see that the proposition asserted in the conclusion is true. Such an epistemic necessity comes close to Aristotle's definition of a syllogism as an argument where "certain things being laid down something follows of necessity from them, i.e., because of them without any further term being needed to justify the conclusion."[4] It is of course right to say that there is an epistemic tie between premisses and conclusion in a valid inference — some kind of thought necessity, we could say, thanks to which the conclusion can become justified. But to say this is not to go much beyond our starting point. It still remains to say how the justification comes about.

Although the idea of Bolzano and Tarski to vary the content of non-logical terms does not help us in defining the validity of inference, this idea is still useful for defining *logical* consequence and the *logical* validity of inference. One important ingredient in our intuitive idea of logical consequence and logical validity of inference is, I think, that they are topic neutral, and one natural way to express this is to say that they are invariant under variations of non-logical notions.

I suggest that we distinguish between *logical consequence* and *deductive* (or *analytic*) *consequence* and similarly between an inference being *logically valid* and it being only (*deductively*) *valid*. Given the latter concept we can easily define logical validity in the style of Bolzano and Tarski:

> *An inference J is* logically valid, *if and only if, for any variation of the content of the non-logical terms occurring in the premisses and conclusion of J, the resulting inference is valid.*

The variation of content may be produced by making substitutions for the non-logical terms in the manner of Bolzano or by considering assignments of values to them in the manner of Tarski; we do not need to go into these technical details here.

This definition makes justice to the idea that whether an inference is logically valid is independent of the meaning of the non-logical terms. However, the logical validity of an inference is now not reduced to truth preservation or to the truth of a generalized material implication but to the validity of an inference under variations of the contents of non-logical terms.

Some inferences are logically valid, in addition to being deductively valid, and this is an interesting feature of them, but it is not a feature on which the conclusiveness of the inference hinges. It thus remains to analyse deductive validity and bring out how such an inference may deliver a ground for its conclusion.

6 Grounds

Rational judgements and sincere assertions are supposed to be made on good grounds. It is not that an assertion is usually accompanied by the

[4] See (Ross, 1949, p. 287).

statement of a ground for it; in other words, the speaker often keeps her ground for herself. But if the assertion is challenged, the speaker is expected to be able to state a ground for it. To have a ground is thus to be in a state of mind that can manifest itself verbally.

I am here interested in grounds that are obtained by making inferences; all grounds can of course not be obtained in this way, and I shall soon return to some examples. When an assertion is justified by way of an inference, it is common to indicate this by simply stating the inference in the way discussed above, and the premisses of the inferences are then often called the ground for the assertion. This way of speaking may be acceptable in an everyday context, but it conceals the problem that we are dealing with, which is probably one reason why the problem has been so neglected. It makes it seem as if one automatically has a ground for a conclusion by just stating an inference that in fact happens to be valid — in effect, it seems as one may get a ground by simply stating that one has one. We have discussed above (Section 3) why a ground for a conclusion is not forthcoming by "inferring" it in this superficial sense.

But there are also other reasons why it is not a good terminology to use the term "ground" for the premisses of an inference. The premisses are judgements or assertions affirming propositions, and the fact that one has judged or asserted them as true cannot constitute a ground for the conclusion, nor can the truth of the propositions affirmed constitute such a ground; at least not in the sense of something that an agent is in possession of, thereby becoming justified in making the assertion expressed in the conclusion. It is rather the fact, if it is a fact, that the agent has *grounds* for the premisses that is relevant for her having a ground for the assertion made in the conclusion. But the grounds for the premisses are grounds for *them*, not for the conclusion. The question that I have posed is therefore put in the form: given the grounds for the premisses, how does one get from them a ground for the conclusion?

We are used to meet challenges of an inference by breaking it down into simpler steps, and when one succeeds to replace the inference by a chain of sufficiently simple inferences there is in practice no more challenges. But the philosophically interesting question is how one can meet a challenge of a simple inference that is not possible to break down into simpler steps. It is tempting to fall back at that point on what our expressions mean or in other words on what propositions it is that we affirm to be true. However, to my mind, it would be dubious to say of all these inferences that we want to defend but cannot break down into simpler inferences that their validity is just constitutive for the meaning of the sentences involved.[5] The line that

[5] In some previous works (e.g., (Prawitz, 1977, 1973); cf also footnote 6) I have identified a ground for a judgement with a proof of the judgement, or I have spoken of grounds for

I shall take is instead roughly that the meaning of a sentence is determined by what counts as a ground for the judgement expressed by the sentence. Or expressed less linguistically: it is constitutive for a proposition what can serve as a ground for judging the proposition to be true. From this point of view I shall specify for each compound form of proposition expressible in first order languages what constitutes a ground for an affirmation of a proposition of that form. If one does not like this line of approach, one may anyway agree with my specification of what constitutes a ground for various judgements, which is what matters here.

For instance, a conjunction $p\&q$ will here be understood as a proposition such that a ground for judging it to be true is formed by bringing together two grounds for affirming the two propositions p and q. We may put the name *conjunction grounding*, abbreviated $\&G$, on this operation of bringing together two such grounds so as to get a ground for affirming a conjunction. If we do not want to take this view of conjunctions, we may still agree for other reasons that there is an operation $\&G$ such that if β is a ground for affirming p and γ is a ground for affirming q, then $\&G(\beta, \gamma)$ is a ground for affirming $p\&q$. Similarly, we may take it as a further constitutive fact about conjunction that conversely any ground for judging it to be true is formed by the operation of conjunction grounding or just agree to that for other reasons. What matters here is that there is such an operation $\&G$ such that something is a ground for judging $p\&q$ to be true if and only if it can be formed by applying $\&G$ to two grounds for judging p to be true and judging q to be true, respectively.

I have spoken primarily of grounds for judgements or assertions. But for brevity, we may also speak derivatively of a *ground for a proposition p* meaning a ground for the judgement or assertion that p is true. We can thus state the equivalence

> α is a ground for the conjunction $p\&q$ if and only if $\alpha = \&G(\beta, \gamma)$
> for some β and γ such β is a ground for p and γ is a ground for q.

Inferences are made not only from premisses that have been established as holding but also from assumptions and premisses that are established under assumptions. To cover such cases I shall introduce what I shall call *open* or *unsaturated* grounds besides the grounds that we have talked about so far and that I shall call *closed* grounds. Both closed grounds and unsaturated grounds will be said to be *grounds*.

An unsaturated ground is like a function and is given with a number of open argument places that have to be filled in or saturated by closed

sentences and have taken them to be valid arguments. I prefer not to use that terminology now, because I want to take proofs to be built up by inferences, and I do not want to say that an inference constitutes a ground for its conclusion — the question is instead how an inference can deliver a ground for the conclusion.

grounds so as to become a closed ground. Something is a ground for an assertion of A under the assumptions $A_1, A_2, \ldots, A_n$ if and only if it is an n-ary unsaturated ground that becomes a closed ground for A when saturated by closed grounds for $A_1, A_2, \ldots, A_n$. Writing $\alpha(\xi_1, \xi_2, \ldots, \xi_n)$ for the unsaturated ground and $\alpha(\beta_1, \beta_2, \ldots, \beta_n)$ for the result of saturating it by closed grounds β_i for A_i, the condition for $\alpha(\xi_1, \xi_2, \ldots, \xi_n)$ to be a ground for A under the assumptions $A_1, A_2, \ldots, A_n$ is thus that $\alpha(\beta_1, \beta_2, \ldots, \beta_n)$ is a closed ground for A.

Grounds are naturally typed by the propositions they are grounds for. The open places in an unsaturated ground, in other words, the variables used in displaying the unsaturated ground, may then also be typed to indicate the type of the grounds that can saturate them at that place, in other words, that can replace the variables. I shall usually supply variables with types but shall otherwise omit type indications.

With these notions at hand, we can specify that a closed ground for an implication $p \to q$ is something that is formed by an operation that we can call *implication grounding*, $\to G$, applied to a 1-ary unsaturated ground $\beta(\xi^p)$ for judging q to be true under the assumption that p is true. The result of applying this operation to the open ground $\beta(\xi^p)$, which I shall write $\to G\xi^p(\beta(\xi^p))$, yields thus a closed ground for $p \to q$; it corresponds on the syntactical level to a variable binding operator, and I indicate this by writing the variable ξ^p behind the operator. If it is applied to an n-ary unsaturated ground for A under the assumptions $A_1, A_2, \ldots, A_n$ written $\alpha(\xi_1, \xi_2, \ldots, \xi_n)$, I shall write $\to G\xi_i(\alpha(\xi_1, \xi_2, \ldots, \xi_n))$ to indicate that it is the i-th place in the unsaturated ground that becomes bound, which then denotes an unsaturated ground for A under the assumptions $A_1, A_2, \ldots, A_{i-1}, A_{i+1}, \ldots, A_n$.

We have thus the equivalence

α is a ground for $p \to q$ if and only if $\alpha = \ \to G\xi^p(\beta(\xi^p))$
where $\beta(\xi^p)$ is an unsaturated ground for judging that q is true
under the assumption that p is true.

Finally we have to pay attention to the fact that the premisses of an inference may be an *open judgement* $A(x_1, x_2, \ldots, x_m)$ (possibly under some open assumptions), by which I mean that its kernel is not a proposition, but a propositional function $p(x_1, x_2, \ldots, x_m)$ defined for a domain of individuals such that for any n-tuple of individuals $a_1, a_2, \ldots, a_m$, $A(a_1, a_2, \ldots, a_m)$ is the judgement that affirms $p(a_1, a_2, \ldots, a_m)$. We must therefore consider unsaturated grounds that are unsaturated not only with respect to grounds but also with respect to individuals that can appear as arguments in propositional functions. Let $A(x_1, x_2, \ldots, x_m)$ and $A_i(x_1, x_2, \ldots, x_m)$ be assertions whose propositional kernels are propositional functions over $x_1, x_2, \ldots, x_m$, and let $A(a_1, a_2, \ldots, a_m)$ and $A_i(a_1, a_2, \ldots, a_m)$ be the assertions that arise when we apply the corresponding propositional functions to the individuals $a_1, a_2, \ldots, a_m$. Then I shall say that something is an unsaturated ground for the open judgement $A(x_1, x_2, \ldots, x_m)$ under the assumptions $A_1(x_1, x_2, \ldots, x_m)$, $A_2(x_1, x_2, \ldots, x_m), \ldots, A_n(x_1, x_2, \ldots, x_m)$ if and only if it is an unsaturated ground $\alpha(\xi_1, \xi_2, \ldots, \xi_n, x_1, x_2, \ldots, x_m)$ with respect to n closed grounds and m individuals such that when saturated by the individuals $a_1, a_2, \ldots, a_m$ it becomes an unsaturated ground for $A(a_1, a_2, \ldots, a_m)$ under the assumptions $A_1(a_1, a_2, \ldots, a_m), A_2(a_1, a_2, \ldots, a_m), \ldots, A_n(a_1, a_2, \ldots, a_m)$.

We can then specify that a closed ground for a generalized proposition $\forall x p(x)$ is something that is formed by an operation that I shall call *universal grounding*, $\forall G$, applied to an unsaturated ground $\alpha(x)$ for the propositional function $p(x)$. The result of applying this operation to the open ground $\alpha(x)$, which I shall write $\forall G x(\alpha(x))$, again indicating that x becomes bound by writing it behind the operator, is thus a closed ground for $\forall x p(x)$. We have thus the equivalence

$$\alpha \text{ is a ground for } \forall x p(x) \text{ if and only if } \alpha = \forall G x(\beta(x))$$
$$\text{where } \beta(x) \text{ is an unsaturated ground for } p(x).$$

If we identify negated propositions, $\neg p$, with $(p \rightarrow \bot)$ where $\bot$ is a constant for falsehood, for which it is specified that there is no ground for $\bot$, we have specified by recursion what can be a ground for sufficiently many forms of propositions expressible in classical first order languages, except that we have said nothing about grounds for atomic propositions. What they are will of course vary with the content of the atomic propositions.

In the language of first order Peano arithmetic we may take a ground for an identity between two numerical terms $t = u$ to be a calculation of the value of t and u showing that they are the same. Alternatively, if we want to analyse a calculation as consisting of steps each of which has a ground, we need to start from more basic grounds. As already said, all grounds cannot be obtained by inferences. There must in other words be some propositions like $t = t$ or '0 is a natural number' for which it is constitutive that there are specific grounds for them that are not derived or built up from something else.

Outside of mathematics, we may consider observation statements, and for them, I suggest, we take relevant verifying observations to constitute grounds. For instance, a ground for a proposition 'it is raining' is taken to consist in seeing that it rains; taking "seeing" in a veridical sense, it constitutes a conclusive ground. It does not seem unreasonable to say that to know what proposition is expressed by "it is raining" is to know, or at least implies that one knows, how it looks when it is raining, and hence that one knows what constitutes a ground for the statement.

In the case of intuitionistic predicate logic we have to say in addition what counts as a ground for disjunctions and existential propositions, which can be done in an obvious way analogously to the case of conjunction (but which becomes too restrictive when disjunctions and existential propositions are understood classically — these forms have then instead to be defined in terms of other logical constants in the usual way).

The grounds that I have described are abstract entities that can be constructed in the mind and that we can become in possession of in that way. Alternatively, we may think of a ground for a judgement as just a representation of the state of our mind when we have justified a judgement.

The possession of a ground for a judgement can manifest itself in the naming of that object, and I have introduced a notation for doing so. An alternative way of defining the grounds would have been to lay down these ways of denoting grounds as the canonical notation for grounds, making a distinction between *canonical* and *non-canonical* forms since the same ground may be denoted by different expressions. To be in possession of a ground for a judgement could then be identified with having constructed a term that denotes a ground for that judgement.

7 Inferences

As the reader has already realized, the primitive operations introduced above to specify the grounds for the affirmation of propositions of various forms correspond to certain inference rules, namely Gentzen's introduction rules in the system of natural deduction for first order languages. For instance, conjunction grounding corresponds to the schema for conjunction introduction. Gerhard Gentzen saw the introduction rules as determining the meaning of the corresponding logical constants. I have not followed that idea here,[6] but have instead seen the specifications of what constitutes grounds for affirming propositions of a certain form to be constitutive for propositions of that form. One can say that I have carried over Gentzen's idea to the domain of grounds, since the grounds are built up by primitive grounding operations that closely correspond to his introduction rules. More precisely, it holds for every such grounding operation Φ that if we form an inference according to the corresponding inference rule, then we can apply Φ to grounds for the premisses of that inference and shall get as a result a ground for the conclusion of the inference. Furthermore, having defined a domain of grounds by presuming these primitive grounding operations, we can define other operations on the grounds of this domain that will have

[6] In some other works (see, e.g., (Prawitz, 1973) and cf. footnote 5) I have used Gentzen's idea more directly in a definition of valid argument, saying that an argument whose last inference is an introduction is valid if and only if the immediate subarguments are valid.

similar property with respect to other inferences. This makes it possible to give some substance to the idea that an inference is something more than just stating a conclusion and reasons for it, an idea that we described above (end of Section 3) in metaphorical terms as "seeing" that the proposition affirmed in the conclusion is true given that the propositions affirmed in the premisses are true. The mental act that is performed in an inference may be represented, I suggest, as an operation performed on the given grounds for the premisses that results in a ground for the conclusion, whereby we see that the proposition affirmed is true.

To illustrate the idea let us consider an inference that is valid but not logically valid, say a case of mathematical induction. How do we see that its conclusion, the induction statement, $A(x)$ say, is true for any natural number n? Is it not reasonable to say that we see this by operating on the given grounds for the induction base and the induction step? We start with the given ground for the induction base $A(0)$ and then successively apply the ground for the induction step. In the induction step we arrive at asserting $A(n+1)$ under the induction assumption $A(n)$, and its ground is thus an unsaturated ground that becomes a closed ground for $A(n+1)$ when saturated with n and a closed ground for $A(n)$. We realize that by applying or saturating this ground n times by the natural numbers $0, 1, \ldots, n-1$, and the grounds that we successively obtain for $A(0), A(1), \ldots, A(n-1)$, we finally get in possession of a ground for $A(n)$, which statement is thus seen to hold.

In accordance with this idea, I shall see an individual inference act as individuated by at least the following five items (for brevity I leave out other additional items that may be needed to individuate an inference such as how hypotheses are discharged):

1. a number of premisses $A_1, A_2, \ldots, A_n$,
2. grounds $\alpha_1, \alpha_2, \ldots, \alpha_n$,
3. an operation Φ applicable to such grounds,
4. a conclusion B, and
5. an agent performing the operation at a specific occasion.

In logic we are usually not interested in individual acts of this kind and therefore abstract away from the agent, which leaves the four items 1–4 individuating what I shall refer to as an (*individual*) *inference*. To *make* or *carry out* such an inference is to apply the operation Φ to the grounds $\alpha_1, \alpha_2, \ldots, \alpha_n$.

I define an individual inference individuated by 1–4 to be *valid* if α_1, $\alpha_2, \ldots, \alpha_n$ are grounds for $A_1, A_2, \ldots, A_n$, respectively, and the result of applying the operation Φ to the grounds $\alpha_1, \alpha_2, \ldots, \alpha_n$, that is $\Phi(\alpha_1, \alpha_2, \ldots, \alpha_n)$, is a ground for B.

According to this definition an individual conjunction introduction, given with two premisses affirming the propositions p_1 and p_2, grounds α_1 and α_2 for them, the operation conjunction grounding $\&G$, and the conclusion affirming the proposition $p_1\&p_2$, is trivially a valid inference since $\&G(\alpha_1, \alpha_2)$ is by definition a ground for affirming $p_1\&p_2$, given that α_i is a ground for affirming p_i.

If we introduce two operations $\&R_1$ and $\&R_2$ defined for grounds for affirmations of propositions of conjunctive form by the equations $\&R_i(\&G(\alpha_1, \alpha_2)) = \alpha_i$ ($i = 1$ or 2), then an individual inference of the type conjunction elimination, given by a premiss affirming a conjunction $p_1\&p_2$, a ground α for it, an operation $\&R_i$, and a conclusion affirming p_i, is valid, since the ground α for the premiss must be of the form $\&G(\alpha_1, \alpha_2)$ where α_i is a ground for p_i, and since the ground α_i is by definition the value of the operation $\&R_i$ applied to $\&G(\alpha_1, \alpha_2)$.

Often we also abstract away from the grounds and from any specific premisses and conclusion of an inference, preserving only a certain formal relation between them. We can then speak of an *inference form* determined only by this formal relation and an operation Φ. For instance, modus ponens may now be seen as such an inference form, individuated by giving an operation Φ, namely the operation $\rightarrow R$ defined below, and by saying that one of the premisses is affirming a proposition of the form of an implication $p \rightarrow q$ while the other premiss affirms the proposition p and the conclusion affirms the proposition q. If we also abstract away from the operation Φ, we get what we may call an *inference schema.*

I shall say that such an inference form is valid when it holds for any instance of the form with premisses $A_1, A_2, \ldots, A_n$, and conclusion B and for all grounds $\alpha_1, \alpha_2, \ldots, \alpha_n$ for $A_1, A_2, \ldots, A_n$ that the result $\Phi(\alpha_1, \alpha_2, \ldots, \alpha_n)$ of applying the operation Φ in question to $\alpha_1, \alpha_2, \ldots, \alpha_n$ is a ground for B. An inference schema is *valid* if it can be assigned an operation Φ such that the resulting inference form is valid.

For instance, modus ponens as usually understood without specifying an operation is an inference schema, which is valid, because by assigning to it the operation $\rightarrow R$ defined by the equation

$$\rightarrow R(\rightarrow G\xi^p(\beta(\xi^p), \alpha) = \beta(\alpha),$$

we get a valid inference form. In the equation above $\beta(\alpha)$ is the result of saturating $\beta(\xi^A)$ by α. To see that the resulting inference form is valid, we have to see that the result of applying the operation $\rightarrow R$ to the grounds for the premisses of an inference of this form is a ground for the conclusion of that inference. Suppose that γ and α are grounds for premisses of that inference and that the premisses are affirming that the propositions $p \rightarrow q$ and p are true. Then by how grounds for implications have been specified, γ is of the form $\rightarrow G\xi^p(\beta(\xi^p))$, where $\beta(\xi^p)$ is an unsaturated ground for affirming that q is true under the assumption that p is true. This means that if $\beta(\xi^p)$ is saturated by a ground for affirming that p is true, the result is a ground for the affirmation that q is true. Now α is a ground for affirming that p is true, hence $\beta(\alpha)$ is a ground for affirming that q is true, which affirmation is the conclusion of the inference. And $\beta(\alpha)$ is the result of applying the operation $\rightarrow R$ to the given grounds for the premisses of the inference according to the definition of $\rightarrow R$.

8 Conclusion

It should now be clear that if the concepts of inference, making an inference, validity of inference, and ground are understood in the way developed here, the question that we started with is easily answered. The general question was how and why we acquire knowledge by making inferences, and this was more precisely formulated as the problem to state the conditions under which an agent P gets a ground for a judgement by inferring it from other judgements. Given that

$$J \text{ is a valid inference from judgements } A_1, A_2, \ldots, A_n \text{ to a judgement } B, \tag{a}$$

and that

$$\text{the agent } P \text{ has grounds } \alpha_1, \alpha_2, \ldots, \alpha_n \text{ for } A_1, A_2, \ldots, A_n, \tag{b}$$

the problem was to state a third condition (c), describing what relation P has to have to the inference J in order that it should follow from (a)–(c) that

$$P \text{ has or gets a ground for } B. \tag{d}$$

When an individual inference is individuated not only by its premisses and conclusion but also by grounds for the premisses and an operation applicable to them, and when making an inference is understood as applying this operation to the grounds, in other words, as transforming the given grounds for the premisses to a ground for the conclusion, it becomes possible to state the third condition that we have sought for simply as

$$P \text{ makes the inference } J. \tag{c}$$

I started out from the conviction that the question why an agent gets a ground for a judgement by inferring it from premisses for which she already has a ground should be easy to answer, once the concepts involved are understood in an appropriate way. This is now actually the case. What it means for an inference J to be valid, as it has now been defined, is simply that the operation Φ that comes with the inference J yields a ground for the conclusion B when applied to the grounds $\alpha_1, \alpha_2, \ldots, \alpha_n$ for the premisses $A_1, A_2, \ldots, A_n$ — in short, that $\Phi(\alpha_1, \alpha_2, \ldots, \alpha_n)$ is a ground B. Therefore, by making the inference J, that is, by applying the operation Φ to the given grounds, the agent gets in possession of a ground for the conclusion.

It remains to say something about what it is for an agent to be in possession of a ground for the conclusion. As already said above, it means basically to have made a certain construction in the mind of which the agent is aware, and which she can manifest by naming the construction. Regardless of whether the construction is only made in the mind or is described, it will be present to the agent under some description, which will normally contain descriptions of a number of operations. It is presupposed that the agent knows these operations, which means that she is able to carry them out, which in turn means that she is able to convert the term that describes the ground to canonical form. Furthermore the agent is presupposed to understand the assertion that she makes and hence to know what kind of ground she is supposed to have for it. It follows that when an agent has got in possession of a ground for an judgement by making an inference, she is aware of the fact that she has made a construction that has the right canonical form to be a ground for the assertion that she makes.

However, it does not mean that the agent has proved that the construction she has made is really a ground for her assertion. As we have already discussed (Section 4), this cannot be a requirement for her judgement to be justified. But if the inference she has made is valid, then she is in fact in possession of a ground for her judgement, and this is exactly what is needed to be justified in making the judgement, or to be said to know that the affirmed proposition is true. Furthermore, although it is not required in order for the judgment to be justified, by reflecting on the inference she has made, the agent can prove that the inference is valid, as has been seen in examples above.

Dag Prawitz
Department of Philosophy, Stockholm University
106 91 Stockholm, Sweden
dag.prawitz@philosophy.su.se

References

Bolzano, B. (1837). *Wissenschaftslehre* (Vols. I–IV). Sulzbach: Seidel.

Carroll, L. (1895). What the tortoise said to Achilles. *Mind*, *IV*, 278–280.

Etchemendy, J. (1990). *The concept of logical consequence.* Cambridge, MA: Harvard University Press.

Martin-Löf, P. (1985). On the meaning of the logical constants and the justification of the logical laws. *Nordic Journal of Philosophical Logic*, *1*, 11–60. (Republishing.)

Prawitz, D. (1973). Towards a foundation of a general proof theory. In P. Suppes et al. (Eds.), *Logic, methodology and philosophy of science* (pp. 225–250). Amsterdam: North-Holland.

Prawitz, D. (1977). Meaning and proofs. *Theoria*, *XLIII*, 2–40.

Prawitz, D. (2009). *Validity of inference.* (To appear in *Proceedings from the* 2nd *Launer Symposium on the Occasion of the Presentation of the Launer Prize at Bern 2006.*)

Ross, W. D. (1949). *Aristotle's prior and posterior analytics.* Oxford: Oxford University Press.

Sundholm, G. (1998). Inference versus consequence. In *The Logica Yearbook 1998.* Prague: Czech Acad. Sc.

Tarski, A. (1936). Über den Begriff der logischen Folgerung. *Actes du Congrès International de Philosophie Scientifiques*, *7*, 1–11. (Translated to English in A. Tarski, *Logic, Semantics and Metamathematics*, Oxford 1956.)

A Sound and Complete Axiomatic System of bdi–stit Logic

Caroline Semmling Heinrich Wansing

1 Introduction

In (Semmling & Wansing, 2008), bdi–stit logic has been motivated and introduced semantically. This logic combines the belief, desire, and intention operators from *BDI* logic (Georgeff & Rao, 1998; Wooldridge, 2000) with the action modalities from d stit logic, the modal logic of deliberatively seeing to it that (Belnap, Perloff, & Xu, 2001), (Horty & Belnap, 1995). The multi-modal bdi–stit logic is an expressively rich logic, which allows a formal analysis of, for example, reasoning about doxastic decisions and belief revision, see (Semmling & Wansing, 2009), (Wansing, 2006a).

In (Semmling & Wansing, 2009), we have presented a sound and complete tableau calculus for bdi–stit logic based on the tableau calculus for d stit logic defined in (Wansing, 2006b). In the present paper we introduce a sound and complete axiomatization of bdi–stit logic and prove decidability by establishing the finite model property.

2 Syntax and semantics

The syntax of bdi–stit logic

The language of bdi–stit logic comprises a denumerable set of sentential variables $(p_1, p_2, p_3, \ldots)$, the constants $\bot$, $\top$, the connectives of classical propositional logic $(\neg, \wedge, \vee, \supset, \equiv)$, and the modal necessity and possibility operators $\Box$ and $\Diamond$. We assume that $\Diamond$ is defined as $\neg\Box\neg$. This vocabulary is supplemented by action modalities and operators used to express the beliefs, desires and intensions of arbitrary (rational) agents. Additionally, there is a possibility operator $\diamonddot$ taken over from (Semmling & Wansing, 2008). We also assume a denumberable set of agent variables $(\alpha_1, \alpha_2, \ldots, \alpha_n, \ldots)$.

Definition 1 (bdi–stit syntax). 1. Every sentential variable $p_1, p_2, \ldots$ and each constant $\bot$, $\top$ is a formula.

2. If α_1, α_2 are agent variables, then $(\alpha_1 = \alpha_2)$ is a formula.

3. If φ, ψ are formulas and α is an agent variable, then $\neg\varphi$, $(\varphi \wedge \psi)$, $\Box\varphi$, $\Diamond\varphi$, $\alpha\,\mathrm{c\,stit} : \varphi$, $\alpha\,\mathrm{bel} : \varphi$, $\alpha\,\mathrm{des} : \varphi$ and $\alpha\,\mathrm{int} : \varphi$ are formulas.

4. Nothing else is a formula.

A formula consisting of only one sentential variable or one constant is called an atomic formula. The reading of a formula $\alpha\,\mathrm{c\,stit} : \varphi$ is "agent α sees to it that φ". In (Semmling & Wansing, 2008), instead of the c stit-operator, an operator of deliberatively seeing to it that, d stit:, is used. We introduce the d stit-operator with the following equivalence

$$\alpha\,\mathrm{d\,stit} : \varphi \equiv (\alpha\,\mathrm{c\,stit} : \varphi \wedge \neg\Box\varphi).$$

This is done because it makes the presentation of the completeness proof easier. But nevertheless it is also possible to use d stit: as a primitive operator and to choose the axioms appropriately, cf. (Belnap et al., 2001).

A formula $\alpha\,\mathrm{bel} : \varphi$ is read as "agent α believes that φ" or "agent α has the belief that φ". The readings of the desire operators $\alpha\,\mathrm{des} :$ and the intention operators $\alpha\,\mathrm{int} :$ are conceived in this vein, too.

The semantics of bdi–stit logic

A bdi–stit model consists of a frame $\mathcal{F} = (\mathrm{Tree}, \leq, \mathcal{A}, N, C, B, D, I)$ and a valuation function v. The frame $\mathcal{F}$ is based on a branching temporal structure as in Stit-Theory, (Belnap & Perloff, 1988). The set Tree is a non-empty set (of moments of time) and $\leq$ is a partial order, which is reflexive, transitive but acyclic, such that every moment $m \in \mathrm{Tree}$ has a unique predecessor. Thus, the set of *histories* H, defined as the set of all maximal linearly ordered subsets of Tree, and the set of *situations* $S = \{(m,h) | m \in \mathrm{Tree}, h \in H\}$ of the frame result from the ordered set $(\mathrm{Tree}, \leq)$. The set of histories passing through moment $m \in \mathrm{Tree}$ ($\{h \mid m \in h, h \in H\}$) is denoted by H_m. The denumerable, non-empty set $\mathcal{A}$ is the set of *agents*, and C is a function that maps every pair of $\mathcal{A} \times \mathrm{Tree}$ to a set of disjoint subsets of histories passing through m, such that the union of all subsets is H_m. Thus, $C(\alpha, m) = C^{\alpha}_{m}$[1] defines an equivalence relation on H_m. Histories h and h' are said to be *choice-equivalent* for agent α at moment m, if they belong to the same set in C^{α}_{m}. The equivalence class of an arbitrary

[1] On this account we do not explicitly distinguish between the variables and the agents and denote both by α, $\alpha_1, \ldots, \alpha_n, \ldots$. The context will disambiguate.

history h in moment m is denoted by $C^\alpha_m(h) = C^\alpha_{(m,h)}$. The set of classes $\left\{C^\alpha_{(m,h)} | h \in H_m\right\}$ represents all distinguishable choice cells of agent α at situation (m, h).

The function $N\colon S \to \mathcal{P}(\mathcal{P}(S))$ assigns a set N_s of non-empty subsets of situations to every situation s. The set N_s is called a *neighbourhood system* of s. Its elements are called neighbourhoods of s. We also denote by N the union of all neighbourhood systems, $N = \{U | U \in N_s, s \in S\}$. The context will disambiguate.

The functions B and D are mappings from $\mathcal{A} \times S$ to $\mathcal{P}(N)$. A set $U \in B(\alpha, s) = B^\alpha_s$ can be regarded as a neighbourhood endorsing certain beliefs of agent α at situation s. In the same way, every set $U \in D(\alpha, s) = D^\alpha_s$ is a neighbourhood endorsing certain desires.

To interpret the ascription of intentions to an agent in a situation s, we use the function I, which maps a pair $(\alpha, s) \in \mathcal{A} \times S$ to a neighbourhood $I(\alpha, s) = I^\alpha_s \in N$ representing all situations compatible with what α intends at s.

Let $\mathcal{F} = (\text{Tree}, \leq, \mathcal{A}, N, C, B, D, I)$ be such a frame and let Select_m be the set of all functions σ from $\mathcal{A}$ into subsets of H_m, such that $\sigma(\alpha) \in C^\alpha_m$. $\mathcal{F}$ satisfies the *independence of agents* condition of their actions, if and only if for every $m \in$ Tree,

$$\bigcap_{\alpha \in \text{Agent}} \sigma(\alpha) \neq \varnothing$$

for every $\sigma \in \text{Select}_m$.

A pair $\mathcal{M} = (\mathcal{F}, v)$ is then said to be a bdi–stit model based on the frame $\mathcal{F}$, where v is a valuation function on $\mathcal{F}$, which maps the agent variables into the set $\mathcal{A}$ of $\mathcal{F}$ and the set of atomic formulas into the powerset of situations $\mathcal{P}(S)$ of $\mathcal{F}$ with the constraints that $v(\bot) = \varnothing$ and $v(\top) = S$.

Satisfiability of a formula in a bdi–stit model $\mathcal{M}$ is then defined as follows, where, for abbreviation, we denote for an arbitrary formula φ by $\|\varphi\|$ the set of situations which contain every situation of $\mathcal{M}$ satisfying formula φ; $\|\varphi\| = \{s | \mathcal{M}, s \models \varphi\}$.

Definition 2 (bdi–stit semantics). Let $s = (m, h)$ be a situation in model $\mathcal{M} = (\mathcal{F}, v)$, let α, α_1, α_2 be agent variables, and let φ, ψ be formulas according to Definition 1. Then:

$\mathcal{M}, s \models \varphi$	iff	$s \in v(\varphi)$, if φ is an atomic formula.
$\mathcal{M}, s \models (\alpha_1 = \alpha_2)$	iff	$v(\alpha_1) = v(\alpha_2)$.
$\mathcal{M}, s \models \neg\varphi$	iff	$\mathcal{M}, s \not\models \varphi$.
$\mathcal{M}, s \models (\varphi \wedge \psi)$	iff	$\mathcal{M}, s \models \varphi$ and$\mathcal{M}, s \models \psi$.
$\mathcal{M}, s \models \Box\varphi$	iff	$\mathcal{M}, (m, h') \models \varphi$ for all $h' \in H_m$.

$\mathcal{M}, s \models \diamondsuit\varphi$ iff there exists $U \in N_s$ with $U \subseteq \|\varphi\|$.

$\mathcal{M}, s \models \alpha\,\text{cstit} : \varphi$ iff $\{(m,h')|h' \in C_s^{v(\alpha)}\} \subseteq \{(m,h')|\mathcal{M},(m,h') \models \varphi, h' \in H_m\}$.

$\mathcal{M}, s \models \alpha\,\text{int} : \varphi$ iff $I_s^{v(\alpha)} \subseteq \|\varphi\|$.

$\mathcal{M}, s \models \alpha\,\text{des} : \varphi$ iff there exists $U \in D_s^{v(\alpha)}$ with $U \subseteq \|\varphi\|$.

$\mathcal{M}, s \models \alpha\,\text{bel} : \varphi$ iff there exists $U \in B_s^{v(\alpha)}$ with $U \subseteq \|\varphi\|$.

Obviously, the operators $\diamondsuit$, $\alpha\,\text{des}$:, and $\alpha\,\text{bel}$: are not defined by a relational semantics, but by a monotonic neighbourhood (alias Scott-Montague, alias minimal models) semantics, cp. (Chellas, 1980; Montague, 1970; Scott, 1970). For such an operator op, a formula $\text{op}\,\varphi \wedge \text{op}\,\neg\varphi$ is satisfiable for any contingent formula φ. Note that it is not possible to satisfy $\text{op}\,\varphi$ for an inconsistent φ in a bdi–stit model, because every neighbourhood is non-empty. A neighbourhood semantics of an operator nop is in use, if $\mathcal{M}, s \models \text{nop}\,\varphi$ iff $\|\varphi\| \in N_s$. Usually, in this semantics nop is interpreted as a kind of necessity-operator. Since for such an operator it is also possible to satisfy formulas $\text{nop}\,\varphi \wedge \text{nop}\,\neg\varphi$, we, however, prefer to read the operator $\diamondsuit$ as a kind of possibility-operator, and call it neighbourhood possibility. A formula $\diamondsuit\varphi$ is true at a situation, if φ is cognitively possible at situation s.

One may wonder about the meaning of the dual operator $\neg\,\text{op}\,\neg\varphi$. The semantics tells us that a formula $\neg\,\text{op}\,\neg\varphi$ is satisfied at a situation s, if each neighbourhood of s necessarily contains a situation s' satisfying φ. But formulas such as $\neg\,\text{op}\,\neg\varphi \wedge \neg\,\text{op}\,\varphi$ are also satisfiable, for example, if every neighbourhood contains at least two situations, one satisfying φ and another satisfying $\neg\varphi$, so that the dual operator does not express a kind of neighbourhood necessity, too. How can we express necessity in a neighbourhood semantics? Our proposal is: $s \models \boxdot\varphi$ iff for every $U \in N_s$, $U \subseteq \|\varphi\|$ (iff φ is a cognitive necessity at situation s). Then obviously it holds that the implications $\diamondsuit\varphi \supset \neg \boxdot \neg\varphi$ and $\boxdot\varphi \supset \neg\diamondsuit\neg\varphi$ are valid, but the implications in the other direction fail. Thus, it is possible to satisfy formulas of the form $\neg \boxdot \neg\varphi \wedge \neg\diamondsuit\varphi$.

In addition to neighbourhood necessity and possibility, there are modal operators not related to the cognitive propositional attitudes of agents: $\Box$ and $\Diamond$. These operators can be read as operators of historical necessity and possibility, respectively, where historical possibility and necessity are defined as dual operators: $\Diamond\varphi \equiv \neg\Box\neg\varphi$. They are adopted from (Belnap & Perloff, 1988; Belnap et al., 2001).

3 Axiomatization

Since the bdi–stit logic is constructed on d stit frames, cf. (Belnap & Perloff, 1988; Belnap et al., 2001), with the addition of some functions as in Scott–Montague models, cf. (Chellas, 1980; Montague, 1970; Scott, 1970), the axiomatization is not too difficult. We assume a complete axiomatization of the non-modal propositional logic and add the following axioms:

(A1) $\Box\varphi \supset \varphi$, $\neg\Box\varphi \supset \Box\neg\Box\varphi$, $\Box(\varphi \supset \psi) \supset (\Box\varphi \supset \Box\psi)$.

(A2) $\alpha\,\text{c stit} : \varphi \supset \varphi$, $\neg\alpha\,\text{c stit} : \varphi \supset \alpha\,\text{c stit} : \neg\alpha\,\text{c stit} : \varphi$,
$\alpha\,\text{c stit} : (\varphi \supset \psi) \supset (\alpha\,\text{c stit} : \varphi \supset \alpha\,\text{c stit} : \psi)$.

(A3) $\Box\varphi \supset \alpha\,\text{c stit} : \varphi$.

(A4) $\alpha = \alpha$, $(\alpha = \beta) \supset (\beta = \alpha)$, $((\alpha = \beta) \wedge (\beta = \gamma)) \supset (\alpha = \gamma)$.

(A5) $(\alpha = \beta) \supset (\varphi \supset \varphi[\alpha/\beta])$.[2]

(AIA_k) $(\Delta(\beta_0, \ldots, \beta_k) \wedge \Diamond\beta_0\,\text{c stit} : \psi_0 \wedge \ldots \wedge \Diamond\beta_k\,\text{c stit} : \psi_k) \supset$
$\supset \Diamond\,(\beta_0\,\text{c stit} : \psi_0 \wedge \ldots \wedge \beta_k\,\text{c stit} : \psi_k)$.

The axioms (AIA_k) represent the independence of agents condition for $k \in \mathbb{N}$ agents. The formula $\Delta(\beta_0, \ldots, \beta_k)$ states that $\beta_0, \ldots, \beta_k$ are pairwise distinct. We also have several derivation rules, cf. (Belnap et al., 2001):

(RN) $\varphi/\Box\varphi$,

(MP) $\varphi, \varphi \supset \psi/\psi$,,

(APC_n) $[\Diamond\alpha\,\text{c stit} : \varphi_1 \wedge \Diamond(\alpha\,\text{c stit} : \varphi_2 \wedge \neg\varphi_1) \wedge \ldots \wedge$
$\Diamond(\alpha\,\text{c stit} : \varphi_n \wedge \neg\varphi_1 \wedge \ldots \wedge \neg\varphi_{n-1})] \supset (\varphi_1 \vee \ldots \vee \varphi_n)$.

By the axiom of n possible choices (APC_n), it is assured that every agent has at most n different alternatives to act. If the axiom (APC_n) is accepted, the resulting logic is denoted by $\mathcal{L}_n$. Evidently, it holds that $\mathcal{L}_{n+1} \subseteq \mathcal{L}_n$.

The new bdi operators are axiomatized by the following axioms and derivation rules taken over from (Chellas, 1980).

(Di) $\alpha\,\text{int} : \varphi \supset \neg\alpha\,\text{int} : \neg\varphi$,

(F,Fb,Fd,Fi) $\neg\diamond\bot$, $\neg\alpha\,\text{bel} : \bot$, $\neg\alpha\,\text{des} : \bot$, $\neg\alpha\,\text{int} : \bot$,

(RM) $(\varphi \supset \psi)/(\diamond\varphi \supset \diamond\psi)$,

[2] Note, that the substitution does not have to be uniform. It is possible, to replace some or all occurrences of α with β.

(RMi) $(\varphi \supset \psi)/(\alpha\,\mathrm{int} : \varphi \supset \alpha\,\mathrm{int} : \psi)$,

(RMb) $(\varphi \supset \psi)/(\alpha\,\mathrm{bel} : \varphi \supset \alpha\,\mathrm{bel} : \psi)$,

(RMd) $(\varphi \supset \psi)/(\alpha\,\mathrm{des} : \varphi \supset \alpha\,\mathrm{des} : \psi)$.

From these axioms, which are proven to be complete in combination with the axioms of the d stit logic in Section 4, we can derive the following theorems, which state the monotony of the neighbourhood operators, and form the typical axioms of the relationally defined ones.

(Ni) $\alpha\,\mathrm{int} : \top$,

(Tc) $\alpha\,\mathrm{c\,stit} : (\varphi \wedge \psi) \equiv (\alpha\,\mathrm{c\,stit} : \varphi \wedge \alpha\,\mathrm{c\,stit} : \psi)$,

(Ti) $\alpha\,\mathrm{int} : (\varphi \wedge \psi) \equiv (\alpha\,\mathrm{int} : \varphi \wedge \alpha\,\mathrm{int} : \psi)$, $\alpha\,\mathrm{int} : \varphi \vee \alpha\,\mathrm{int} : \psi) \supset \alpha\,\mathrm{int} : (\varphi \vee \psi)$,

(T) $\diamondsuit(\varphi \wedge \psi) \supset (\diamondsuit\varphi \wedge \diamondsuit\psi)$,

(Tb) $\alpha\,\mathrm{bel} : (\varphi \wedge \psi) \supset (\alpha\,\mathrm{bel} : \varphi \wedge \alpha\,\mathrm{bel} : \psi)$,

(Td) $\alpha\,\mathrm{des} : (\varphi \wedge \psi) \supset (\alpha\,\mathrm{des} : \varphi \wedge \alpha\,\mathrm{des} : \psi)$

4 Completeness and decidability

Since bdi–stit logic is based on d stit logic, which is decidable, and since it is supplemented with some operators, which are interpreted as in decidable classical modal logics with a neighbourhood semantics, it is not surprising that also bdi–stit logic is decidable. We first show the completeness of the axiomatization presented in Section 3, by extending the construction of a canonical BT + AC (agents and choices in branching time) structure of d stit logic, presented, for example, in (Belnap et al., 2001), to the construction of a frame of a canonical bdi–stit model. Subsequently, we show that bdi–stit logic has the finite model property, i.e., each non-theorem of $\mathcal{L}_n$ is falsifiable in a finite bdi–stit model, by doing the same as in (Belnap et al., 2001) for d stit logic. Since the number of axiom schemes and derivation rules is also finite, the decidability of bdi–stit logic ensues.

Completeness

The soundness of the system of axioms and derivation rules of Section 3 is straightforward and for (APC_n) and (AIA_k) as adduced in (Belnap et al., 2001). Thus, this section deals only with the completeness of the axiomatization. Therefore, we intend to combine the construction of a canonical model for d stit logic, represented in (Belnap et al., 2001), and the construction of a canonical model of a classical monotonic modal logic, cf. (Chellas, 1980). Since the construction of the canonical d stit model has a more complicated structure, this construction constitutes the basis and we expand it appropriately to comprise the interpretation of the belief, desire, and intention operators.

We will present the construction of the canonical BT + AC structure which was defined by Ming Xu, cf. (Belnap et al., 2001; Xu, 1994, 1998). But first, properties of the set $W_{\mathcal{L}_n}$ of maximal $\mathcal{L}_n$-consistent sets of formulas, relations on this set $W_{\mathcal{L}_n}$ as well as on subsets X, W of it and on sets of agent variables are stated in almost the same manner as for the d stit logic Ldm_n. The subscript $_n$ indicates that the axiom (APC_n) is included.

For a given subset $W \subseteq W_{\mathcal{L}_n}$ we define a relation $\cong_W$ on a set of agent variables, by stipulating that $\alpha_1 \cong_W \alpha_2$, if only if, $\alpha_1 = \alpha_2 \in w$ for all $w \in W$.

Lemma 3. *The relation $\cong_W$ is an equivalence relation for any $W \subseteq W_{\mathcal{L}_n}$.*

Proof. Cf. (Belnap et al., 2001). The property of being an equivalence relation results from the axioms (A4), which correspond to reflexivity, symmetry and transitivity. □

The other way around, we define a relation on $X \subseteq W_{\mathcal{L}_n}$ by a set of agent variables A: $w \cong_A w'$, if and only if, $\alpha_1 = \alpha_2 \in w$ iff $\alpha_1 = \alpha_2 \in w'$ for all $\alpha_1, \alpha_2 \in A$.

Lemma 4. *The relation $\cong_A$ is an equivalence relation on an arbitrary set $W \subseteq W_{\mathcal{L}_n}$ for any set A of agent variables.*

Proof. It is self-evident. □

For the next two lemmas we fix an arbitrary subset $W \subseteq W_{\mathcal{L}_n}$. Then, the relation $R \subseteq W \times W$ is defined by wRw' iff $\{\phi | \Box\phi \in w\} \subseteq w'$.

Lemma 5. *The relation $R \subseteq W \times W$ is an equivalence relation.*

Proof. Cf. (Belnap et al., 2001). The property of being an equivalence relation results from axioms (A1), since $\Box\phi \supset \phi$ corresponds to reflexivity and $\neg\Box\phi \supset \Box\neg\Box\phi$ to euclidity. □

Therefore, we can partition W into equivalence classes $\{X_i\}_{i \in I}$ with respect to R. Let X be an arbitrary element of $\{X_i\}_{i \in I}$. For such a subset $X \subseteq W$ it holds, that if $(\alpha = \beta) \in w$ for some $w \in X$, it follows by Rule (RN) and Axiom (A5) that $\Box(\alpha = \beta) \in w$, such that $(\alpha = \beta) \in w'$ for all $w' \in X$. That warrants the use of the equivalence classes $\{\beta_j\}_{j \in J} = \{[\alpha]_X\}$ of $\equiv_X$

instead of agent variables to define the following relations $R_{\beta_j} \subseteq X \times X$ for all $j \in J$;

$$wR_{\beta_j}w' \quad \text{iff} \quad \{\phi | \beta_j \,\text{c stit} : \phi \in w\} \subseteq w'.$$[3]

Lemma 6. *For $X \in \{X_i\}_{i \in I}$ the relation $R_{\beta_j} \subseteq X \times X$ is an equivalence relation for all $j \in J$.*

Proof. Cf. (Belnap et al., 2001). The first and the second axiom of (A2) express reflexivity and euclidity, respectively. □

This definition of the relations R_{β_j} depends on the set X. In the following, this X will be an arbitrary but fixed equivalence class of relation R. We denote by E_{β_j} the set of all equivalence classes of relation R_{β_j} on X.

Lemma 7. *Let X, $\{\beta_j\}_{j \in J}$, R, R_{β_j}, E_{β_j} for all $j \in \mathbb{N}$ be given by the definitions above. Then it holds for all $w \in X$, ϕ:*

(i) $\Box\phi \in w$ *iff* $\phi \in w'$ *for all* $w' \in X$ *iff* $\Box\phi \in w'$ *for all* $w' \in X$.

(ii) $\beta_j \,\text{c stit} : \phi \in w$ *iff* $\phi \in w'$ *for all* w' *with* $wR_{\beta_j}w'$ *iff* $\beta_j \,\text{c stit} : \phi \in w'$ *for all* w' *with* $wR_{\beta_j}w'$.

(iii) $\beta_j \,\text{d stit}\, \phi \in w$ *iff* $\phi \in w'$ *for all* w' *with* $wR_{\beta_j}w'$ *and* $\neg\phi \in w''$ *for some* $w'' \in X$.

(iv) *Assume f to be an arbitrary function from $\{\beta_j\}_{j \in J}$ into the union of E_{β_j} for all $j \in J$ such that $f(\beta_j) \in E_{\beta_j}$. This entails*

$$\bigcap_{j \in J} f(\beta_j) \neq \varnothing.$$

(v) *Let $\mathcal{L}_n$ with $n \geq 1$ and let X, $\{\beta_j\}_{j \in J}$, $\{R_{\beta_j}\}_{j \in J}$, $\{E_{\beta_j}\}_{j \in J}$ be defined with respect to $\mathcal{L}_n$. Then there are at most n different equivalence classes R_{β_j} for every $j \in J$, i.e.*

$$\left|E_{\beta_j}\right| \leq n.$$

Proof. Cf. (Belnap et al., 2001). For (i), the claim follows by Axiom (A1) and Rule (RN). For (ii) Axioms (A2) and (A3) and Rule (RN) are needed. Assertion (iii) results from the definition of d stit and (i), (ii). Clearly, (iv) is backed up by (AIA_k) and (v) by (APC_n) for appropriate k, n. □

[3] The abbreviation $\beta_j \,\text{c stit} : \phi$ means that for some $\alpha \in \beta_j$, $\alpha \,\text{c stit} : \phi \in w$, because of Axiom (A5) it follows for all $\tilde{\alpha} \in \beta_j$, $\tilde{\alpha} \,\text{c stit} : \phi \in w$.

Theorem 1 (completeness). *Each $\mathcal{L}_n$-consistent set Φ of bdi–stit formulas is satisfiable by a bdi–stit model.*

Proof. Let $W_{\mathcal{L}_n}$ be the set of all maximal $\mathcal{L}_n$-consistent sets, let

$$A = \{\alpha | \text{the agent variable } \alpha \text{ occurs in } \Phi\}.$$

Then $\cong_A$ is an equivalence relation on $W_{\mathcal{L}_n}$. We denote by W the equivalence class, such that for all agent variables $\alpha, \tilde{\alpha} \in A$, $\beta, \tilde{\beta} \notin A$ it holds that $\alpha = \tilde{\alpha} \notin \Phi$ iff $\alpha = \tilde{\alpha} \notin w$ and $\beta = \tilde{\beta} \in w$ and $\alpha = \beta \notin w$ for all $w \in W$. Let $\{X_i\}_{i \in I}$ be the set of all equivalence classes of relation R on W.

The basis of our bdi–stit model is a frame $\mathcal{F} = (\text{Tree}, \leq, \mathcal{A}, N, C, B, D, I)$ defined on a Branching-time structure $(\text{Tree}, \leq)$. We define it as follows:

- Tree $:= \{w | w \in W\} \cup \{X_i | i \in I\} \cup \{W\}$;
- $\leq := \text{trcl}(\{(w, w) | w \in W\} \cup \{(W, X_i), (X_i, X_i), (X_i, w) | w \in X_i, i \in I\} \cup \{(W, W)\})$;[4]
- $\mathcal{A} := \{\alpha | \alpha$ belongs to an arbitrary but fixed set of class representatives of $\cong_W$ on all agent variables.$\}$;[5]
- for all $i \in I$ we define the choice equivalence classes of any agent $\alpha \in \mathcal{A}$ at any moment in Tree:

  $C(\alpha, w) := \{\{h_w\}\}$, where h_w is the unique history passing through moment w with $h_w = \{w, X_i, W\}$, where X_i is the equivalence class containing w.

  $C(\alpha, W) := \{\{h_w | w \in W\}\}$,

  According to Lemma 6, there is an equivalence class β_j with $\alpha \in \beta_j$ and an equivalence relation $R^i_{\beta_j}$ on X_i. We denote the classes of $R^i_{\beta_j}$ on X_i by $E^i_{\beta_j}$ and then we can define:

  $$C(\alpha, X_i) := \{H \,|\exists e : e \in E^i_{\beta_j} \text{ and } H = \{h_w | w \in e\}\}.$$

Since there is a one-to-one correspondence between all $w \in W$ and all histories of $(\text{Tree}, \leq)$, the following concepts are well-defined. For all $m \in$ Tree, $w \in W$ and $\alpha \in \mathcal{A}$, we have:

- $|\varphi| := \bigcup_{i \in I} \{(X_i, h_{w'}) | \varphi \in w', X_i \in h_{w'}\} \in N_{(m, h_w)}$ iff $\diamond\varphi \in w$;

[4] Here trcl stands for the transitive closure of a binary relation.

[5] Recall that we use the same α for agent variables and agents. Since we now interpret the agent variable by the agent variable itself, this naming was just a kind of forestalling. But note that $\mathcal{A} \neq A$ in general.

- $|\varphi| \in B^{\alpha}_{(m,h_w)}$ iff there is $\alpha\,\text{bel} : \varphi \in w$;[6]

- $|\varphi| \in D^{\alpha}_{(m,h_w)}$ iff there is $\alpha\,\text{des} : \varphi \in w$;

- We define a relation $S_\alpha \subseteq W \times W$ for all $\alpha \in \mathcal{A}$, by stipulating that $wS_\alpha w'$ iff $\{\varphi | \alpha\,\text{int} : \varphi \in w\} \subseteq w'$. Then we choose the sets I^{α}_{s} for every situation $s = (m, h_w)$ in $\mathcal{M}$:

$$I^{\alpha}_{(m,h_w)} = \{(w', h_{w'}) | wS_\alpha w'\}.$$

  We have to show that for all $w \in W$ there is a $w' \in W$ with $wS_\alpha w'$, which means that $I^{\alpha}_{(m,h_w)} \neq \varnothing$ for any $m \in \text{Tree}$. From axiom (Di), derivation rule (RMi) and theorems (Ti), it is evident that for any $w \in W$ the set $S = \{\varphi | \alpha\,\text{int} : \varphi \in w\}$ is consistent, thus there is a maximal consistent set w' with $S \subseteq w'$.

In analogy to (Belnap et al., 2001), we claim that the frame $\mathcal{F}$ satisfies the independence of agents condition. For a given moment $m \in \text{Tree}$ let Select_m be the set of all functions from $\mathcal{A}$ into subsets of H_m, the set of histories passing through moment m, where for all $\sigma \in \text{Select}_m$ it holds that $\sigma(\alpha) \in C^{\alpha}_{m}$. Then $\mathcal{F}$ satisfies this condition if and only if for every moment m and any $\sigma \in \text{Select}_m$

$$\bigcap_{\alpha \in \mathcal{A}} \sigma(\alpha) \neq \varnothing.$$

Let $m = w$ for an arbitrary $w \in W$ or $m = W$, then the condition is evidently satisfied. Now, let for an arbitrary $i \in I$, σ_{X_i} be any function from $\mathcal{A}$ into $\mathcal{P}(H_{X_i})$, such that $\sigma_{X_i}(\alpha) \in C^{\alpha}_{X_i}$ for all $\alpha \in \mathcal{A}$. By the above definition of $C^{\alpha}_{X_i}$, there is an equivalence class $e_j \in E^{i}_{\beta_j}$ with $\sigma_{X_i}(\alpha) = \{h_w | w \in e_j\}$. Define a function f_i by $f_i(\alpha) = e_j \in E^{i}_{\beta_j}$, where $\alpha \in \beta_j$ for $j \in J$. Then for all $\alpha, \tilde{\alpha} \in \beta_j$, $f_i(\alpha) = f_i(\tilde{\alpha})$, there is a well defined corresponding function $\tilde{f}_i \colon \{\beta_j\}_{j \in J} \to \bigcup_{j \in J} E_{\beta_j}$. As well, $w \in f_i(\alpha)$ iff $h_w \in \sigma_{X_i}(\alpha)$ and by Lemma 7 (iv), it holds that

$$\bigcap_{\alpha \in \mathcal{A}} f_i(\alpha) = \bigcap_{j \in J} \tilde{f}_i(\beta_j) \neq \varnothing, \text{ such that } \bigcap_{\alpha \in \mathcal{A}} \sigma_{X_i}(\alpha) \neq \varnothing.$$

Since $|C^{\alpha}_{w}| = |C^{\alpha}_{W}| = 1$ for all $w \in W$ and $\alpha \in \mathcal{A}$, and for all $i \in I$ it holds that $|C^{\alpha}_{X_i}| = |E^{i}_{\beta_j}|$, it obviously follows by Lemma 7 (v) that for any $\alpha \in \mathcal{A}$ and $m \in \text{Tree}$, $C^{\alpha}_{m} \leq n$, cf. (Belnap et al., 2001). So any agent α has at most n possible choices in the frame $\mathcal{F}$.

[6] Because of Axiom (A5) for all $\beta \in [\alpha]_W$ it follows: $\beta\,\text{op} : \phi \in w$ iff $\alpha\,\text{op} : \phi \in w$ for $\text{op} \in \{\text{c stit}, \text{d stit}, \text{bel}, \text{des}, \text{int}\}$ and for all $w \in W$.

Now, we define a canonical model on that frame $\mathcal{M} = (\mathcal{F}, v)$, where v is an interpretation function, which maps each agent variable β on $v(\beta) = \alpha \in \mathcal{A}$ with $\beta \in [\alpha]_W$ and every atomic formula p on a subset of S containing all situations $s = (m, h_w)$ with $p \in w$ for all $m \in$ Tree. Evidently, $v(\top) = S$ and $v(\bot) = \varnothing$. For any agent variable α, $h \in H$, $w \in W$ it holds that $h \in C^{\alpha}_{X_i}(h_w)$ iff there is $e \in E^i_{[\alpha]X_i}$ and $w, w' \in e$, where $h = h_{w'}$. Furthermore, $w, w' \in e \in E^i_{[\alpha]X_i}$ iff $wR^i_{[\alpha]X_i}w'$, such that $h \in C^{\alpha}_{X_i}(h_w)$ iff $wR^i_{[\alpha]X_i}w'$.

We show by induction that $\mathcal{M}, (X_i, h_w) \models \varphi$ iff $\varphi \in w$ for every bdi–stit formula φ and $w \in X_i$, for all $i \in I$.

$$\mathcal{M}, (X_i, h_w) \models p \Leftrightarrow (X_i, h_w) \in v(p) \overset{\text{by definition}}{\Leftrightarrow} p \in w.$$

$$\mathcal{M}, (X_i, h_w) \models (\alpha = \beta) \Leftrightarrow v(\alpha) = v(\beta) \Leftrightarrow \alpha \cong_W \beta \Leftrightarrow \alpha = \beta \in w.$$

$$\mathcal{M}, (X_i, h_w) \models \neg\varphi \Leftrightarrow (X_i, h_w) \not\models \varphi \overset{\text{by induction}}{\Leftrightarrow} \varphi \notin w \Leftrightarrow \neg\varphi \in w.$$

$$\begin{aligned}\mathcal{M}, (X_i, h_w) \models \varphi \wedge \psi &\Leftrightarrow (X_i, h_w) \models \varphi \text{ and } (X_i, h_w) \models \psi \overset{\text{by induction}}{\Leftrightarrow} \\ &\varphi \in w \text{ and } \psi \in w \Leftrightarrow (\varphi \wedge \psi) \in w.\end{aligned}$$

$$\begin{aligned}\mathcal{M}, (X_i, h_w) \models \Box\varphi &\Leftrightarrow \text{for all } h \in H_{X_i} \text{ it holds that } (X_i, h) \models \varphi \\ &\overset{\text{by induction}}{\Leftrightarrow} \text{for all } h \in H_{X_i} \text{ there is } w' \in X_i \\ &\qquad \text{with } h = h_{w'} \text{ and } \varphi \in w' \\ &\Leftrightarrow \text{for all } w' \in X_i, \varphi \in w' \overset{\text{by Lemma 7 (i)}}{\Leftrightarrow} \Box\varphi \in w.\end{aligned}$$

$$\begin{aligned}\mathcal{M}, (X_i, h_w) \models \beta\,\text{c stit} : \varphi &\Leftrightarrow \text{for all } h \in C^{\alpha}_{X_i}(h_w) \text{ with } \beta \in [\alpha]_W \\ &\quad \text{it holds that } (X_i, h) \models \varphi \\ &\Leftrightarrow \text{for all } h \in C^{\alpha}_{X_i}(h_w) \text{ there is } w' \text{ with } h = h_{w'} \\ &\quad \text{and it holds that } (X_i, h_{w'}) \models \varphi \overset{\text{by induction}}{\Leftrightarrow} \\ &\quad \text{for all } w' \in X_i, \text{ if } wR^i_{[\alpha]X_i}w' \text{ then } \varphi \in w' \\ &\overset{\text{by Lemma 7 (ii)}}{\Leftrightarrow} \alpha\,\text{c stit} : \varphi \in w \\ &\overset{\text{by Axiom (A5)}}{\Leftrightarrow} \beta\,\text{c stit} : \varphi \in w.\end{aligned}$$

$$\begin{aligned}\mathcal{M}, (X_i, h_w) \models \Diamond\varphi &\Leftrightarrow \text{there is } U \in N_{(X_i, h_w)} \varnothing \neq U \subseteq \|\varphi\| \\ &\Leftrightarrow \text{there is } \psi \text{ with } \varnothing \neq |\psi| \subseteq \|\varphi\| \text{ and } \Diamond\psi \in w \\ &\overset{(*)}{\Leftrightarrow} \Diamond\varphi \in w.\end{aligned}$$

We want to show the equivalence $\overset{(*)}{\Leftrightarrow}$.

$\Leftarrow$: If $\Diamond\varphi \in w$, then $|\varphi| \in N_{(X_i, h_w)}$ with $w \in X_i$. That means there is $\psi = \varphi$ with $|\varphi| \subseteq \|\varphi\|$ by induction.

⇒: There exists ψ with $\varnothing \neq |\psi| \subseteq \|\varphi\|$ and $\diamond\psi \in w$. Since

$$|\psi| = \bigcup_{i\in I}\{(X_i, h_{w'})|\psi \in w', X_i \in h_{w'}\},\ W = \bigcup_{i\in I}\{w|(X_i, h_w) \text{ is a situation}\},$$

and $|\psi| \subseteq \|\varphi\|$ it follows for all $w' \in W$ and $i \in I$: if $\mathcal{M}, (X_i, h_{w'}) \models \psi$ then $\mathcal{M}, (X_i, h_{w'}) \models \varphi$. By induction we have for all $w' \in W$: if $\psi \in w'$ then $\varphi \in w'$. Thus it holds for all $w' \in W$ that $(\psi \supset \varphi) \in w'$. By rule (RM) it holds that $(\diamond\psi \supset \diamond\phi) \in w'$ for all $w' \in W$. Since $\diamond\psi \in w$, we have $\diamond\varphi \in w$.

$$\begin{aligned}
\mathcal{M}, (X_i, h_w) \models \beta\,\text{bel} : \varphi &\Leftrightarrow \text{there is } U \in B^{\alpha}_{(X_i,h_w)} \text{ with } \beta \in [\alpha]_W \text{ and} \\
&\qquad \varnothing \neq U \subseteq \|\varphi\| \\
&\Leftrightarrow \text{there is } \psi \text{ with } \varnothing \neq |\psi| \subseteq \|\varphi\| \text{ and } \alpha\,\text{bel} : \psi \in w \\
&\overset{(**)}{\Leftrightarrow} \beta\,\text{bel} : \varphi \in w. \\
\mathcal{M}, (X_i, h_w) \models \beta\,\text{des} : \varphi &\Leftrightarrow \text{there is } U \in D^{\alpha}_{(X_i,h_w)} \text{ with } \beta \in [\alpha]_W \text{ and} \\
&\qquad \varnothing \neq U \subseteq \|\varphi\| \\
&\Leftrightarrow \text{there is } \psi \text{ with } \varnothing \neq |\psi| \subseteq \|\varphi\| \text{ and } \alpha\,\text{des} : \psi \in w \\
&\Leftrightarrow \beta\,\text{des} : \varphi \in w.
\end{aligned}$$

We only show $(**)$, which is similar to the argument for the $\diamond$-operator. The corresponding equivalence for the desire operator is shown analogously.

⇐: If $\beta\,\text{bel} : \varphi \in w$, then, by Axiom (A5), $\alpha\,\text{bel} : \varphi \in w$, such that $|\varphi| \in B^{\alpha}_{(X_i,h_w)}$ with $w \in X_i$. That means there is $\psi = \varphi$ with $|\varphi| \subseteq \|\varphi\|$ by induction.

⇒: There is ψ with $\varnothing \neq |\psi| \subseteq \|\varphi\|$ and $\alpha\,\text{bel} : \psi \in w$. Since

$$|\psi| = \bigcup_{i\in I}\{(X_i, h_{w'})|\psi \in w', X_i \in h_{w'}\},\ W = \bigcup_{i\in I}\{w|(X_i, h_w) \text{ is a situation}\},$$

and $|\psi| \subseteq \|\varphi\|$ it follows for all $w' \in W$ and $i \in I$: if $\mathcal{M}, (X_i, h_{w'}) \models \psi$ then $\mathcal{M}, (X_i, h_{w'}) \models \varphi$. By induction we have for all $w' \in W$: if $\psi \in w'$ then $\varphi \in w'$. Thus it holds for all $w' \in W$ that $(\psi \supset \varphi) \in w'$. By rule (RMb) it holds that $(\alpha\,\text{bel} : \psi \supset \alpha\,\text{bel} : \phi) \in w'$ for all $w' \in W$. Since $\alpha\,\text{bel} : \psi \in w$, we have $\alpha\,\text{bel} : \varphi \in w$. Again because of (A5), it follows $\beta\,\text{bel} : \varphi \in w$.

$$\begin{aligned}
\mathcal{M}, (X_i, h_w) \models \beta\,\text{int} : \varphi &\Leftrightarrow I^{\alpha}_{(X_i,h_w)} \subseteq \|\varphi\| \text{ with } \beta \in [\alpha]_W \\
&\Leftrightarrow \text{for all } s = (w', h'_w) : \text{ if } wS_\alpha w', \text{ then } \mathcal{M}, s \models \varphi \\
&\overset{\text{by induction}}{\Leftrightarrow} \text{for all } s = (w', h'_w) : \text{if } wS_\alpha w', \text{ then } \varphi \in w' \\
&\overset{(***)}{\Leftrightarrow} \beta\,\text{int} : \varphi \in w.
\end{aligned}$$

At last we have to show the equivalence $(***)$.

$\Leftarrow$: If $\beta\,\text{int} : \varphi \in w$, then $\alpha\,\text{int} : \varphi \in w$ and for all $w' \in W$ with $wS_\alpha w'$ it follows that $\varphi \in w'$.

$\Rightarrow$: For all $s = (w', h'_w)$: if $wS_\alpha w'$, then $\varphi \in w'$. Because of axiom (Di) and maximality for any φ and $w \in W$ it holds that either $\alpha\,\text{int} : \varphi \in w$ or $\alpha\,\text{int} : \neg\varphi \in w$ or both is not the case. The assumption $\alpha\,\text{int} : \neg\varphi \in w$ is contradictory, since $\neg\varphi \in \{\psi|\alpha\,\text{int} : \psi \in w\} \subseteq w'$ for any w' with $wS_\alpha w'$. Assume $\neg\alpha\,\text{int} : \varphi \in w$ and $\neg\alpha\,\text{int} : \neg\varphi \in w$. Then $\varphi, \neg\varphi \notin \{\psi|\alpha\,\text{int} : \psi \in w\}$. Since w is maximal, the set $\{\psi|\alpha\,\text{int} : \psi \in w\}$ is closed under implication by (RMi) and (Ti), such that the sets $\{\varphi\} \cup \{\psi|\alpha\,\text{int} : \psi \in w\}$ and $\{\neg\varphi\} \cup \{\psi|\alpha\,\text{int} : \psi \in w\}$ are both consistent. But then there is a maximal world w'' with $\{\neg\varphi\}\cup\{\psi|\alpha\,\text{int} : \psi \in w\} \subseteq w''$ and $wS_\alpha w''$. But that conflicts with $\varphi \in w'$ for all $wS_\alpha w'$. Thus, $\alpha\,\text{int} : \varphi \in w$, resp. $\beta\,\text{int} : \varphi \in w$.

At any rate, there is one (maybe more than one, then choose one) maximal consistent set $w_0 \in W$ with $\Phi \subseteq w_0$. This w_0 belongs to an equivalence class X_{i_0} of R. Then, $\mathcal{M}, (X_{i_0}, h_{w_0}) \models \varphi$ for any $\varphi \in \Phi$. Thus, any consistent set Φ is satisfiable. □

Finite model property

We construct to a given sentence φ according to Definition 1 a finite frame $\mathcal{F}_{\text{fin}}$ and add a special interpretation v, such that for every subsentence of φ it is decidable whether the subsentence is satisfiable by $(\mathcal{F}_{\text{fin}}, v)$. We adopt the filtration method as used in (Belnap et al., 2001). We take the canonical frame $\mathcal{F} = (\text{Tree}, \leq, \mathcal{A}, N, C, B, D, I)$ of the previous section and define a filtration first over all worlds by the set of all subformulas of the given formula φ including all formulas derived by Axioms (AIA$_k$) and (A$_i$) for all $1 \leq i \leq 5$, cf. (Belnap et al., 2001). Then again, we filtrate the equivalence classes of relation R by a set of formulas implied by subformulas of φ prefixed by d stit or c stit operators, such that we can define choice-equivalent histories. To begin with, we define the sets of subformulas:

$$\begin{aligned}
\Sigma_\varphi &= \{\psi|\psi \text{ is a subsentence of } \varphi\},\\
\Sigma_i &= \Sigma_\varphi \cup \{\neg\beta\,\text{int} : \neg\psi|\beta\,\text{int} : \psi \in \Sigma_\varphi\},\\
\Sigma_d &= \Sigma_\varphi \cup \{\beta\,\text{c stit} : \psi, \neg\Box\psi|\beta\,\text{d stit} : \psi \in \Sigma_\varphi\} \cup \{\neg\Box\neg\psi|\Diamond\psi \in \Sigma_\varphi\},\\
\Sigma_e &= \{\psi|\psi \text{ is a subsentence of a formula of } \Sigma_d \text{ or of}\\
&\qquad \{\beta\,\text{c stit} : \neg\beta\,\text{c stit} : \psi|\beta\,\text{c stit} : \psi \in \Sigma_d\}\},\\
\Sigma_p &= \{\Diamond(\beta_0\,\text{c stit} : \phi_0 \wedge \cdots \wedge \beta_n\,\text{c stit} : \phi_n)|n \geq 0, \beta_0, \ldots \beta_n \text{ differ}\\
&\qquad \text{pairwisely, occur in } \varphi, \text{ and for all } 0 \leq i \leq n, \phi_i = \psi_0 \wedge \cdots \wedge \psi_{m_i},\\
&\qquad 0 \leq j \leq m_i, \text{there is } \beta_i\,\text{c stit} : \psi_j \text{ in } \Sigma_e\},\\
\Sigma_a &= \{\psi|\psi \text{ is a subsentence of a formula of } \Sigma_p \cup \Sigma_e \cup \Sigma_i\}.
\end{aligned}$$

For all $w, w' \in W$, we define the equivalence relation $\equiv_{\Sigma_a}$ by setting $w \equiv_{\Sigma_a} w'$ iff for all $\psi \in \Sigma_a$: $(m, h_w) \models \psi$ iff $(m, h_{w'}) \models \psi$. By $\tilde{W}$ we denote a chosen set of representatives of all equivalence classes. By $\tilde{X}_i$ we denote the subset of $\tilde{W}$ consisting of all representatives, which belong to the equivalence class X_i for all $i \in I$. Note that if the relation R is first applied to the set W and then relation Σ_a is implemented on the equivalence classes, one gets an isomorphic Branching Time Structure in the end. Since Σ_a is finite, so it is $\tilde{W}$ and therewith every $\tilde{X}_i$. Thus, we define a finite frame $\mathcal{F}_{\text{fin}}$:

- Tree $:=\{w|w \in \tilde{W}\} \cup \{\tilde{X}_i|i \in I\} \cup \{\tilde{W}\}$ is finite.
- $\leq :=$ trcl$(\{(w,w)|w \in \tilde{W}\} \cup \{(\tilde{W}, \tilde{X}_i), (\tilde{X}_i, \tilde{X}_i), (\tilde{X}_i, w)|w \in X_i, i \in I\} \cup \{(\tilde{W}, \tilde{W})\})$, such that there is again an one-to-one corresponding relation between the histories $\tilde{H}$ and the set $\tilde{W}$, $\tilde{H} = \{h_w|w \in \tilde{W}\}$.
- the set of agents is chosen as $\tilde{\mathcal{A}} = \{\alpha|$ there is α occurring in $\varphi\}$, thus $\tilde{\mathcal{A}}$ is also finite.[7]
- for all $\alpha \in \tilde{\mathcal{A}}$ we define relations $\equiv^{\alpha}_{\Sigma_e}$ on every set $\tilde{X}_i$, by $w \equiv^{\alpha}_{\Sigma_e} w'$ iff for all $\alpha\,\text{cstit} : \psi \in \Sigma_e$ it holds that $\alpha\,\text{cstit} : \psi \in w$ iff $\alpha\,\text{cstit} : \psi \in w'$. By $\tilde{U}^i_{[\alpha]_{\tilde{X}_i}}$ we denote the set of all equivalence classes on $\tilde{X}_i$. With this definition it is possible to define the choice equivalent function $\tilde{C}$ in the finite frame:

$$\begin{aligned}
\tilde{C}(\alpha, w) &:= \{\{h_w\}\}, \text{ where } h_w \text{ is the unique history passing through} \\
&\qquad\qquad \text{moment } w \text{ with } h_w = \{w, \tilde{X}_i, \tilde{W}\}, \\
\tilde{C}(\alpha, \tilde{W}) &:= \{\{h_w|w \in \tilde{W}\}\}, \\
\tilde{C}(\alpha, \tilde{X}_i) &:= \left\{H|\exists e : e \in U^i_{[\alpha]_{\tilde{X}_i}} \text{ and } H = \{h_w|w \in e\}\right\}.
\end{aligned}$$

Since there is a one-to-one correspondence between all $w \in \tilde{W}$ and all histories of (Tree, $\leq$), the following notions are well-defined. For all $\diamond\phi, \alpha\,\text{bel} : \phi, \alpha\,\text{des} : \phi \in \Sigma_\varphi$, $m \in$ Tree, $w \in \tilde{W}$ and $\alpha \in \tilde{\mathcal{A}}$, we have:

- $|\phi| := \bigcup_{i \in I} \{(w', h_{w'})|\phi \in w'\} \in N_{(m,h_w)}$ iff $\diamond\phi \in w$,
- $|\phi| \in \tilde{B}^{\alpha}_{(m,h_w)}$ iff $\alpha\,\text{bel} : \phi \in w$,
- $|\phi| \in \tilde{D}^{\alpha}_{(m,h_w)}$ iff $\alpha\,\text{des} : \phi \in w$,

[7] We neglect the problem of identity statements of agents, since it can be handled as above.

- For all $w \in \tilde{W}$ and each $\alpha \in \tilde{\mathcal{A}}$ set $t_w^\alpha = \{\phi | \alpha\, \text{int} : \phi \in w\}$. Then, there is at least one $\tilde{w}^\alpha \in \tilde{W}$ with $t_w^\alpha \cap \Sigma_a \subseteq \tilde{w}^\alpha$, since t_w^α is consistent by (Ti), (Di). We define for all $m \in$ Tree:

$$\tilde{I}^\alpha_{(m,h_w)} = \{(\tilde{w}^\alpha, h_{\tilde{w}^\alpha}) | \tilde{w}^\alpha \in \tilde{W}, t_w^\alpha \cap \Sigma_a \subseteq \tilde{w}^\alpha\}.$$

By construction, these sets are not empty.

Lemma 8. *Let $\mathcal{F} = (\text{Tree}, \leq, \mathcal{A}, N, C, B, D, I)$ be the canonical frame, let i be fixed, X_i the corresponding equivalence class $X_i \in$ Tree, $\tilde{X}_i$ the corresponding class in the finite frame $\mathcal{F}_{fin} = (\text{Tree}, \leq, \tilde{\mathcal{A}}, \tilde{N}, \tilde{C}, \tilde{B}, \tilde{D}, \tilde{I})$ filtrated by the sets of subformulas of φ and $U^i_{[\alpha]_{\tilde{X}_i}}$ the sets of equivalence classes of $\equiv_{\Sigma_e}$ on $\tilde{X}_i$ for all $\alpha \in \tilde{\mathcal{A}}$.*

(i) *If $\Box\psi \in \Sigma_a$, $w \in \tilde{X}_i$, then*

$$\Box\psi \in w \quad \textit{iff} \quad \psi \in w' \textit{ for all } w' \in \tilde{X}_i.$$

(ii) *If $\alpha\,\text{cstit} : \psi \in \Sigma_e$, $w \in \tilde{X}_i$, then*

$$\alpha\,\text{cstit} : \psi \in w \quad \textit{iff} \quad \psi \in w' \textit{ for all } w' \in \tilde{X}_i \textit{ with } w \equiv^\alpha_{\Sigma_e} w'.$$

(iii) *For all equivalence classes $e_\alpha \in \tilde{U}^i_{[\alpha]_{\tilde{X}_i}}$ it holds that*

$$\bigcap_{\alpha \in \tilde{A}} e_\alpha \neq \varnothing.$$

(iv) *Let $\varphi \in \mathcal{L}_n$ with $n \geq 1$, then for any $j \in [0, |\tilde{\mathcal{A}}|]$ it holds that*

$$\left| U^i_{[\alpha]_{\tilde{X}_i}} \right| \leq n.$$

Proof. Cf. (Belnap et al., 2001). □

This frame satisfies the independence of agents condition. For all moments $m \in \{\tilde{W}, w | w \in \tilde{W}\}$ it is evident that for an arbitrary function $\sigma_m : \tilde{A} \to \tilde{C}_m$ the intersection $\bigcap\{\sigma_m(\alpha) | \alpha \in \tilde{\mathcal{A}}\}$ is not empty. If $m = \tilde{X}_i$, then for all $\alpha \in \tilde{\mathcal{A}}$ there is $e_\alpha \in U^i_{[\alpha]_{\tilde{X}_i}}$ with $\sigma_m(\alpha) = e_\alpha$, such that with the frame property 8 (iii) the set $\bigcap\{\sigma_{\tilde{X}_i}(\alpha) | \alpha \in \tilde{\mathcal{A}}\}$ is also not empty. Assuming $\mathcal{L}_n$, the corresponding axiom of possible choices is also fulfilled, since for all $\alpha \in \tilde{\mathcal{A}}$, $\left|\tilde{C}^\alpha_{\tilde{W}}\right| = \left|\tilde{C}^\alpha_w\right| = 1$ and $\left|\tilde{C}^\alpha_{\tilde{X}_i}\right| = \left|U^i_{[\alpha]_{\tilde{X}_i}}\right| \leq n$ by Lemma 8 (iv).

For any $\psi \in \Sigma_\varphi$ we can show that for any equivalence class $\tilde{X}_i$ it holds that

$$\mathcal{M}_{\text{fin}}(\tilde{X}_i, h_w) \models \psi \quad \text{iff} \quad \psi \in w,$$

where $\mathcal{M}_{\text{fin}} = (\mathcal{F}_{\text{fin}}, v)$ and v is the valuation function defined as for the canonical model, but restricted to $\tilde{W}$. The proof is by induction.

$$\mathcal{M}_{\text{fin}}, (\tilde{X}_i, h_w) \models p \Leftrightarrow \tilde{X}_i, h_w) \in v(p) \Leftrightarrow p \in w.$$

$$\mathcal{M}_{\text{fin}}, (\tilde{X}_i, h_w) \models \neg\psi \Leftrightarrow (\tilde{X}_i, h_w) \not\models \psi \Leftrightarrow \psi \notin w \Leftrightarrow \neg\psi \in w.$$

$$\begin{aligned}\mathcal{M}_{\text{fin}}, (\tilde{X}_i, h_w) \models \phi \wedge \psi &\Leftrightarrow (\tilde{X}_i, h_w) \models \phi \text{ and } (\tilde{X}_i, h_w) \models \psi \\ &\Leftrightarrow \phi \in w \text{ and } \psi \in w \Leftrightarrow (\phi \wedge \psi) \in w.\end{aligned}$$

$$\begin{aligned}\mathcal{M}_{\text{fin}}, (\tilde{X}_i, h_w) \models \Box\psi &\Leftrightarrow \text{for all } h \in H_{\tilde{X}_i} \text{ it holds that } (\tilde{X}_i, h) \models \psi \\ &\Leftrightarrow \text{for all } h \in H_{\tilde{X}_i} \text{ there is } w' \in \tilde{X}_i \text{ with} \\ &\quad\; h = h_{w'} \text{ and } \psi \in w' \\ &\Leftrightarrow \text{for all } w' \in \tilde{X}_i, \psi \in w' \overset{\text{by Lemma 8 (i)}}{\Leftrightarrow} \Box\psi \in w.\end{aligned}$$

$$\begin{aligned}\mathcal{M}_{\text{fin}}, (\tilde{X}_i, h_w) \models \alpha\,\text{cstit} : \psi &\Leftrightarrow \text{ for all } h \in \tilde{C}^{\alpha}_{\tilde{X}_i}(h_w) \text{ it holds that} \\ &\quad\; (\tilde{X}_i, h) \models \psi \\ &\Leftrightarrow e_\alpha \in U^i_{[\alpha]_{\tilde{X}_i}} \text{ with } w \in e_\alpha : \\ &\Leftrightarrow \text{for all } w' \in \tilde{X}_i, \text{ if } w, w' \in e_j \text{ then } \psi \in w' \\ &\overset{\text{by Lemma 8 (ii)}}{\Leftrightarrow} \alpha\,\text{cstit} : \psi \in w.\end{aligned}$$

$$\begin{aligned}\mathcal{M}_{\text{fin}}, (\tilde{X}_i, h_w) \models \diamondsuit\psi &\Leftrightarrow \text{ there is } U \in \tilde{N}_{(\tilde{X}_i, h_w)} \varnothing \neq U \subseteq \|\psi\| \\ &\Leftrightarrow \text{ there is } \phi \in \Sigma_\varphi \text{ with } \varnothing \neq |\phi| \subseteq \|\psi\| \text{ and} \\ &\quad\; \diamondsuit\phi \in w \\ &\Leftrightarrow \diamondsuit\psi \in w.\end{aligned}$$

If $|\phi| \subseteq \|\psi\|$, then $|\phi| \subseteq |\psi|$, i.e. for all $w' \in \tilde{W} : (\psi \supset \phi) \in w'$. Assume there is $w \in W$ with $(\psi \supset \phi) \notin w$. Since $\tilde{W}$ is a complete set of representatives of all $\equiv_{\Sigma_a}$-equivalence classes, there is $\tilde{w} \in \tilde{W}$ with $w \equiv_{\Sigma_a} \tilde{w}$. For the canonical model $\mathcal{M}$ it holds that $(\psi \supset \phi) \notin w$. Then $\mathcal{M}, (X_i, h_w) \models \neg(\psi \supset \phi)$ and $\mathcal{M}, (X_i, h_{\tilde{w}}) \models (\psi \supset \phi)$. Thus, $\mathcal{M}, (X_i, h_w) \not\models \psi$ or $\mathcal{M}, (X_i, h_w) \models \phi$ and $\mathcal{M}, (X_i, h_{\tilde{w}}) \models \psi$ and $\mathcal{M}, (X_i, h_{\tilde{w}}) \not\models \phi$, but this conflicts with $w \equiv_{\Sigma_a} \tilde{w}$, as $\phi, \psi \in \Sigma_a$. Therefore, for all $w \in W : \phi \supset \psi \in w$, and so by (RM) $\diamondsuit\phi \supset \diamondsuit\psi \in w$. Consequently, $\diamondsuit\psi \in w$. The other direction is obvious with $\psi = \phi$. Similar considerations give:

$$\begin{aligned}\mathcal{M}_{\text{fin}}, (X_i, h_w) \models \alpha\,\text{bel} : \psi &\Leftrightarrow \text{there is } U \in \tilde{B}^{\alpha}_{(X_i, h_w)} \varnothing \neq U \subseteq \|\psi\| \\ &\Leftrightarrow \text{there is } \phi \in \Sigma_\varphi \text{ with } \varnothing \neq |\phi| \subseteq \|\psi\| \text{ and} \\ &\quad\; \alpha\,\text{bel} : \phi \in w \Leftrightarrow \alpha\,\text{bel} : \psi \in w.\end{aligned}$$

$$\begin{aligned}
\mathcal{M}_{\text{fin}}, (X_i, h_w) \models \alpha\,\text{des} : \psi &\Leftrightarrow \text{there is } U \in \tilde{D}^{\alpha}_{(X_i,h_w)} \varnothing \neq U \subseteq \|\psi\| \\
&\Leftrightarrow \text{there is } \phi \in \Sigma_{\varphi} \text{ with } \varnothing \neq |\phi| \subseteq \|\psi\| \text{ and} \\
&\qquad \alpha\,\text{des} : \phi \in w \Leftrightarrow \alpha\,\text{des} : \psi \in w. \\
\mathcal{M}_{\text{fin}}, (X_i, h_w) \models \alpha\,\text{int} : \psi &\Leftrightarrow \tilde{I}^{\alpha}_{(X_i,h_w)} \subseteq \|\psi\| \\
&\Leftrightarrow \text{for all } w' \in \tilde{W} : \text{if } t^{\alpha}_{w} \cap \Sigma_a \subseteq w', \text{ then } \psi \in w'. \\
&\Leftrightarrow \alpha\,\text{int} : \psi \in w.
\end{aligned}$$

Assume $\alpha\,\text{int} : \psi \notin w$, then $\neg\alpha\,\text{int} : \psi \in w$. Because of (Di) there are two different cases possible, (i) $\alpha\,\text{int} : \neg\psi \in w$ or (ii) $\neg\alpha\,\text{int} : \neg\psi \in w$. If (i), then $\neg\psi \in t^{\alpha}_{w}$ and, since $\Sigma_i \subseteq \Sigma_a$, $\neg\psi \in t^{\alpha}_{w} \cap \Sigma_a \subseteq w'$. Or (ii) $\neg\alpha\,\text{int} : \psi \in w$ and $\neg\alpha\,\text{int} : \neg\psi \in w$; then $\psi, \neg\psi \notin t^{\alpha}_{w}$. But then there is a w'' with $t^{\alpha}_{w} \cap \Sigma_a \subseteq w''$ and $\neg\psi \in w''$. These contradictions imply $\alpha\,\text{int} : \psi \in w$.

To sum up, like d stit logic, bdi–stit logic is finitely axiomatizable and has the finite model property. Therefore, it is decidable.

Caroline Semmling
Institute of Philosophy, Dresden University of Technology
Dresden, Germany
Caroline.Semmling@gmx.de

Heinrich Wansing
Institute of Philosophy, Dresden University of Technology
Dresden, Germany
Heinrich.Wansing@tu-dresden.de

References

Belnap, N. D., & Perloff, M. (1988). Seeing to it that: a canonical form for agentives. *Theoria*, *54*, 175–199.

Belnap, N. D., Perloff, M., & Xu, M. (2001). *Facing the future: Agents and choices in our indeterminist world.* New York: Oxford University Press.

Chellas, B. (1980). *Modal logic: An introduction.* Cambridge: Cambridge University Press.

Fagin, R., & Halpern, J. Y. (1988). Belief, awareness, and limited reasoning. *Artificial Intelligence*, *34*, 39-76.

Georgeff, M. P., & Rao, A. S. (1998). Decision procedures for BDI logics. *Journal of Logic and Computation*, *8*, 293–342.

Horty, J. F., & Belnap, N. D. (1995). The deliberative stit: A study of action, omission, ability and obligation. *Journal of Philosophical Logic*, *24*, 583–644.

Montague, R. (1970). Universal grammar. *Theoria*, *36*, 373–398.

Scott, D. (1970). Advice in modal logic. In K. Lambert (Ed.), *Philosophical problems in logic* (pp. 143–173). Dordrecht: Reidel.

Semmling, C., & Wansing, H. (2008). From *bdi* and *stit* to *bdi-stit* logic. *Logic and Logical Philosophy*, 185–207.

Semmling, C., & Wansing, H. (2009). *Reasoning about belief revision.* (To appear in: E. J. Olsson, S. Rahman, and T. Tulenheimo (eds.), *Science in Flux*, Springer-Verlag, Berlin.)

Wansing, H. (2006a). Doxastic decisions, epistemic justification, and the logic of agency. *Philosophical Studies*, *128*.

Wansing, H. (2006b). Tableaux for multi-agent deliberative-stit logic. In G. Governatori, I. Hodkinson, & Y. Venema (Eds.), *Advances in modal logic* (Vol. 6, pp. 503–520). London: College Publications.

Wooldridge, M. (2000). *Reasoning about rational agents.* Cambridge MA: MIT Press.

Xu, M. (1994). Decidability of deliberative stit theories with multiple agents. In D. M. Gabbay & H. J. Olbach (Eds.), *Temporal logic, first international conference, ICTL '94* (pp. 332–348). Berlin: Springer-Verlag.

Xu, M. (1998). Axioms for deliberative STIT. *Journal of Philosophical Logic*, *27*, 505–552.

A Procedural Interpretation of Split Negation

Sebastian Sequoiah-Grayson*

1 Introduction

Taking the procedural/dynamic turn in the study of information seriously means that we need to make the transition from the study of bodies of information, to the study of the manipulations *of* such bodies of information. In this case, we will not be able to carry out the study of informational dynamics by restricting our attention to bodies of information, or even to the structure of the bodies of information, although this is an important component. We will also need to pay attention to the procedures via which such bodies of information are combined and developed, and processed.

One restriction that we might place on a particular study of informational dynamics is that we examine only *positive information*. That is, we might restrict our attention to the positive fragments of various logics used to underpin logics of information flow. Restricting ourselves to the study of positive information is justifiable on several counts, not the least of which is that it makes perfect sense to restrict ourselves to simpler cases, as even these may turn out to be surprisingly complicated. However, to do justice to the phenomena of information flow, any adequate theory of information processing will have to allow for the representation of both positive, and negative information. In this case, attention will not be restricted to the positive fragment of the various logics used to underpin logics of information flow.

* Many thanks to Vladimír Svoboda, Michal Peliš, and all behind Logica 2008! This research was made possible by the generous support of the Harold Hyam Wingate Foundation. This research was carried out whilst undertaking a Visiting Research Fellowship at the Tilburg Institute of Logic and Philosophy of Science (TiLPS), at Tilburg University, The Netherlands. I am extremely grateful to Stephan Hartmann and everyone at TiLPS for the vibrant, research-griendly atmosphere provided. I am also extremely grateful to Edgar Andrade, Johan van Benthem, Francesco Berto, Catarina Dutilh, Volker Halbach, Christian Kissig, Greg Restall, Heinrich Wansing, and Tim Williamson for many invaluable suggestions. Any errors that remain are my own.

This essay is an argument for a particular procedural interpretation of negative information. In particular, it is an argument for a procedural interpretation of split negation. A split negation pair $\langle\sim,\neg\rangle$ is definable in any non-commutative logic. As such, a procedural interpretation of split negation should adoptable in principle for any non-commutative logic, be this a non-commutative linear logic, or a variant of the Lambek calculus or whatever. Accordingly, we will be abstracting across non-commutative logics in general as opposed to looking in detail at any one non-commutative logic in particular. However, an information–processing application will be the general motivation. From a philosophical standpoint, the closest analogues are the non-commutative linear logics, albeit under procedural interpretations. Linear logics were developed in order to track resource-use: formula are understood as resources, and in this case number of times they occur becomes relevant. As such, the mark of linear logics in general is the rejection of contraction. However, if the formula are taken to be concrete data, then the accessibility of these resources also becomes relevant. It is often the case that data have spatiotemporal locations, such as in the memories of agents or computers, and remote data will be less easy to access than adjacent data. In a more sensitive logic of resources then, it is not only the multiplicity of data, but also their order that is important. Spatiotemporal obstacles often need to be circumvented so that data may be accessed, hence commutation is inappropriate by virtue of its destroying the very ordering that we would like to preserve. In situations where actual information processing is being carried out, the arrangement of the data is crucial (Paoli, 2002, 28-9). For recent work on non-commutative linear logics, see (Abrusci & Ruet, 2000), and for an explicitly procedural examination of commutation in the context of agent-based information processing, see (Sequoiah-Grayson, 2009).

Interpreting split negation is a known difficulty (Dosen, 1993, 20). For any negation type there will be more than one way of defining it. Given a definition, we then need to provide an interpretation of the resulting negation in terms commensurate with the intended application. In our case, the intended application is the area of dynamic information processing. Given the procedural aspect, we will define the negation of A in terms of A implying bottom ($\mathbf{0}$). This is commensurate with information processing due to the implication doing the work being analysed in procedural terms. Sans the procedural aspect, an interpretation of the negation of A in terms of A implying $\mathbf{0}$ goes back at least to (Kripke, 1965).

This essay develops and proposes a particular interpretation of the negation of A in terms of A implying $\mathbf{0}$ in information processing terms: In section 2, information frames and information models are introduced. We also introduce the definition of split negation. The informational reading of the ternary relation R of frame semantics is introduced. In section 3, a procedural interpretation of split negation under the definition given in section 2 is proposed. Up to this point our exploration will have been conducted in purely model-theoretic terms. It is in section 3 that we touch on to proof-theoretical matters. This is essentially to check the procedural interpretation against a series of universally valid proof-theoretical properties of split negation. Put simply, the proposal is that we interpret the negation of A in terms of the ruling out of particular procedures, with these procedures being any procedure that involves combining the negation of A with A itself.

The first step is to introduce the notion of an information frame and model, so that we may specify our definition of split negation.

2 Information frames and models

Take a non-commutative information frame $\mathbf{F}\ \langle S, \sqsubseteq, \bullet, \otimes, \rightarrow, \leftarrow, \mathbf{0}\rangle$ where S is a set of information states $x, y, \ldots$ that may be inconsistent, incomplete, or both, the binary relation $\sqsubseteq$ is a partial order on S of informational development/inclusion, $\bullet$ is the binary composition operator on information states such that due to commutation failure we have it that $x \bullet y \neq y \bullet x$, $\otimes$ is (non-commutative) fusion, $\rightarrow$ and $\leftarrow$ are right and left implication respectively, and $\mathbf{0}$ is bottom.[1] Making all of this clear is easier once we have a model.

A model $\mathbf{M} := \langle \mathbf{F}, \Vdash \rangle$ is an ordered pair $\mathbf{F}\ \langle S, \sqsubseteq, \bullet, \otimes, \rightarrow, \leftarrow, \mathbf{0}\rangle$ and $\Vdash$ such that $\Vdash$ is an evaluation relation that holds between members of S and formula. Where A is a propositional formula, and $x, y, z \in \mathbf{F}$, $\Vdash$ obeys the heredity condition:

$$\text{For all } A, \text{ if } x \Vdash A \text{ and } x \sqsubseteq y, \text{ then } y \Vdash A, \tag{1}$$

And also obeys the following conditions for each of our connectives:

$$x \Vdash A \otimes B \textit{ iff } \text{for some } y, z, \in \mathbf{F} \text{ s.t. } y \bullet z \sqsubseteq x, y \Vdash A \text{ and } z \Vdash B. \tag{2}$$

$$x \Vdash A \rightarrow B \textit{ iff } \text{for all } y, z \in \mathbf{F} \text{ s.t. } x \bullet y \sqsubseteq z, \text{if } y \Vdash A \text{ then } z \Vdash B. \tag{3}$$

$$x \Vdash A \leftarrow B \textit{ iff } \text{for all } y, z \in \mathbf{F} \text{ s.t. } y \bullet x \sqsubseteq z, \text{if } y \Vdash A \text{ then } z \Vdash B. \tag{4}$$

$$x \Vdash \mathbf{0} \text{ for no } x \in \mathbf{F}. \tag{5}$$

[1] A notational note: **0** is commonly written as $\bot$. The difference in notation is to ensure that no confusion is made between bottom, and the perp relation of incompatibility (Dunn, 1993), (Dunn, 1994), (Dunn, 1996), written as $\bot$. In the recent literature on negation, $\bot$ is so often used for the perp relation that using it for bottom creates too great a risk for misunderstanding. Hence, we follow (Girard, 1987) in the use of **0** for bottom.

Many non-commutative logics are also non-associative, such as the non-associative Lambek calculus among others. However, since nothing that follows depends on either the presence or absence of associativity, we should be able to safely ignore this issue for our purposes.

The evaluation relation $\Vdash$ may be understood in different ways, depending on the context of application. For example, if we were to be working with language frames and syntactically categorising particular alphabetical strings, we would understand $x \Vdash A$ to mean *string x is of type A*. We might instead consider a scientific research project with its various developmental phases. In this case the development relation $\sqsubseteq$ will order different states of a research project over time (with the idealisation that there is no information-loss). Here we would understand $x \Vdash A$ to mean that the proposition A is known at state x, and that this particular state of development in the project supports A. We will in fact return to this very idea in 5 below. For now however, we need something a little more general. Along with (Mares, 2009) we will understand $x \Vdash A$ to mean that *the information in state x carries the information that A*. Hence, we may also say that x *supports the information that A*. This is very similar to the familiar semantic entailment relation $\models$. The difference is that we want to allow for the information at x being incomplete and/or inconsistent. There are many applications where we might want to do this. Taking inconsistency as the running example, consider various states of an agent as the agent reasons deductively. In this case, x may support A where A is '*p and not p*', but this is different from x making A true, at least in the usual sense of "making true", as there is no possible way that the world can be such that x could be true of it. One might wish to understand 'supports' as 'makes true' if one holds to a *dialethic paraconsistentism* whereby at least some contradictions are taken to be true. However, we will sidestep this particular debate and stay with the interpretation of 'supports' that takes it to be the subtler relative of 'makes true' in the manner stipulated above.

The reader familiar with the ternary relation R of frame semantics will recognise (2)–(4) as the ternary conditions for $\otimes$, $\rightarrow$, and $\leftarrow$ respectively, under an explicitly informational reading. R may be parsed in terms of the two binary relations $\bullet$ and $\sqsubseteq$ and such that $Rxyz$ comes out as $x \bullet y \sqsubseteq z$. How should we read formulas containing the binary composition relation $\bullet$? A common and traditional way of understanding binary composition is simply to take $x \bullet y$ as x together with y. In this case $\bullet$ will behave much the same as set union such that x together with y is no different to y together with x, and x together with itself is no different from x and so forth. However, we are not restricted to such a reading of $\bullet$. There is in general no canonical reading of $Rxyz$. That is to say that there is no canonical interpretation of the model theory. Although this is frustrating when one encounters the ternary relation for the first time, it is a key point with respect to the flexibility of ternary semantics. In our case, we have it that $\bullet$ is non-commutative, so $x \bullet y \neq y \bullet x$. In ternary terms, non-commutation comes out as $Rxyz \neq Ryxz$. Hence, simply interpreting $x \bullet y$ as x together with y will blur the very ordering that non-commutation is trying to preserve. We could stipulate that "x composed with y" differs from "y composed with x", however this is slightly strained and does not read straight off a casual use of 'composed'. A better way to keep this distinction robust is to read $x \bullet y$ as x *applied to* y. "Applying" is an order-sensitive notion, and one that fits comfortably with dynamic/procedural operations. So we can think of $x \bullet y$ as the composition of x with y, where this composition is order-sensitive, and we will mark this order-sensitivity by speaking of "application" instead of "composition".

Of course we are not merely concerned with syntactic constructions, as $x, y, z \in S$, and S is a set of information states. We are concerned with the application of the information in one state to the information in another. One way then, of reading $Rxyz$, is that $Rxyz$ holds iff the result of applying the information in x to the information in y is contained in the information in z, and this is precisely what $x \bullet y \sqsubseteq z$ tells us. Another way of putting this is to say that the information in z is a development of the information resulting from the application of the information in x to the information in y.

The role that *information application* plays here is not redundant, and neither is it *merely* to mark order-sensitivity. We are not simply concerned with ordered sequences of information states — something like an order-sensitive conjunction where we would have one piece of information, then another and then another etc. We are concerned with something much more subtle. We are concerned with the *interaction between* information states. This concern with interaction, or *process*, is precisely why it is that we are concerned with order-sensitivity in the first place. Order-sensitivity is in this sense a means to an end, with this end being the individuation of procedures of dynamic information processing. This sense of "applied" carries over in a natural way from the information states themselves, to the propositions supported by the information states. It is easiest to see this with an example.

Take fusion, and its frame conditions given in (2). (2) can be interpreted to state that an information state x carries the information resulting from the application of the information that A to the information that B if and only if x is itself a development of the application of the information in state y to the information in state z, where y carries the information that A and z carries the information that B. This is a little longwinded, and going in the right to left hand direction is a little more straightforward: for two states y and z that carry the information that A and that B respectively, the application of y to z will result in a new information state, x, such that x carries the information that results from the application of the information that A to the information that B. The analogous interpretations of the frame conditions for right and left implication ((3), and (4), respectively) unpack in a similar manner.

The fusion connective and the implication connectives are not independent; they form a family of sorts. Our fusion and implication connectives interrelate in the following manner:

$$A \otimes B \vdash C \textit{ iff } B \vdash A \rightarrow C. \tag{6}$$

$$A \otimes B \vdash C \textit{ iff } A \vdash C \leftarrow B. \tag{7}$$

In deductive information processing, we understand the premises as databases and the consequence relation '$\vdash$' as the information processing mechanism, a more brutally syntactic operation that the information carrying/supporting of $\Vdash$. In informational terms, we may read $A \vdash B$ as information of type B follows from information of type A, or the information in B follows from the information in A etc. We can think of typing as encoding, in which case we might also read $A \vdash B$ as the information encoded by B follows from the information encoded by A. (6) and (7) make sense.

Take (6), starting with the left-to-right-hand direction: If the information in C follows from the information resulting from the application of the information in A to the information in B, then from the information in B alone it follows that we have the information in C conditional on the information in A. The right-to-left-hand direction works out similarly: If from the information in B alone we can get the information in C conditional on the information in A, then we can get the information in C via the application of the information A to the information in B. Now take the left-to-right-hand direction of (7): If, again, the information in C follows from the information resulting from the application of the information in A to the information in B, then from the information in A alone it follows that we have the information in C, this time conditional on the information in B. The right-to-left-hand direction works on similarly here too: If from the information in A alone we can get the information in C conditional on the information in B, then we can get the information in C via the application of the information in A to the information in B. (6) and (7) are informational processing versions of the deduction theorem.

With regards to (5), no information, in any context whatsoever, is of type $\mathbf{0}$. There is nothing that we can do to get $\mathbf{0}$, and $\mathbf{0}$ is not supported by any information state in our frame $\mathbf{F}$.

Now we have the logical tools that we need in order to begin looking at negative information. We can define a split negation pair in terms of double implication:

$$\sim A := A \rightarrow \mathbf{0}, \tag{8}$$

$$\neg A := \mathbf{0} \leftarrow A. \tag{9}$$

In this case, the frame conditions for $\sim A$ and $\neg A$ are cashed out in explicit informational terms as follows:

$$x \Vdash {\sim} A[A \to \mathbf{0}] \textit{ iff } \text{for each } y, z \text{ s.t. } x \bullet y \sqsubseteq z, \text{ if } y \Vdash A \text{ then } z \Vdash \mathbf{0}, \quad (10)$$

$$x \Vdash \neg A[\mathbf{0} \leftarrow A] \textit{ iff } \text{for each } y, z \text{ s.t. } y \bullet x \sqsubseteq z, \text{ if } y \Vdash A \text{ then } z \Vdash \mathbf{0}. \quad (11)$$

The major points so far have been the informational translation of the ternary relation R, such that $Rxyz$ comes out as as $x \bullet y \sqsubseteq z$, and the definition of split negation in terms of double implication, such that ${\sim} A :=$ $A \to \mathbf{0}$ and $\neg A := \mathbf{0} \leftarrow A$. The definitional component here is important. Our double implication connectives $\to$ and $\leftarrow$ have their conditions given by R, albeit under an informational reading, in (3) and (4) respectively. This means that our split negation connectives $\sim$ and $\neg$ ultimately have their definitions in terms of the ternary relation also.

3 A procedural interpretation of split negation

How should we interpret ${\sim} A$ and $\neg A$ given their respective definitions, $A \to \mathbf{0}$ and $\mathbf{0} \leftarrow A$? The type of answer we give here will depend on the domain. For example, if we were working with actions, then we could interpret ${\sim} A$ as the type of action that cannot be followed by an action of type A, and we could interpret $\neg A$ as the type of action that cannot follow an action of type A etc. For any interpretation that we give to split negation, the interpretation has to be compatible with certain properties that hold universally for any split negation. The purpose of this section is to check the presently proposed procedural interpretation of split negation against these properties, which are listed as (12)–(18) below.

We are working with information. This is still fairly general though, and various informational applications will likely influence our choice of interpretation. By taking the procedural/dynamic turn and working with information *flow*, the *application* aspect at work in both fusion and the binary combination operator get taken very seriously. In this case, the suggestion is that we interpret ${\sim} A$ as the body of information that cannot be applied to bodies of information of type A, and that we interpret $\neg A$ as the body of information that cannot have bodies of information of type A applied to it. The interpretation is supported by the model theory; by the information states supporting ${\sim} A$, $\neg A$, and A. If x supports ${\sim} A$ and y supports A, x cannot be applied to y. Similarly, if x supports $\neg A$ and y supports A, then y cannot be applied to x. This is not because such an application would cause an explosion of information, but because it could never generate any information. The interpretation of split negation in terms of ruling out particular informational applications is not gerrymandered. It is directly supported by the frame conditions for ${\sim} A$ and $\neg A$.

To see this, note that the frame conditions in informational terms for $\sim A$ laid out in (10) above tell us that there is no information resulting from the application of x to y where $x \Vdash \sim A$ and $y \Vdash A$, since $x \bullet y \sqsubseteq z$ and $z \Vdash \mathbf{0}$ (and $z \Vdash \mathbf{0}$ nowhere). Suppose though that we were to attempt to apply $\sim A$ to A, in other words to attempt $\sim A \otimes A$. In informational terms, the frame conditions for fusion (2) tell us that an information state x will support $\sim A \otimes A$ iff for some information states y and z such that x is an informational development of the application of the information in y to the information in z, y supports $\sim A$ and z supports A. However, we know from our definition of $\sim A$ in terms of $A \rightarrow \mathbf{0}$, that there is no state x such that it supports the application of $\sim A$ to A, this is simply what (10) tells us.

Support for the ruling out conditions on $\neg A$ from the frame conditions for $\neg A$ works similarly, and involves only a directional change. The frame conditions for $\neg A$ laid out in (11) above tell us that there is no information resulting from the application of the information state y to the information state x where $y \Vdash A$ and $x \Vdash \neg A$ since $y \bullet x \sqsubseteq z$ and $z \Vdash \mathbf{0}$ (and $z \Vdash \mathbf{0}$ nowhere). If we were to attempt to apply A to $\neg A$, in other words attempt $A \otimes \neg A$, then there would need to be an information state x that supported $A \otimes \neg A$, and this would be the case iff there were some information states y and z such that y supported A and z supported $\neg A$ and x was an informational development of the application of y to z. From our definition of $\neg A$ in terms of $\mathbf{0} \leftarrow A$ however, we know that there is no state x such that it supports the application of A to $\neg A$, this is marked out by (11).

Given the non-gerrymandered nature of the interpretation of split negation in terms of procedural prohibition, we should be able to give a natural interpretation of general proof-theoretic, hence information–processing properties of split negation in such terms. For any split negation, independently of which structural rules are present, the following (12)–(18) hold:

$$\frac{A \vdash B}{\sim B \vdash \sim A} \tag{12}$$

(12) makes sense in terms of the ruling out of information processing procedures. Given a split negation, and given also that information of type B follows from information of type A, then ruling out the procedure $\sim A \otimes A$ follows from ruling out the procedure $\sim B \otimes B$. This is just to say that given that we can get information of type B from information of type A, then from the body of information that can never be applied to bodies of type B, we can get the body of information that can never be applied to bodies of information of type A.

$$\frac{A \vdash B}{\neg B \vdash \neg A} \tag{13}$$

The reasoning with regards to (13) is directly analogous to that surrounding (12): Again given a split negation, and again given that information of type

B follows from the information of type A, then ruling out the procedure $A \otimes \neg A$ follows from ruling out the procedure $B \otimes \neg B$. This is just to say that given that we can get information of type B from information of type A, then from the body of information that can never have bodies of type B applied to it, we can get the body of information that can never have bodies of information of type A applied to it. The reasoning surrounding (14) and(15) is slightly more involved than in (12) and (13).

$$\frac{A \vdash {\sim}B}{B \vdash \neg A} \tag{14}$$

$$\frac{B \vdash \neg A}{A \vdash {\sim}B} \tag{15}$$

We can take (14) and (15) together, getting the *split negation property*:

$$A \vdash {\sim}B \textit{ iff } B \vdash \neg A. \tag{16}$$

Starting with the left-to-right-hand direction: If we can, on the basis of information of type A alone, get the body of information that can never be applied to bodies of information of type B, then on the basis of information of type B alone, we can get the body of information that can never have bodies of information of type A applied to it. The intermediate step is this: If we were to apply A to B (i.e. $A \otimes B$) then we would get nothing, viz. $\mathbf{0}$, since $A \otimes B \vdash \mathbf{0}$, since if $A \vdash {\sim}B$ then $A \otimes B \vdash \mathbf{0}$. As such, from information of type B alone we can get the body of information that can never have bodies of information of type A applied to it. The right-to-to-left-hand direction is similar: If we can, on the basis of information of type B alone, get the body of information that can never have bodies of information of type A applied to it, then were to apply B to A (i.e. $B \otimes A$) then we would get nothing, viz. $\mathbf{0}$, since $B \otimes A \vdash \mathbf{0}$, since if $B \vdash \neg A$ then $B \otimes A \vdash \mathbf{0}$. As such, then from information of type A alone we can get the body of information that can never be applied to bodies of information of type B. (17) and (18) are more straightforward:

$$A \vdash \neg{\sim}A, \tag{17}$$

$$A \vdash {\sim}\neg A. \tag{18}$$

On the basis of information of type A alone, we can rule out the body of information that can never be composed with bodies of information of type A. This is just to say that we can rule out $A \to \mathbf{0}$. This is just what (17) states under a procedurally focused informational interpretation. Similarly, on the basis of information of type A alone, we can also rule out the body of information that can never have bodies of information of type A applied to it. In this case we are ruling out $\mathbf{0} \leftarrow A$. In this context, "ruling out" is a form of procedural prohibition.

The proposal for a procedural interpretation of split negation has been that we interpret $\sim A$ (that is $A \rightarrow \mathbf{0}$) as the body of information that cannot be applied to bodies of information of type A, and that we interpret $\neg A$ (that is $\mathbf{0} \leftarrow A$) as the body of information that cannot have bodies of information of type A applied to it. We have seen that this interpretation of split negation is entirely natural once we translate the ternary relation R into its informational form, in which case the interpretation is directly supported by the frame conditions for $\sim A$ and $\neg A$, (10) and (11) respectively. We have also seen that the interpretation is compatible with the universally valid split negation properties (12)–(18).

4 Conclusion

We have seen how it is that we may reconstruct the ternary relation of frame semantics, $Rxyz$ in explicitly dynamic informational terms, as $x \bullet y \sqsubseteq z$. This dynamic informational reconstruction carries over to any connective defined in terms of the ternary relation, allowing us to give explicitly procedural accounts of double implication and fusion. Since we have used double implication to define a split negation, $\sim A := A \rightarrow \mathbf{0}$, and $\neg A := \mathbf{0} \leftarrow A$, we have a procedural definition of split negation.

Given the definition of split negation in these dynamic informational terms, we have been able to "read off" a natural procedural interpretation of split negation. This interpretation has been shown to be compatible with the universally valid properties of a split negation.

Sebastian Sequoiah-Grayson
Formal Epistemology Project
IEG – Computing Laboratory – University of Oxford
sebsequoiahgrayson@hiw.kuleuven.be
http://users.ox.ac.uk/~ball1834/index.shtml

References

Abrusci, V. M., & Ruet, P. (2000). Non-commutative logic, I: the multiplicative fragment. *Ann. Pure Appl. Logic*, *101*(1), 29–64.

Dosen, K. (1993). A historical introduction to substructural logics. In P. Schroeder-Heister & K. Dosen (Eds.), *Substructural logics, studies in logic and computation no. 2* (pp. 1–30). Oxford: Clarendon Press.

Dunn, J. M. (1993). Partial gaggles applied to logics with restricted structural rules. In P. Schroeder-Heister & K. Dosen (Eds.), *Substructural logics, studies in logic and computation no. 2* (pp. 63–108). Oxford: Clarendon Press.

Dunn, J. M. (1994). Star and perp: Two treatments of negation. In J. E. Tomberlin (Ed.), *Philosophical perspectives* (Vol. 7, pp. 331–357). Atascadero, CA: Ridgeview.

Dunn, J. M. (1996). Generalised ortho negation. In H. Wansing (Ed.), *Negation: A notion in focus* (pp. 3–26). Berlin: de Gruyter.

Girard, J.-Y. (1987). Linear logic. *Theoretical Computer Science*, *50*, 1–101.

Kripke, S. A. (1965). Semantical analysis of intuitionistic logic I. In J. Crossley & M. Dummett (Eds.), *Formal systems and recursive functions* (pp. 92–129). Amsterdam: North-Holland.

Mares, E. (2009). *General information in relevant logic.* (Forthcoming in *Synthese*, section: *Knowledge, Rationality, and Action*, L. Floridi and S. Sequoiah-Grayson (eds.): *The Philosophy of Information and Logic, Synthese, KRA*, Proceedings of PIL–07, The First Workshop on the Philosophy of Information and Logic, University of Oxford, November 3–4, 2007.)

Paoli, F. (2002). *Substructural logics: A primer.* Berlin–New York: Spinger Verlag.

Sequoiah-Grayson, S. (2009). *A positive information logic for inferential information.* (Forthcoming in *Synthese*, section: *Knowledge, Rationality, and Action*, L. Floridi and S. Sequoiah-Grayson (eds.): *The Philosophy of Information and Logic, Synthese, KRA*, Proceedings of PIL–07, The First Workshop on the Philosophy of Information and Logic, University of Oxford, November 3–4, 2007.)

Reference to Indiscernible Objects

Stewart Shapiro*

1 The problem

Some critics of my ante rem structuralism (Shapiro, 1997) argue that I have an issue with structures that have indiscernible places.[1] A structure is said to be "rigid" if its only automorphism is the trivial one based on the identity mapping. The main exemplars of the alleged problem are non-rigid structures. It is an easy theorem that isomorphic structures are equivalent: Let f be an automorphism on a given structure M and let $\Phi(x_1, \ldots, x_n)$ be any formula in the language of the structure. Then for any objects $a_1, \ldots, a_n$ in the domain of M, M satisfies $\Phi(a_1, \ldots, a_n)$ if and only if M satisfies $\Phi(fa_1, \ldots, fa_n)$. If f is a non-trivial automorphism, then there is an object a such that $fa \neq a$. In this case, a and fa are indiscernible, at least concerning the language of the structure: anything true of one of them will be true of the other. So non-rigid structures have indiscernible objects.

The most-cited example is that of complex analysis. Start with the language of fields, and consider the algebraic closure of the reals, which is unique up to isomorphism (in its second-order formulation). The complex numbers are the intended model. The function that takes a complex number $a + b\mathrm{i}$ to its conjugate $a - b\mathrm{i}$ is an automorphism. It follows that for any formula $\Phi(x)$, with only x free, $\Phi(a + b\mathrm{i})$ if and only if $\Phi(a - b\mathrm{i})$. In particular, $\Phi(\mathrm{i})$ if and only if $\Phi(-\mathrm{i})$. So i and −i are indiscernible; they

* I gave early versions of parts of this paper at the philosophy of mathematics workshop at Oxford, the Arché Research Centre at the University of St. Andrews, Ohio State University and the University of Minnesota. Thanks to all of the audiences there. I am indebted to Craige Roberts, Gabriel Uzquiano, Ofra Magidor, Cathy Müller, Dan Isaacson, Graham Priest, Kevin Scharp, Robert Kraut, and Jason Stanley.

[1] The early critics include (Burgess, 1999, pp. 287–288), (Hellman, 2001, pp. 192–196), and (Keränen, 2001, 2006). More recent participants in the debate include (Ladyman, 2005), (Button, 2006), (Ketland, 2006), (MacBride, 2005, § 3), (MacBride, 2006b), and (Leitgeb & Ladyman, 2008). My own contributions include (Shapiro, 2006b), (Shapiro, 2006a), and (Shapiro, 2008).

have the same properties, at least among those that can be expressed in the language. Another oft-cited example is Euclidean space, where things are even worse. Any two points in Euclidean space can be connected with a linear translation, which is an automorphism. So, it seems, *all* of the points in Euclidean space are indiscernible, at least with respect to properties that can be expressed in the language of geometry. Hannes Leitgeb and James Ladyman (2008) point out that since some (simple) graphs have no relations, *any* bijection on them is an isomorphism. So with those graphs, every point is indiscernible from every other. The simplest of these simple graphs are isomorphic to the finite cardinal structures introduced in my chapter on epistemology in (Shapiro, 1997, Ch. 4).

Why is this a problem for ante rem structuralism? Some ill-chosen remarks in my book at least suggest a principle of the identity of indiscernibles, which, in light of examples like these, would reduce the view to absurdity. I'd be committed to saying that $\mathrm{i} = -\mathrm{i}$, and that there is only one point in Euclidean space. But there is little point in trying to figure out what my view was. I reject the identity of indiscernibles now.

Much of the discussion of this issue is metaphysical. John Burgess (1999, p. 288) points out that although the two complex roots of -1 are distinct, on my view "there seems to be *nothing* to distinguish them." This seems to invoke something in the neighborhood of the Principle of Sufficient Reason. If something is so, then there must be something that makes it so, or at least something that explains why it is so. Jukka Keranänen (2001) articulates a general metaphysical thesis that anyone who puts forward a theory of a type of object must provide an account of how those objects are to be "individuated". According to Keränen, for each object a in the purview of a proposed theory, we have to be told "the fact of the matter that makes a the object it is, distinct from any other object" of the theory, by "providing a *unique* characterization thereof."

Some authors entered the discussion, on my behalf, by suggesting metaphysical principles that are weaker than Keränen's individuation requirement but still meet Burgess's demand that the theorist find something that distinguishes distinct objects. The idea, it seems, is that one can distinguish objects without individuating them. The weakest of these requirements is a thesis that for any a, b, if $a \neq b$ then there is an irreflexive relation R such that Rab (Ladyman, 2005). Complex analysis and Euclidean geometry easily pass this test. For example, i and $-$i satisfy the irreflexive relation of being additive inverse to each other and distinct from 0, and any pair of distinct points in Euclidean space satisfy the irreflexive relation of determining exactly one straight line. Nevertheless, the finite cardinal structures and some graphs still fail the test, unless non-identity counts as an irreflexive relation (in which case, of course, we do not have a substantial test).

I wish to put aside these metaphysical matters here, at least as far as possible. There are some related and, I think, more interesting issues concerning the semantics and pragmatics of mathematical languages, and perhaps languages generally. These issues also bear on logic, and they go well beyond local disputes concerning ante rem structuralism. How do we manage to talk about, and thus, in some sense, refer to indiscernible objects? I do not intend to offer a detailed solution to this semantic problem here, just to highlight it and indicate its generality. A solution, I believe, would involve an extended foray into linguistics, the philosophy of language, and the philosophy of logic.

To be sure, the whole project presupposes that there are indiscernible objects, and this presupposes that the metaphysical principles adopted by some of my opponents are false. I also do not intend to argue for that in detail here (but see (Shapiro, 2008) and related work cited there). The informal language of complex analysis has a term i which is supposed to denote one of the square roots of -1. At least grammatically, i is a constant, a proper name. And the role of a constant is to denote a single object — at least in a sufficiently regimented language. But which of the square roots does i denote? Is it not as if the mathematical community has managed to single out one of the roots, in order to baptize it with the name "i". They cannot do so, as the two roots are indiscernible.

Gottlob Frege (1884) seems to have noted our problem:

> We speak of 'the number 1', where the definite article serves to class it as an object (§ 57). If, however, we wished to use [a] concept for defining an object falling under it [by a definite description], it would, of course, be necessary first to show two distinct things:
>
> that some object falls under the concept;
> that only one object falls under it (§ 74n).
>
> Nothing prevents us from using the concept 'square root of -1'; but we are not entitled to put the definite article in front of it without more ado and take the expression 'the square root of -1' as having a sense. (§ 97)

Complex analysis is perhaps the most salient example of the logical-semantic phenomenon in question, but there are a some others, at least if we go with a straightforward reading of various mathematical languages (see (Brandom, 1996)). Consider, for example, the integers, with addition as the only operation. It is, of course, an Abelian group, whose elements are:

$$\ldots, -3, -2, -1, 0, 1, 2, 3, \ldots$$

In the relevant language, the operation that takes any integer a to $-a$ is an automorphism. So anything in the relevant language that holds of an integer a also holds of $-a$. In this structure, then, 1 is indiscernible from -1, but, of course, 1 is distinct from -1. Another example is the Klein group. It has four elements, which are usually called e, a, b, and c, and there is one operation, given by the following table:

$$\begin{array}{cccc} e & a & b & c \\ a & e & c & b \\ b & c & e & a \\ c & b & a & e \end{array}$$

It is easy to verify that any function f that is a permutation on the four elements such that $fe = e$ is an automorphism. The three non-identity elements are thus indiscernible, in the language of groups, and yet there are three such elements and not just one. But which one is a?

Category theory is rampant with examples of the phenomenon under study here. The main reason for this, it seems, is that *every* categorical notion is preserved under isomorphism. To take one instance, an object O in a category is called *terminal*, if for any object A in the category, there is exactly one map from A to O. In categories with a terminal object, it is common to introduce a term "1" for such an object. It is trivial to show that any two terminal objects are isomorphic. Indeed, if 1 and $1'$ are both terminal, then there is exactly one map from 1 to $1'$ and exactly one map from $1'$ to 1. These two maps must compose to an identity map — either the unique map from 1 to 1, or the unique map from $1'$ to $1'$, depending on the order of composition. Moreover, any object isomorphic to a terminal object is itself terminal. So if a category has a terminal object, it usually has many. In the category of sets, for example, any singleton is terminal. Which of them is *the* terminal object of the category, the one picked out by the term "1"? The answer, of course, is that it does not matter — just as it does not matter which square root of -1 is i. And, here too, "1" functions as a singular term, at least as far as syntax goes.

In a category with a terminal object, it is common to define an *element* of an object A to be a map from 1 to A. So, it would seem, to know which maps are the elements of A, we have to know which object is 1. In a sense, we can't know this, but, again, it does not matter. In like manner, a *product* of two objects A, B is an object, usually written $A \times B$, and a pair of maps, one from $A \times B$ to A and one from $A \times B$ to B, that satisfies a certain universal property. Again, products are not usually unique: any object isomorphic to a product of A and B can itself be shown to be a product of A and B. Nevertheless, the "$\times$" symbol seems to be a function symbol, and it is common to talk about $A \times B$ as *the* product of A and B — just as it is common to talk about i as *the* square root of -1.

There is a related phenomenon concerning the use parameters or free variables in mathematical discourse. Those act like singular terms in context, but often fail uniqueness. Suppose that in the course of a demonstration, a geometer says "let $ABCD$ be any parallelogram, with the line AB congruent and parallel to the line CD." It follows that the pair of points A, B, (and the line segment AB) are indiscernible from the pair C, D (and the segment CD). Anything one can say about one of the pairs (and one of the segments) will be true of the other pair (and the other segment). So which one is AB and which one is CD?

2 Indiscernibility, semantics, and expressive resources

To say that two objects are indiscernible is to say that they cannot be told apart. This brief characterization suggests that indiscernibility is relative. Two balls may be visually indiscernible to someone who is color blind while being discernible to someone with more normal vision. In the present context, indiscernibility is relative to expressive resources. Two mathematical objects may be indiscernible with some batch of resources, but discernible once expressive resources are added. In the case of the integers, for example, 1 and -1 are discernible if one can invoke multiplication: 1 is the multiplicative identity and -1 is not.

At the outset, I formulated the issue in terms of rigid structures, with "rigidity" defined in terms of automorphisms. This registers the relativity to expressive resources as well, since "automorphism" is itself defined in terms of the background language: all of the primitive relations must be preserved.

Suppose, then, that we just add a constant i to the official language of complex analysis, with the obvious axiom $i^2 = -1$. Then, technically, the structure becomes rigid: there are no non-trivial automorphisms. The reason is that isomorphisms must preserve all of the structure in the language, and, in particular, it must preserve the denotations of the constants. If f is an isomorphism between M_1 and M_2, in the language of arithmetic, for example, then if the constant "0" denotes a in M_1, then "0" must denote fa in M_2. Similarly, let N be any model of complex analysis, in the envisioned language with a constant "i", and let f be an automorphism. If "i" denotes a in N, then "i" denotes fa in N. That is, $a = fa$. It follows that for each element b in the domain, $fb = b$. So f is trivial.

Nevertheless, it seems to me that, in the relevant intuitive sense, the two square roots of -1 are still indiscernible, even in this language. Let N' be a model of the theory that is just like N, except that in N', "i" denotes $-a$ (and thus "$-$i" denotes a). Technically, N is *not* isomorphic to N', for the above reason. However, it seems to me that the two models are equivalent, in the intuitive sense. Both have the same domain and they agree on the operations. In particular, in each model, the same two objects are the roots of -1. The only difference between them is that N *calls* one of them "i": and N' *calls* the other one "i". It seems to me that this is not a significant difference — not unless we add some structure to the naming relation.

To develop this point a bit, let us go up a level, so to speak, and think of the semantic relations themselves in formal, or structural terms. Begin with a simple graph that has two nodes and no edges. As noted above, this structure is completely homogeneous. Now add two new objects, a, b, and a relation R to the structure. The new item a bears R to one of the nodes in the original graph and b bears R to the other node. This is the structure of some very simple semantic relations on the graph: think of a and b as names, and R as the reference relation. *This* mathematical-cum-semantic structure is not rigid. If we modify it by switching the "referents" of a and b, we get an automorphism. And, intuitively, we have not really changed the extended structure with this switch. It is still the same simple graph with the same two new objects, the same relation R, and the same structural–semantic relations.

We can do the same with our more standard mathematical example. Consider a structure M that includes the places and relations of our model N, of complex analysis. In addition, M has nodes representing the primitive *terms* of the language of complex analysis ("0", "i", "+", "A"), and a relation R representing reference. So, for example, Rix would be an atomic formula in the envisioned object language, saying that the constant "i" refers to, or denotes, x. The theory would include the axioms of complex analysis (over N) and the Tarskian satisfaction clauses between the "terms" ("0", "i", "+", "A") and the relevant items constructed from N. So, for example, our structure would satisfy $\forall x \forall y((Rix \& Riy) \rightarrow x = y)$ and Ria (recalling that a is one of the square roots of -1 in N).

This mathematical-cum-semantic structure is not rigid. The function that takes $x+ay$ to $x-ay$ (within N), takes each "term" ("0", "1", "i", "+", "A") to itself, and adjusts the relation R accordingly, is an automorphism. We still have a model of complex analysis, as above, and all of the Tarskian satisfaction clauses are still satisfied (see (Leitgeb, 2007, pp. 133–134)). The problem, here, is to say something about the semantics and logic of the languages of mathematics, so construed.

3 The identity of indiscernibles

The issues here are related to those in Max Black's (1952) celebrated discussion of the identity of indiscernibles. The paper is in the form of a dialogue. One character, A, takes the identity of indiscernibles to be "obviously true", while the opponent, B, takes it to be "obviously false". The latter gives a thought experiment meant to refute the principle in question:

> Isn't it logically possible that the universe should have contained nothing but two exactly similar spheres? We might suppose that each was made of chemically pure iron, had a diameter of one mile, that they had the same temperature, color, and so on, and that nothing else existed. Then every quality and relational characteristic of the one would also be a property of the other. (Black, 1952, p. 156)

Black's two spheres are analogous to the two square roots of -1. Of course, I am not claiming that there is a possible world which consists of just these two complex numbers. But the rest of the analogy holds, at least in the language of complex analysis.

The main thrust of (Black, 1952) is metaphysical and, as noted above, such matters are being put aside here as much as possible. Along the way, however, the article broaches the logico-semantic issues of present concern. The defender of the indiscernibility of identicals, A, first denies that B has described a coherent possibility, and then continues, "But supposing that you *have* described a possible world, I still don't see that you have refuted the principle. Consider one of the spheres, a." At this point, B interrupts, protesting: "How can I, since there is no way of telling them apart? Which one do you want me to consider?". That is, B refuses to let his opponent use a variable, a parameter, or a singular term for one of the spheres. Character A responds: "This is very foolish. I mean either of the two spheres, leaving you to decide which one you wished to consider." In our case, it strikes me as eminently reasonable to say, "let i designate one of the square roots of -1. I don't care which." The present problem is to make sense of this locution. Character B denies that it can be made sense of.

Robert Brandom (1996, p. 298) puts our problem in similar terms:

> Now if we ask a mathematician '*Which* square root of -1 is i?,' she will say 'It doesn't matter: pick one.' And from a *mathematical* point of view this is exactly right. But from the *semantic* point of view, we have the right to ask how this trick is done — how is it that I *can* 'pick one' if I can't tell them apart? What must I do in order to be *picking* one, and picking *one*? For we really *cannot* tell them apart — and... not just because of some lamentable incapacity of ours.

The next exchange in Black's dialogue puts some detail to the differing semantic presuppositions. Character A, the proponent of the (obvious truth of) the identity of indiscernibles, continues, "If I were to say to you 'Take any book off the shelf' it would be foolish on your part to reply 'Which?'." B retorts: "It's a poor analogy. I know how to take a book off a shelf, but I don't know how to identify one of the two spheres supposed to be alone in space..." It seems that, for the purposes of this argument, B claims that one cannot use a singular term to designate an object without first "identifying" it, or at least knowing how to identify one of the objects in question. That, I take it, is the matter at hand here, whether it is possible to introduce a lexical item to refer to an indiscernible object. The character A takes the bait: "Can't you imagine that one sphere has been designated as 'a'?" The dialogue continues:

> B. I can imagine only what is logically possible. Now it is logically possible that somebody should enter the universe I have described, see one of the spheres on his left hand and proceed to call it 'a'...
>
> A. Very well, now let me try to finish what I began to say about a... [ellipsis in original]
>
> B. I still can't let you, because you, in your present situation, have no right to talk about a. All I have conceded is that if something were to happen to introduce a change in my universe, so that an observer entered and could see the two spheres, one of them could then have a name. But this would be a different supposition from the one I wanted to consider. My spheres don't yet have names... You might just as well ask me to consider the first daisy in my lawn that would be picked by a child, if a child were to come along and do the picking. This doesn't now distinguish any daisy from the others. You are just pretending to use a name.
>
> A. And I think you are just pretending not to understand me.

4 Is this a problem?

What seems to matter here, at a minimum, is one's philosophy of mathematics, and one's account of reference. If someone has a philosophy of mathematics that accepts a principle of the identity of indiscernibles, and also accepts a certain naive account of what indiscernibility comes to, then he will not allow the foregoing, implicit characterization of the complex numbers as the algebraic closure of the reals (or as the structure characterized by the standard axiomatization). That very characterization violates the identity of indiscernibles, since it introduces *two* distinct objects that cannot be told apart. The same goes for the other examples, the integers under addition, the Klein group, and all of category theory. The philosopher who accepts these strictures will not face the foregoing problem of reference — since there is no such problem on that view — but she will need some other account of the various structures and theories.

One way to avoid the problem is to break the symmetry. Whether one accepts the identity of indiscernibles or not, one can think of complex numbers as pairs of reals, following a now common mathematical technique. In that case, i is the pair $\langle 0, 1\rangle$ and $-$i is $\langle 0, -1\rangle$. Since one can distinguish 1 from -1 in the reals, and thus in the second coordinate, there is no problem distinguishing those two pairs.

The problem with complex analysis, however, at least appears to be robust — liable to reappear. The custom nowadays is to use polar coordinates, in which case i is $\left\langle 1, \frac{\pi}{2}\right\rangle$ and $-$i is $\left\langle 1, \frac{3\pi}{2}\right\rangle$. So it now becomes a matter of distinguishing those pairs from each other. This, in turn, becomes a matter of distinguishing the *angle* $\frac{\pi}{2}$ from the *angle* $\frac{3\pi}{2}$. And how does one do that? Well, we can say that the first is positive and the second is negative; or that the first goes counterclockwise and the second clockwise; or that the first is above the x-axis and the second below it. But when it comes to angles, "positive–negative", "counterclockwise–clockwise", and "above–below" all point to symmetries — automorphisms of the plane. So it seems that the indiscernibility has returned. To break the symmetry in the complex numbers, we need to break the symmetry in the plane.

As noted above, one can break the symmetry in the integers under addition by thinking of that as a substructure of the integers under addition and multiplication. There is no problem distinguishing 1 from -1 in *that* structure. And perhaps one can deny that there is such a thing as *the* Klein group. Instead, there are a number of Klein groups. In each such group, the four elements are properly individuated. One would have to give a similar interpretation of the languages invoked in category theory. One might just break the symmetry globally — once and for all — by insisting that the ontology of all of mathematics is the iterative hierarchy, or some other rigid structure.

All this could be done, of course. The technical resources required are well-known. Notice, however, that the need to break the symmetry involves reinterpreting the languages of mathematics. One question would concern how natural the reinterpretations are, as readings of the original languages of mathematics.

I suspect that there would not be a problem for Frege concerning our issue. As noted above, he demanded two things before one could use the definite article in a properly rigorous mathematical treatment. One has to show "that some object falls under" the concept in question, and the other is "that only one object falls under it" (Frege, 1884, § 74n). It would not do, for

Frege, to simply declare that the complex numbers are the algebraic closure of the reals, or even to say that we are interested in *an* algebraic closure of the reals. In doing this, we would fail the *first* requirement, of showing that *some* object falls under the concept "square root of -1". How do we know that there are any algebraic closures of the real numbers? Presumably, Frege would have given an explicit definition of the complex numbers, perhaps as pairs of reals (which, in turn, would be defined in terms of certain courses-of-values).[2] This explicit definition would presumably break the symmetry between i and −i. If it didn't, then the account would fail Frege's second requirement, "that only one object falls under" the concept. In that case, Frege would not allow the use of the definite article.

Consider, next, a nominalist, a philosopher who denies the existence of mathematical objects. On such views, mathematics has no distinct ontology. Then the devil is in the details of the view. I do not see an issue here for a fictionalist, one who likens mathematics to make-believe. One can surely tell a coherent *story* about objects that are indiscernible as far as the details provided by the story go (Black's character B notwithstanding). Consider, for example, the following short story: "One day, two people met, fell in love, and lived together, happily ever after." Whatever its literary merits, this is surely a coherent piece of fiction. In both cases, there is nothing *in the story* to distinguish the characters. Anything in the language of the story that holds of one also holds of the other. And, of course, we have nothing to go on besides the details of the story, plus common knowledge of human psychology, naming conventions, and the like. Consider this variation on our story: "One day, Chris met Kelly. They fell in love, and lived together, happily ever after." One might claim that, here, Chris is distinguished from Kelly because he or she is the one who is *called* "Chris" in that story. But, as with complex analysis, this does not seem like a distinction that matters. To be sure, there are interesting issues concerning the semantics and, perhaps, the ontology and metaphysics of fiction, but I do not propose to enter that realm here.

Reconstructive nominalists provide translations of mathematical languages into vocabulary that does not commit the mathematician to the existence of mathematical objects. Typically, singular terms and bound variables in mathematical languages are rendered as bound variables within the scope of modal operators. Instead of speaking of what exists, the reconstructive nominalist speaks of what might exist (as in Geoffrey Hellman (1989)), what can be constructed (ala Charles Chihara (1990)), or what follows from axioms, or whatever. Whether an issue analogous to the present

[2] Thanks to Michael Dummett for some key insights. One must be speculative here, since Frege only gave the barest hints at how *real* analysis is to fit into his logicist program (see, for example, (Simons, 1987)).

one arises depends on the details of the translation, and I propose to avoid that as well.[3]

Suppose, finally, that someone does accept the existence of mathematical objects — contra nominalism — and agrees that in some cases, distinct objects are indiscernible — contra the program of symmetry-breaking-via-reinterpretation. For present purposes, it does not seem to matter what the metaphysical nature of these mathematical objects may be. Our philosopher may be a traditional platonist, or she may hold that mathematical objects are somehow mental constructions, or that they are social constructs, or whatever else the philosopher dreams up. Our philosopher may even be a quietist about mathematical ontology, insisting that the only things to say about them are what follows from the mathematical theories. All that matters, for now, is that the languages be understood literally, and that some numerically distinct objects are indistinguishable.

Without much loss of generality, we might as well keep on with our standard example: our philosopher holds that the complex numbers exist, that the square roots of -1 are indiscernible, and that there is a role played by the term "i". Then our problem arises. He must either come up somehow with a referent for 'i', which would be to break the symmetry, or else he must describe the logico-semantic role of that term.

This is not the place to attempt to solve the present problem. That is a matter for future work which, I believe, involves substantial themes in semantics, pragmatics, and logic, both for the languages of mathematics and for natural languages generally. What is the role and function of singular terms (or linguistic items that look and function like singular terms), and how do such terms get introduced into the language? The purpose of the present article is to articulate the issue, and to delimit the range of philosophers of mathematics for whom it is a substantial issue. At the very least, I hope to have convinced the gentle reader that it is not a problem local to ante rem structuralism.

Stewart Shapiro
Department of Philosophy, The Ohio State University
350 University Hall, 230 North Oval Mall, Columbus, Ohio 43210, USA
`shapiro.4@osu.edu`

[3] Chihara's (1990) modal constructivism is an interesting case here. Accordingly, a singular term, such as a numeral, represents the possibility of constructing an open sentence with certain semantic features. So one can wonder which open sentence would correspond to "i, as opposed to the open sentence that corresponds with "$-$i". I presume that Chihara would liken complex numbers to pairs of reals, as above. This would avoid the (analogue of) the present problem, by breaking the symmetry.

References

Black, M. (1952). The identity of indiscernibles. *Mind*, *61*, 153–164.

Brandom, R. (1996). The significance of complex numbers for Frege's philosophy of mathematics. *Proceedings of the Aristotelian Society*, *96*, 293–315.

Burgess, J. (1999). Review of (Shapiro, 1997). *Notre Dame Journal of Formal Logic*, *40*, 283–291.

Button, T. (2006). Realist structuralism's identity crisis: a hybrid solution. *Analysis*, *66*, 216–222.

Chihara, C. (1990). *Constructibility and mathematical existence.* Oxford: Oxford University Press.

Frege, G. (1884). *Die Grundlagen der Arithmetik.* Breslau: Koebner. (The Foundations of Arithmetic, translated by J. Austin, second edition, New York, Harper, 1960.)

Hellman, G. (1989). *Mathematics without numbers.* Oxford: Oxford University Press.

Hellman, G. (2001). Three varieties of mathematical structuralism. *Philosophia Mathematica*, *9*(III), 184–211.

Keränen, J. (2001). The identity problem for realist structuralism. *Philosophia Mathematica*, *3*(3), 308–330.

Keränen, J. (2006). The identity problem for realist structuralism II: a reply to Shapiro. In F. MacBride (Ed.), *Identity and modality* (pp. 146–163). Oxford: Oxford University Press.

Ketland, J. (2006). Structuralism and the identity of indiscernibles. *Analysis*, *66*, 303–315.

Ladyman, J. (2005). Mathematical structuralism and the identity of indiscernibles. *Analysis*, *65*, 218–221.

Leitgeb, H. (2007). Struktur und Symbol. In H. Schmidinger & C. Sedmak (Eds.), *Der Mensch: Ein "animal Symbolicum"?* (Vol. 4, pp. 131–147). Darmstadt: Wissenschaftliche Buchgesellschaft.

Leitgeb, H., & Ladyman, J. (2008). Criteria of identity and structuralist ontology. *Philosophia Mathematica*, *16*(3), 388–396.

MacBride, F. (2005). Structuralism reconsidered. In S. Shapiro (Ed.), *Oxford handbook of philosophy of mathematics and logic* (pp. 563–589). Oxford: Oxford University Press.

MacBride, F. (2006a). *Identity and modality.* Oxford: Oxford University Press.

MacBride, F. (2006b). What constitutes the numerical diversity of mathematical objects? *Analysis*, *66*, 63–69.

Shapiro, S. (1997). *Philosophy of mathematics: Structure and ontology.* New York–Oxford: Oxford University Press.

Shapiro, S. (2006a). The governance of identity. In F. MacBride (Ed.), *Identity and modality* (pp. 164–173). Oxford: Oxford University Press.

Shapiro, S. (2006b). Structure and identity. In F. MacBride (Ed.), *Identity and modality* (pp. 109–145). Oxford: Oxford University Press.

Shapiro, S. (2008). Identity, indiscernibility, and ante rem structuralism: the tale of i and −i. *Philosophia Mathematica*, *16*(3), 285–309.

Simons, P. (1987). Frege's theory of real numbers. *History and Philosophy of Logic*, *8*, 25–44.

Sequent Calculi and Bidirectional Natural Deduction: On the Proper Basis of Proof-theoretic Semantics

Peter Schroeder-Heister*

Philosophical theories of logical reasoning are intrinsically related to formal models. This holds in particular of Dummett–Prawitz-style proof-theoretic semantics and calculi of natural deduction. Basic philosophical ideas of this semantic approach have a counterpart in the theory of natural deduction. For example, the "fundamental assumption" in Dummett's theory of meaning (Dummett, 1991, p. 254 and Ch. 12) corresponds to Prawitz's formal result that every closed derivation can be transformed into introduction form (Prawitz, 1965, p. 53). Examples from other areas in the philosophy of logic support this claim.

If conceptual considerations are genetically dependent on formal ones, we may ask whether the formal model chosen is appropriate to the intended conceptual application, and, if this is not the case, whether an inappropriate choice of a formal model motivated the wrong conceptual conclusions. We will pose this question with respect to the paradigm of natural deduction and proof-theoretic semantics, and plead for Gentzen's sequent calculus as a more adequate formal model of hypothetical reasoning. Our main argument is that the sequent calculus, when philosophically re-interpreted, does more justice to the notion of *assumption* than does natural deduction. This is particularly important when it is extended to a wider field of reasoning than just that based on logical constants.

To avoid confusion, a terminological *caveat* must be put in place: When we talk of the sequent calculus and the reasoning paradigm it represents, we mean, as its characteristic feature, its symmetry or bidirectionality, i.e.,

* This work has been supported by the ESF EUROCORES programme "LogiCCC — Modelling Intelligent Interaction" (DFG grant Schr 275/15–1) and by the joint German-French DFG/ANR project "Hypothetical Reasoning: Logical and Semantical Perspectives" (DFG grant Schr 275/16–1). I would like to thank Luca Tranchini and Bartosz Więckowski for helpful comments and suggestions.

the fact that it uses introduction rules for formulas occurring in different positions. We do not assume that these positions are syntactically represented by the left and right sides of a sequent, i.e., we do not stick to the *sequent format* which gave the calculus its name. In particular, we propose a natural-deduction variant of the sequent calculus called *bidirectional natural deduction*, which embodies the basic conceptual features of the sequent calculus.[1] Conversely, the natural-deduction paradigm to be criticized is the reasoning based on (conventional) introduction and elimination inferences, even though it can be given a sequent-calculus format as in so-called "sequent-style natural deduction".[2] The conceptual meaning of *natural deduction vs. sequent calculus*, which we try to capture by the notions of *unidirectionality vs. bidirectionality*, is to be distinguished from the particular syntax of these systems. We hope it will always be clear from the context whether a conceptual model or a specific syntactic format is meant.

We do not claim originality for the translation of the sequent calculus into bidirectional natural deduction. This translation is spelled out in detail in (von Plato, 2001). The system itself has been known much longer.[3] Here we want to make a philosophical point concerning the proper concept of hypothetical reasoning that also pertains to applications beyond logic and logical constants. The term "bidirectional natural deduction" seems to us to be a very appropriate characterization of the system considered. To our knowledge, it has not been used before.[4]

1 Assumptions in natural deduction

In a natural-deduction framework, there are essentially two things that can be done with assumptions: introducing and discharging. If we introduce an assumption

$$\begin{array}{c} A \\ \vdots \end{array}$$

[1] Other variants would be Schütte-style systems with metalinguistically specified right and left parts of formulas (Schütte, 1960) or even Frege-style systems, see (Schroeder-Heister, 1999).

[2] First suggested by Gentzen in (Gentzen, 1935), though not under that name.

[3] See, for example, (Tennant, 1992), (Tennant, 2002). For a brief history see (Schroeder-Heister, 2004, p. 33) (footnote). Unfortunately, the earliest proposal of this system (Dyckhoff, 1988) was accidentally omitted there.

[4] Von Plato (2001) simply speaks of "natural deduction with general elimination rules", which can also be understood in the unidirectional way (depending on the treatment of major premisses of elimination inferences). — The term "bidirectional" came up in personal discussions with Luca Tranchini on the proper treatment of negation in proof-theoretic semantics, a topic which is closely related to bidirectional reasoning. See his contribution to this volume (Tranchini, 2009).

then we make the derivation below A dependent on that assumption, and if we discharge it at an application of an inference

$$\begin{array}{c} (n) \\ A \\ \vdots \\ \dfrac{B}{C}\,(n) \end{array}$$

we retract this dependency, so that the conclusion of that inference is not longer dependent on A. As proposed in (Prawitz, 1965), the numeral n indicates the link between assumptions and inferences at which they are discharged. Other notations are Fitch's (Fitch, 1952) explicit notation of subproofs, which goes back to (Jaśkowski, 1934), where the idea of discharging assumptions was developed even before (Gentzen, 1934/35).

Introducing and discharging assumptions is not very much one can do. Especially, there are no operations that change the form of an assumption and therefore have to do with its meaning. In this sense, they are purely structural operations. However, it is definitely more than can be done in Hilbert-type calculi, where we have at best the introduction of assumptions but never their discharging. In Hilbert-type systems assumptions can never disappear by means of a formal step. However, we can metalinguistically prove that we can work without assumptions by using them as the left side of a conditional statement. This is the content of the deduction theorem: If, in a Hilbert-type system, we have derived B from A, we can instead derive $A \to B$ by an appropriate transformation of the derivation of B from A.

Since in natural deduction we have the discharging of assumptions as a formal operation at the object level, we can express the content of the deduction theorem as a formal rule of implication introduction:

$$\begin{array}{c} (n) \\ A \\ \vdots \\ \dfrac{B}{A \to B}\,(n) \end{array}$$

Although this is an important step beyond Hilbert-type calculi, it is not all that can possibly be done in extending the expressive power of formal systems. Our claim is that a genuinely semantic treatment of assumptions is more appropriate than a purely structural one as in natural deduction.

In natural deduction, assumptions have a close affinity to free variables: Assumptions which are not discharged are called *open*, whereas discharged assumptions are called *closed*. This terminology is justified since undischarged assumptions are open for the substitution of derivations whose end formula is the assumption in question, whereas closed assumptions are not.

Given a derivation $\begin{matrix} A \\ \mathcal{D} \\ B \end{matrix}$ with the open assumption A and a derivation $\begin{matrix} \mathcal{D}_1 \\ A \end{matrix}$ of A, then

$$\begin{matrix} \mathcal{D}_1 \\ A \\ \mathcal{D} \\ B \end{matrix}$$

is a derivation of B which may be considered a substitution instance of the original derivation. In this sense an open derivation corresponds to an open term, and a closed derivation, i.e. a derivation without open assumptions, corresponds to a closed term. This relationship between open and closed proofs and open and closed terms can be made formally explicit by a Curry–Howard-style association between terms and proofs, where the discharging of assumptions becomes a formal binding operation.

In our example, the derivation $\begin{matrix} \mathcal{D}_1 \\ A \end{matrix}$ can itself be open, just like a variable which is substituted with an open term. So in the formal concept of natural deduction and the composition of derivations there is no primacy of closed derivations over open ones. However, this primacy enters with the philosophical interpretation of natural deduction in the tradition of Dummett and Prawitz. There open assumptions are interpreted as placeholders for closed proofs.[5]

2 Assumptions in Dummett-Prawitz-style proof-theoretic semantics

Proof-theoretic semantics as advanced by Dummett and Prawitz[6] was framed by Prawitz in the form of a definition of validity of proofs, where a proof corresponds to a derivation in natural-deduction form. According to this definition, closed proofs in introduction form are primary as based on "self-justifying" steps, whereas the validity of closed proofs not in introduction form as well as the validity of open proofs is reduced to that of closed proofs using certain transformation procedures on proofs, called "justifications". Given a notion of validity for atomic proofs (i.e. proofs of atomic sentences), the definition of validity for the case of conjunction and implication formulas (to take two elementary cases) can be sketched as follows:

[5] Here we switch terminology from "derivation" to "proof", as in the semantical interpretation we are no longer dealing with purely formal objects, for which we reserve the term "derivation". Prawitz himself often speaks of "arguments" to avoid formalistic connotations still present with "proof".

[6] For an overview of this sort of semantics see (Schroeder-Heister, 2006) and the references therein.

- A closed proof of an atomic formula A is valid if there is a valid atomic proof of A.

- A closed proof of $A \wedge B$ in the introduction form

$$\frac{\begin{array}{cc}\mathcal{D}_1 & \mathcal{D}_2 \\ A & B\end{array}}{A \wedge B}$$

  is valid if the subproofs $\mathcal{D}_1$ and $\mathcal{D}_2$ are valid closed proofs of A and B, respectively.

- A closed proof of $A \rightarrow B$ in the introduction form

$$\frac{\begin{array}{c}(n) \\ A \\ \mathcal{D} \\ B\end{array}}{\mathrm{A} \rightarrow \mathrm{B}}\,(n)$$

  is valid if for every closed proof $\begin{array}{c}\mathcal{D}_1 \\ A\end{array}$ of A, the closed proof

$$\begin{array}{c}\mathcal{D}_1 \\ A \\ \mathcal{D} \\ B\end{array}$$

  of B is valid.

- A closed proof of A not in an introduction form is valid if it reduces, by means of the given justifications, to a valid closed proof of A in an introduction form.

If we are only interested in closed proofs, this definition is sufficient. In view of the last clause, it is a generalized inductive definition proceeding on the complexity of end formulas and the reduction sequences generated by justifications. If we also want to consider open proofs, we would have to define:

- An open proof $\begin{array}{c}A_1, \ldots, A_n \\ \mathcal{D} \\ B\end{array}$ is valid if for all closed valid proofs $\begin{array}{c}\mathcal{D}_1 \\ A_1\end{array}, \ldots, \begin{array}{c}\mathcal{D}_n \\ A_n\end{array}$, the proof $\begin{array}{c}\mathcal{D}_1 \quad\quad \mathcal{D}_n \\ A_1, \ldots, A_n \\ \mathcal{D} \\ B\end{array}$ is a valid closed proof.

Given this clause for open proofs, the defining clause for the validity of a closed proof of $A \rightarrow B$ in introduction form might be replaced with

- A closed proof of $A \to B$ in the introduction form

$$\cfrac{\begin{matrix}(n)\\ A\\ \mathcal{D}\\ B\end{matrix}}{A \to B}\,(n)$$

is valid if its immediate open subproof

$$\begin{matrix}A\\ \mathcal{D}\\ B\end{matrix}$$

is valid,

yielding a uniform clause for all closed proofs in introduction form. However, this way of proceeding makes the definitions of validity of open and closed proofs intertwined, which obscures the fact that there is an independent definition of validity for closed proofs.

According to this definition, closed proofs are conceptually prior to open proofs. Furthermore, assumptions in open proofs are considered to be placeholders for closed proofs, as the validity of open proofs is defined by the validity of their closed instances obtained by substituting a free assumption with a closed proof of it. So we have identified two central features of standard proof-theoretic semantics:

$$\textit{The primacy of closed over open proofs} \qquad (\alpha)$$

$$\textit{The placeholder view of assumptions} \qquad (\beta)$$

The definition of validity shows a further feature which is connected to (α) and (β). The fact that in an open proof $\begin{matrix}A\\ \mathcal{D}\\ B\end{matrix}$ the open assumption A is a placeholder for closed proofs $\begin{matrix}\mathcal{D}_1\\ A\end{matrix}$ of A, yielding a closed proof

$$\begin{matrix}\mathcal{D}_1\\ A\\ \mathcal{D}\\ B\end{matrix}$$

means that the validity of $\begin{matrix}A\\ \mathcal{D}\\ B\end{matrix}$ is expressed as the transmission of validity from [the closed proof] $\mathcal{D}_1$ to [the closed proof]

$$\begin{matrix}\mathcal{D}_1\\ A\\ \mathcal{D}\\ B\end{matrix}$$

If one considers an open proof $\begin{matrix} A \\ \mathcal{D} \\ B \end{matrix}$ to be a proof of the consequence statement that B holds under the hypothesis A, this expresses

The transmission view of consequence (γ)

i.e., the idea that the validity of a consequence statement is based on the transmission of the validity of closed proofs from the premisses to the conclusion. This idea is closely related to the classical approach according to which hypothetical consequence is defined as the transmission of categorical truth (in a model) from the premisses to the conclusion. In that respect, Dummett–Prawitz-style proof-theoretic semantics does not depart from the classical view present in truth-condition semantics (see (Schroeder-Heister, 2008b)). Of course, there are fundamental differences between the classical and constructive approaches, which must not be blurred by this similarity, in particular with respect to epistemological issues (see (Prawitz, 2009)).[7]

A further point showing up in the definition of validity is the assumption of global reduction procedures for proofs (called "justifications"). This is what makes the (generalized) induction on the reduction sequence for proofs possible. It is assumed that it is not individual valid proof steps that generate a valid proof, but the overall proof which may *reduce* to a proof of a particular form (viz., a proof in introduction form). We call this

The global view of proofs (δ)

These four features are intimately connected to the model of natural deduction as its formal background. This holds especially for (β) and (δ), which specify (α) and (γ), respectively. Natural deduction permits to place a derivation on top of another one, and it is natural deduction where we have the notion of proof reduction. In the sequent calculus, this sort of connection is not present.

In the sequent calculus, logical inferences not only concern the right side of a sequent (corresponding to the end formula in natural deduction) but the right and left sides of sequents likewise. In this sense the sequent calculus is

---

[7] It might be mentioned that the definition of validity for a closed proof of $A \rightarrow B$ is closely related to Lorenzen's admissibility interpretation of implication. According to (Lorenzen, 1955), $A \rightarrow B$ expresses the admissibility of the rule $\frac{A}{B}$. The claim that every closed proof of A can be transformed into a closed proof of B can be viewed upon as expressing admissibility. At first glance, this contradicts the fact that in natural deduction an open proof $\begin{matrix} A \\ \vdots \\ B \end{matrix}$ is a proof of B *from* A and should as such be distinguished from an admissibility statement. However, even if, in the formal system, we are dealing with proofs *from* assumptions rather than admissibility statements, the semantic interpretation in terms of validity comes very close to the admissibility view. See (Schroeder-Heister, 2008a).

inherently *bidirectional* compared to the *unidirectional* formalism of natural deduction that underlies Dummett–Prawitz-style proof-theoretic semantics. In the following we will make a case for the bidirectional framework.

3 The sequent calculus and bidirectional natural deduction

According to the traditional, i.e. pre-natural-deduction reasoning model, we start with true sentences and proceed by inferences which lead from true sentences to true sentences. This guarantees that we always stay in the realm of truth.[8] Alternatively, we could start with assumptions and assert sentences under hypotheses. This is the background of natural deduction. Natural deduction adds the feature of discharging assumptions, i.e., the dependency on assumptions may disappear in the course of an argument. In this way the dynamics of reasoning not only affects assertions but at the same time the hypotheses assumed. However, this dynamics is very limited as the only options are introducing and discharging, so there is no more than a yes/no attribution to hypotheses. We cannot introduce and eliminate assumptions according to their specific meaning, which would be a more sophisticated dynamics. In this sense reasoning in standard natural deduction is assertion centred and unidirectional. This is even more so, as the hypotheses assumed are placeholders for closed proofs.[9]

A genuinely different model is given by the sequent calculus. The particular feature of this system, i.e. introduction rules on the left side of the sequent sign, can be philosophically understood as the meaning-specific introduction of assumptions. Consider conjunction with left sequent rules

$$\frac{\Gamma, A \vdash C}{\Gamma, A \wedge B \vdash C} \qquad \frac{\Gamma, B \vdash C}{\Gamma, A \wedge B \vdash C}$$

These rules can be interpreted as follows: Suppose we have asserted C under the hypotheses Γ and A. Then we may claim C by assuming $A \wedge B$ as an assumption and discharging A as an assumption, and similarly for B. Written in natural-deduction style this corresponds to the general elimination rules for conjunction

$$\frac{\overline{A \wedge B} \quad \begin{matrix} \overset{(n)}{A} \\ \vdots \\ C \end{matrix}}{\mathrm{C}}\,(n) \qquad \frac{\overline{A \wedge B} \quad \begin{matrix} \overset{(n)}{B} \\ \vdots \\ C \end{matrix}}{\mathrm{C}}\,(n)$$

[8] This was, for example, the picture drawn by Bolzano and Frege.

[9] This is not essentially changed if we replace assertion with denial and in this sense dualize natural deduction. Unidirectionality would just point into the opposite direction. See (Tranchini, 2009).

but with the crucial modification that the major premiss must now be an assumption, i.e., must occur in top position[10] (this is here indicated by the line over the major premiss). Similarly, the left implication rule

$$\frac{\Gamma \vdash A \quad \Gamma, B \vdash C}{\Gamma, A \to B \vdash C}$$

is interpreted as follows: Suppose we have asserted both A under the hypotheses Γ, and C under the hypotheses Γ and B. Then we may claim C under the assumption $A \to B$ instead of B, i.e., discharge B and assert $A \to B$ instead. Written in natural-deduction style, this yields the general $\to$-elimination rule

$$\frac{\overline{A \to B} \quad A \quad \begin{matrix} (n) \\ B \\ \vdots \\ C \end{matrix}}{\mathrm{C}}\,(n)$$

again with the crucial difference to the standard general elimination rule that the major premiss occurs in top position.[11]

By presenting the sequent-calculus rules in a natural-deduction framework we are no longer working in "standard" or "genuine" natural deduction but in the reasoning model suggested by the sequent calculus, as the restriction on major premisses of elimination rules runs counter to the way premisses are treated in standard natural deduction. We call this modified system *bidirectional natural deduction* as it acts on both the assertion and the assumption side, with rules that depend on the forms of the formulas assumed or asserted. So the possible operations on assumptions are no longer merely structural.[12]

In proposing bidirectional natural deduction, as a natural-deduction-style variant of the sequent calculus, as our model of reasoning, we establish a symmetry between assertions and assumptions. Like assertions, assumptions can be introduced according to their meaning, namely as major premisses of elimination inferences. By imposing the restriction that major premisses must always be assumptions, elimination inferences receive an

[10] In Tennant's (Tennant, 1992) terminology, the major premiss "stands proud".

[11] A translation between sequent calculus and natural deduction with general elimination rules is carried out in full detail in (von Plato, 2001). Note that for implication, we are here considering the general elimination rule used by von Plato, as they correspond to the left sequent calculus rule, rather than the more powerful one proposed in (Schroeder-Heister, 1984), which extends the standard framework of natural deduction with rules as assumptions.

[12] We also call it "natural-deduction-style sequent calculus", as it is conceptually a sequent calculus which is presented in the form of a natural deduction system (Schroeder-Heister, 2004). In (Negri & von Plato, 2001), this term is used in a different sense, meaning a specific form of the sequent calculus.

entirely different reading. They are no longer justified by reference to the way the major premiss can be (canonically) derived. They are rather viewed as ways of introducing complex assumptions, given the derivations of the minor premisses. Elimination inferences in bidirectional natural deduction combine the introduction of an assumption with an elimination step and can thus be viewed as a special form of assumption introduction. Therefore we also call them "upward introductions", as opposed to "downward introductions" which are the common introduction rules.

Assumptions which are major premisses of elimination inferences are no longer placeholders for closed proofs as they cannot be inferred by means of an inference. They are always starting points of elimination inferences. Of course, it might be possible to show that given a proof $\begin{matrix}\mathcal{D}_1\\ A\end{matrix}$ of A and a proof $\begin{matrix}A\\ \mathcal{D}_2\\ B\end{matrix}$ of B from A, we can obtain a proof $\begin{matrix}\mathcal{D}\\ B\end{matrix}$ of B. However, this would have to be established as a theorem corresponding to cut elimination for the sequent calculus. It is no longer a trivial matter as in standard (unidirectional) natural deduction, since

$$\begin{matrix}\mathcal{D}_1\\ A\\ \mathcal{D}_2\\ B\end{matrix}$$

is no longer a well-formed proof if A is a major premiss of an elimination inference. Therefore bidirectionality overcomes the placeholder view of assumptions (β). With this it also overcomes the primacy of closed over open proofs (α) as closed proofs are no longer used to interpret assumptions. Only a premiss of an introduction rule can be viewed as a placeholder for a closed proof, which means that the uniform interpretation of assumptions by reference to closed proofs is given up.

The transmission view of consequence (γ) disappears as well. As assumptions can be introduced in the course of a proof (in the sequent calculus by left introduction, in bidirectional natural deduction as the major premiss of an elimination inference), it is no longer a defining feature of them that they transform closed proofs into closed proofs. If this happens to be the case, then it is "accidental" and has to be proved. The introduction of an assumption is just as primitive as the introduction of an assertion. In the terminology of Dummett–Prawitz-style proof-theoretic semantics, both the introduction of an assertion and the introduction of an assumption is a canonical, i.e. definitional step. More precisely, the distinction between canonical and non-canonical steps disappears. In this sense the concept of validity is much more rule-oriented than proof-oriented: We now consider a proof to be valid if it consists of proper applications of right and left rules (in the sequent calculus) or downward and upward introduction rules (in bidirectional natural deduction) rather than if it reduces to a proof in introduction form for all its closed instances. In this way, the global view of proofs (δ) also disappears, as it is based on the fundamental assumption that proofs are primary to rules and that the validity of rules is based on proofs and proof reduction. The idea of bidirectional reasoning is very much local rather than global.[13]

This does not mean that right and left (sequent calculus), or downward and upward (bidirectional natural deduction) introductions are unrelated to each other. We will still require some notion of harmony between the two sorts of inferences as an adequacy condition. However, this harmony will be local rather than global, and not based on proof reduction. One criterion would be *uniqueness* in the sense of (Belnap, 1961/62), which means that if we duplicate rules for a constant $*$, yielding a constant $*'$ with the same right (or downward) and left (or upward) rules, we can prove $A[*] \dashv\vdash A[*']$ in the combined system. There $A[*]$ is any expression containing $*$, and $A[*']$ is obtained from $A[*]$ by replacing $*$ with $*'$. However, unlike Belnap, we would not rely on conservativeness, as this is a global concept, but rather on local inversion in the sense that the defining conditions for a constant $*$ can be obtained back from this constant. Our main criticism of Belnap's proposal of conservativeness and uniqueness in his discussion of the connective "tonk" is that he mixes a local condition (uniqueness) with a global one (conservativeness).[14]

4 Why going local?

Why should we switch to a concept of hypothetical reasoning which is different from the standard one characterized by (α)–(δ), and which is prevailing both in classical and constructive semantics? The lack of an intuitive justification of the principles (α)–(δ) is no reason for abandoning them, if we cannot also tell why the bidirectional alternative has greater explanatory power. In fact, we gain access to a much wider range of phenomena, if we stick to the bidirectional paradigm. We just mention two points.

Atomic reasoning and inductive definitions

The discussion in proof-theoretic semantics has traditionally focused on logical constants. Logical constants are a particularly well-behaved case where we can apply the global considerations characteristic of the standard approach. Natural-deduction-based proof-theoretic semantics has been devel-

[13] The local approach to hypothetical reasoning put forward here was originally proposed by Hallnäs (Hallnäs, 1991, 2006).

[14] This point will be worked out elsewhere.

oped as a semantics of logical constants. However, this focus is much too narrow. Proof-theoretic definitions of logical constants just feature as particular cases of inductive definitions. Looking at inductive definitions as basic structural entities that confer meaning to objects, the distinction between atomic and non-atomic (i.e. logically compound) objects disappears. Most generally, we would deal with definitional clauses of the form

$$a \Leftarrow C$$

where a is an object to be defined and C is a defining condition. Starting with a definition of this kind, right (downward) and left (upward) introduction rules can be generated from this inductive definition in a canonical way, representing a way of putting inductive definitions into action, and resulting in powerful closure and reflection principles. The form of definitional clauses look like clauses in logic programming, and logic programs can be viewed as particular cases of inductive definitions. We would even generalize the framework set up by logic programming by considering clauses where the body C of a clause may contain hypothetical statements and therefore negative occurrences of defined objects. This goes beyond standard definite Horn clause programming and even transcends the classical field of logic programming with negation (Hallnäs & Schroeder-Heister, 1990/91). It differs from systems investigated in (Martin-Löf, 1971) in that it is not mainly directed at induction principles but rather the local inversion of rules. Systems of this kind have recently been considered by Brotherston and Simpson (Brotherston & Simpson, 2007), where also the relationship between inversion-based reasoning and induction principles for iterated inductive definitions is discussed. Considering inductive definitions in general opens up a wider perspective at hypothetical reasoning which is no longer based on logical constants. It can also integrate subatomic reasoning in the sense of (Więckowski, 2008), where the validity of atomic sentences is reduced to certain assumptions concerning predicates and terms.

Non-wellfounded phenomena

The global reductionist perspective underlying unidirectional natural deduction excludes non-wellfounded cases such as the paradoxes. The inductive definition of validity expects that there is no loop or infinite descent in the reasoning chain. However, in the case of the paradoxes, we have exactly this situation. Our local framework can easily accommodate such phenomena. For example, if we define p by $\neg q$ and q in turn by p, then both p and q are locally defined. The global loop is irrelevant for the local definition. In such a situation we can no longer prove global properties of proofs such as cut elimination, but this we do not require.

As it is now a matter of (mathematical) fact rather than a definitional requirement whether certain global properties hold, we do not rule out non-wellfounded phenomena by definition. This is a great advantage, as it gives us a better chance to understand them. Following (Hallnäs, 1991), we might call this approach a *partial* approach to meaning. According to Hallnäs, this would be in close analogy to recursive function theory, where it is a potential mathematical result that a given partial recursive function is total, rather than something that has to be established for the function definition to make sense.

There are other applications of the local approach that we cannot mention here such as the proper treatment of substructural issues, generalized inversion principles, evaluation strategies in extended logic programs, etc.

5 Final digression: dialogues

We have pleaded for a bidirectional view of reasoning as it is incorporated in Gentzen's sequent calculus and can be given the form of bidirectional natural deduction. As there are certain adequacy conditions governing such a system that relate right/downwards and left/upwards rules with one another, so that they are linked together in a certain way, we might ask of whether it would be possible to obtain them from a single principle. One possible answer might be the dialogical approach proposed by Lorenzen (Lorenzen, 1960) and his followers. If one carries its ideas over to the case of inductive clauses

$$\begin{array}{c} a \Leftarrow C_1 \\ \vdots \\ a \Leftarrow C_n \end{array}$$

one would be lead to an approach where an attack on the defined object a would have to be defended by a choice among the defining conditions C_i, which are themselves attacked by choosing one of its components. The distinction between right and left rules would then be obtained by strategy considerations for and against certain atoms. In this way a more unified approach could be achieved. The dialogical motivation, as based on local attack and defence rules, would not involve global reductive features compared to validity notions in standard proof-theoretic semantics. Therefore, it appears to be more faithful to our local approach, as the global perspective is only introduced at a later stage in terms of strategies and their transformations. In this way the dialogical research programme promises a novel perspective at the local/global distinction.

Peter Schroeder-Heister
Wilhelm–Schickard–Institut für Informatik, Universität Tübingen
Sand 13, 72076 Tübingen, Germany
psh@informatik.uni-tuebingen.de
http://www-ls.informatik.uni-tuebingen.de/psh

References

Belnap, N. D. (1961/62). Tonk, plonk and plink. *Analysis*, *22*, 130–134.

Brotherston, J., & Simpson, A. (2007). Complete sequent calculi for induction and infinite descent. In *Proceedings of the 22nd annual IEEE symposium on Logic in Computer Science (LICS)* (pp. 51–62). Los Alamitos: IEEE Press.

Dummett, M. (1991). *The logical basis of metaphysics.* London: Duckworth.

Dyckhoff, R. (1988). Implementing a simple proof assistant. In *Workshop on programming for logic teaching (Leeds, 6–8 July 1987): Proceedings 23/1988* (pp. 49–59). University of Leeds: Centre for Theoretical Computer Science.

Fitch, F. B. (1952). *Symbolic logic: An introduction.* New York: Ronald Press.

Gentzen, G. (1934/35). Untersuchungen über das logische Schließen. *Mathematische Zeitschrift*, *39*, 176–210, 405–431. (English translation in: *The Collected Papers of Gerhard Gentzen* (ed. M. E. Szabo), Amsterdam: North Holland (1969), pp. 68–131.)

Gentzen, G. (1935). Die Widerspruchsfreiheit der reinen Zahlentheorie. *Mathematische Annalen*, *112*, 493–565. (English translation in: *The Collected Papers of Gerhard Gentzen* (ed. M. E. Szabo), Amsterdam: North Holland (1969), pp. 132–201.)

Hallnäs, L. (1991). Partial inductive definitions. *Theoretical Computer Science*, *87*, 115–142.

Hallnäs, L. (2006). On the proof-theoretic foundation of general definition theory. *Synthese*, *148*, 589–602.

Hallnäs, L., & Schroeder-Heister, P. (1990/91). A proof-theoretic approach to logic programming: I. Clauses as rules. II. Programs as definitions. *Journal of Logic and Computation*, *1*, 261–283, 635–660.

Jaśkowski, S. (1934). On the rules of suppositions in formal logic. *Studia Logica*, *1*, 5–32 (reprinted in: S. McCall (ed.), Polish Logic 1920–1939, Oxford 1967, 232–258).

Lorenzen, P. (1955). *Einführung in die operative Logik und Mathematik.* Berlin: Springer (2nd edition 1969).

Lorenzen, P. (1960). Logik und Agon. In *Atti del XII Congresso Internazionale di Filosofia (Venezia, 1958)* (pp. 187–194). Firenze: Sansoni.

Martin-Löf, P. (1971). Hauptsatz for the intuitionistic theory of iterated inductive definitions. In J. E. Fenstad (Ed.), *Proceedings of the Second Scandinavian Logic Symposium* (pp. 179–216). Amsterdam: North-Holland.

Negri, S., & von Plato, J. (2001). Sequent calculus in natural deduction style. *Journal of Symbolic Logic*, *66*, 1803–1816.

Prawitz, D. (1965). *Natural deduction: A proof-theoretical study.* Stockholm: Almqvist & Wiksell (Reprinted Mineola NY: Dover Publ., 2006).

Prawitz, D. (2009). Inference and knowledge. In M. Peliš (Ed.), *The Logica Yearbook 2008.* London: College Publications [this volume].

Schroeder-Heister, P. (1984). A natural extension of natural deduction. *Journal of Symbolic Logic*, *49*, 1284–1300.

Schroeder-Heister, P. (1999). Gentzen-style features in Frege. In *Abstracts of the 11th International Congress of Logic, Methodology and Philosophy of Science, Cracow, Poland (August 1999)* (p. 449).

Schroeder-Heister, P. (2004). On the notion of *assumption* in logical systems. In R. Bluhm & C. Nimtz (Eds.), *Selected papers contributed to the sections of GAP5, Fifth International Congress of the Society for Analytical Philosophy, Bielefeld, 22–26 September 2003* (pp. 27–48). Paderborn: mentis (Online publication: `http://www.gap5.de/proceedings`).

Schroeder-Heister, P. (2006). Validity concepts in proof-theoretic semantics. *Synthese*, *148*, 525–571.

Schroeder-Heister, P. (2008a). Lorenzen's operative justification of intuitionistic logic. In M. van Atten, P. Boldini, M. Bourdeau, & G. Heinzmann (Eds.), *One hundred years of intuitionism (1907–2007): The Cerisy conference* (pp. 214–240 [References for whole volume: 391–416]). Basel: Birkhäuser.

Schroeder-Heister, P. (2008b). Proof-theoretic versus model-theoretic consequence. In M. Peliš (Ed.), *The Logica Yearbook 2007* (pp. 187–200). Prague: Filosofia.

Schütte, K. (1960). *Beweistheorie (revised: Proof Theory, 1977).* Berlin: Springer.

Tennant, N. (1992). *Autologic.* Edinburgh: Edinburgh University Press.

Tennant, N. (2002). Ultimate normal forms for parallelized natural deductions. *Logic Journal of the IGPL*, *10*, 299–337.

Tranchini, L. (2009). The role of negation in proof-theoretic semantics: A proposal. In M. Peliš (Ed.), *The Logica Yearbook 2008.* London: College Publications [this volume].

von Plato, J. (2001). Natural deduction with general elimination rules. *Archive for Mathematical Logic*, *40*, 541–567.

Więckowski, B. (2008). Predication in fiction. In M. Peliš (Ed.), *The Logica Yearbook 2007* (pp. 267–285). Prague: Filosofia.

Relatives of Robinson Arithmetic

Vítězslav Švejdar*

1 Introduction: numbers, or strings?

Robinson arithmetic Q was introduced in (Tarski, Mostowski, & Robinson, 1953) as a base axiomatic theory for investigating incompleteness and undecidability. It is very weak, but all its recursively axiomatizable consistent extensions are both incomplete and undecidable. In logic textbooks, it often plays the role of the weakest reasonable theory with this property.

A. Grzegorczyk recently proposed to study the *theory of concatenation* as a possible alternative theory for studying incompleteness and undecidability. Unlike Robinson (or Peano) arithmetic, where the individuals are numbers that can be added or multiplied, in the theory of concatenation one has *strings* (or *texts*) that can be concatenated. So in the language of the theory of concatenation there is a binary function symbol $\frown$ for laying two strings end to end to form a new string. Axioms of the theory of concatenation postulate, e.g., associativity of the operation $\frown$, or the existence of irreducible, i.e. single-letter, strings. Particular variants of the theory of concatenation may differ in the number of irreducible strings (with two as the most obvious choice), or in the existence of the empty string.

Before Grzegorczyk, some aspects of concatenation were considered and some axioms were formulated by Quine (1946) and Tarski. One variant of the theory, called theory F, appears already in the book (Tarski et al., 1953), where it is claimed but not proved that F is essentially undecidable.

Grzegorczyk's motivation to study the theory of concatenation is philosophical. When reasoning or when performing a computation, we deal with texts. Our human capacity to perform these intellectual tasks depends on our ability to discern texts. Then it is natural to define notions like undecidability directly in terms of texts, without reference to natural numbers.

* This work is a part of the research plan MSM 0021620839 that is financed by the Ministry of Education of the Czech Republic.

When proving Gödel 1st incompleteness theorem, choosing the theory of concatenation as the base theory could be preferable to choosing Peano or Robinson arithmetic, because then one of the essential steps in the incompleteness proof, formalization of logical syntax, would be practically effortless.

We will discuss properties of two theories of concatenation, theory F defined in Tarski et al. (1953) and theory TC proposed by Grzegorczyk. It appears that an appropriate method of showing undecidability of all consistent extensions is proving mutual interpretability of these theories with Robinson arithmetic Q. We will consider methods of constructing interpretations, one of these being the well known Solovay method of shortening of cuts. We will also discuss the Grzegorczyk's project of replacing Robinson's Q by some version of theory of concatenation in more details. The pros of the project are obvious, but there are also some cons.

2 Some preliminaries

For an axiomatic theory T, let $\mathrm{Thm}(T)$ be the set of all sentences *provable* in T, in symbols $\mathrm{Thm}(T) = \{\varphi; T \vdash \varphi\}$, and let $\mathrm{Ref}(T)$ be the set of all sentences *refutable* in T, in symbols $\mathrm{Ref}(T) = \{\varphi; T \vdash \neg\varphi\}$. A theory T is *consistent* if $\mathrm{Thm}(T) \cap \mathrm{Ref}(T) = \varnothing$, i.e., if no sentence of T is simultaneously provable and refutable in T. A theory T is *complete* if it is consistent and each sentence of T is either provable or refutable in T. A theory T is *recursively axiomatizable* if it is equivalent to a theory T' with an algorithmically decidable set of axioms (i.e. with T' algorithmically decidable). A theory is *decidable* if there exists an algorithm that decides about its provability, i.e., if the set $\mathrm{Thm}(T)$ is algorithmically decidable.

A theory S is an *extension* of a theory T if the language of T (i.e. the set of all non-logical symbols of T) is a subset of the language of S, and each sentence of T provable in T is provable also in S. A theory T is *essentially incomplete* if no recursively axiomatizable extension of T is complete; T is *essentially undecidable* if no consistent extension of T is decidable. It is known that a theory is essentially incomplete if and only if it is essentially undecidable. Thus we use these notions interchangeably or, following Grzegorczyk, we preferably speak about essential undecidability.

An *interpretation* of a theory T in a theory S is a mapping from formulas of T to formulas of S that well-behaves w.r.t. logical symbols and maps all axioms of T to sentences provable in S. A theory T is *interpretable* in S if there exists an interpretation of T in S. The notion of interpretation, as well as the notion of essential undecidability, first appeared in (Tarski et al., 1953). Important facts about interpretability are the following: (i) if T is interpretable in S and S is consistent then T is consistent, too; (ii) if T is interpretable in S and T is essentially undecidable then then S is essentially undecidable, too. The notion of interpretability can be used as a means to measure strength of axiomatic theories: if T is interpretable in S and vice versa, i.e., if T and S are *mutually interpretable*, then we can think that T and S represent the same expressive and deductive strength.

3 The importance of Robinson arithmetic

Robinson arithmetic Q is an axiomatic theory having seven simple axioms formulated in the language $\{+, \cdot, 0, \mathrm{S}\}$ with symbols for addition and multiplication (of natural numbers), a constant for the number zero, and a unary function symbol S for the successor function $x \mapsto x+1$. *Peano arithmetic* PA is obtained from Q by adding the induction schema. The theory $\mathrm{I}\Delta_0$ is like Peano arithmetic, but with the induction schema restricted to Δ_0-formulas (*bounded* formulas) only. The theory $\mathrm{I}\Delta_0+\Omega_1$ is $\mathrm{I}\Delta_0$ enhanced by the axiom asserting the totality of the function $x \mapsto x^{\log x}$. For a non-expert, the properties of natural numbers expressible by Δ_0-formulas constitute a class that is a subclass of all algorithmically decidable properties. An example of a Δ_0-formula is the formula $\exists v(v \cdot x = y)$, i.e. the formula the number x is a divisor of the number y. Two other examples are the number x is prime and the number x is divisible by some prime. An example of a formula that is *not* Δ_0 is there exists a $y > x$ such that $y \neq 0$ and y is divisible by all v such that $v \neq 0$ and $v \leq x$; this formula speaks about a thing similar to the factorial of x. Another example of a non-Δ_0 formula is there exists a y such that $y > x$ and y is prime. In the theory $\mathrm{I}\Delta_0$, one cannot prove that a factorial of x exists for each number x, while provability of the sentence a prime $y > x$ exists for each x is a difficult open problem. Both sentences are easily proved by unrestricted induction, i.e. in Peano arithmetic.

Basic properties of natural numbers, like associativity and commutativity of addition and multiplication, are provable in $\mathrm{I}\Delta_0$, but unprovable in Q. Generally, universal sentences are seldom provable in Q. However, $\mathrm{I}\Delta_0+\Omega_1$ is interpretable in Q. Gödel 1st incompleteness theorem, or better, its Rosser generalization, says that any recursively axiomatizable extension of Q is incomplete. So Q is essentially incomplete (essentially undecidable). The meaning of Gödel 2nd incompleteness theorem is somewhat questionable for Q. However, its usual proof goes through in $\mathrm{I}\Delta_0+\Omega_1$ without any changes.

Thus Robinson arithmetic Q is a very weak but still essentially undecidable theory. It represents a rich "degree of interpretability" because a lot of stronger theories, like $\mathrm{I}\Delta_0+\Omega_1$, are interpretable in it. Since it is finitely axiomatizable, it can be used in a straightforward proof of undecidability of classical predicate logic.

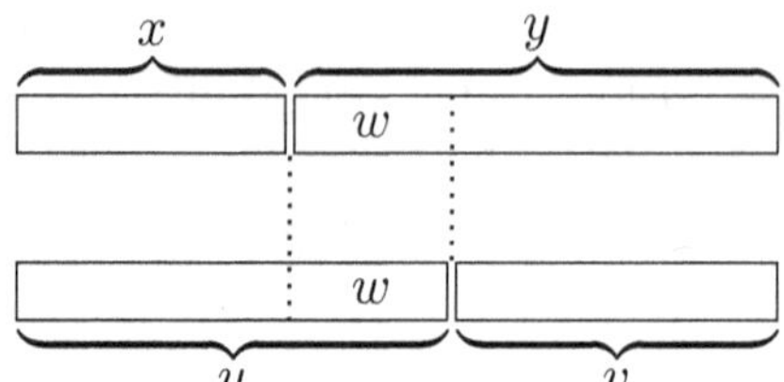

Figure 1. The editors axiom

4 The theory TC

The *theory of concatenation* TC has the language $\{{}^\frown, \varepsilon, \mathtt{a}, \mathtt{b}\}$ with a binary function symbol ${}^\frown$, a constant ε for the empty string, and two other constants $\mathtt{a}$ and $\mathtt{b}$. We usually omit the symbol ${}^\frown$, i.e., write xy for the concatenation $x{}^\frown y$ of the strings x and y. The axioms of TC are the following:

TC1: $\forall x(x\varepsilon = \varepsilon x = x)$,

TC2: $\forall x \forall y \forall z(x(yz) = (xy)z)$,

TC3: $\forall x \forall y \forall u \forall v(xy = uv \rightarrow$
$\rightarrow \exists w((xw = u \ \&\ wv = y) \vee (uw = x \ \&\ wy = v)))$,

TC4: $\mathtt{a} \neq \varepsilon \ \&\ \forall x \forall y(xy = \mathtt{a} \rightarrow x = \varepsilon \vee y = \varepsilon)$,

TC5: $\mathtt{b} \neq \varepsilon \ \&\ \forall x \forall y(xy = \mathtt{b} \rightarrow x = \varepsilon \vee y = \varepsilon)$,

TC6: $\mathtt{a} \neq \mathtt{b}$.

The axioms TC1 and TC2 can be described as axioms of semigroups; by TC2 we can omit parentheses in expressions whenever convenient. The axioms TC4–TC6 postulate that the strings $\mathtt{a}$ and $\mathtt{b}$ are different, and each of them is non-empty and irreducible (cannot be non-trivially decomposed into two strings). The axiom TC3 is called *editors axiom* in (Grzegorczyk, 2005). It describes what happens if two editors of a large work independently suggest splitting the text into two volumes. If their suggestions are x, y and u, v respectively, as shown in the Figure 1, then the first volume of one of the editors consists of two parts: the other editor's first volume, and a text w (possibly empty) that simultaneously occurs as a starting part of the other editor's second volume. In (Ganea, 2007) this text w is called an *interpolant* (of the equation $xy = uv$).

The theory of concatenation TC was defined in (Grzegorczyk, 2005). However, the editors axiom is attributed to Tarski, and the idea about the importance of concatenation in incompleteness proofs can be traced back to Quine, who in (Quine, 1946) cites Tarski and Hermes and says: *Gödel's proof [...] depended on constructing a model of concatenation theory within arithmetic.* Note that Quine does not list any axioms, and thus when he says "concatenation theory", he in fact means its standard model (defined below). Grzegorczyk (2005) proved (mere) undecidability of TC. Later Grzegorczyk and Zdanowski (2008) showed essential undecidability of TC and left open the question whether Robinson arithmetic is interpretable in TC. A. Visser and R. Sterken, see (Visser, 2009), M. Ganea in (Ganea, 2007), and the present author in (Švejdar, 2009) independently gave a positive answer to this question. More about interpretability in (and of) TC is in Section 5 below.

The papers (Grzegorczyk, 2005) and (Grzegorczyk & Zdanowski, 2008) work with a variant of TC having no empty string. Then, for example, the axiom TC4 has the form $\forall x \forall y(xy \neq \mathsf{a})$. The paper (Švejdar, 2009) works with a variant of TC having three instead of two irreducible strings. The exact choice of variant of the theory is a matter of taste because, as shown in (Grzegorczyk & Zdanowski, 2008), all variants of the theory of concatenation are mutually interpretable, provided the irreducible strings are at least two in number.

Let A be the set $\{\mathsf{a}, \mathsf{b}\}^*$ of all strings in the two-letter alphabet $\{\mathsf{a}, \mathsf{b}\}$, and let $\mathbb{A}$ be the structure with A as a universe, with concatenation defined "normally" and with constants a and b realized by a and b, respectively. Then $\mathbb{A}$ is the *standard model of* TC. The structure $\mathbb{B}$ having the set $B = \{\mathsf{a}, \mathsf{b}, \mathsf{e}\}^*$ as its universe and with all symbols also defined normally is another example of a model of TC. Let $x \sqsubseteq y$ mean $\exists u \exists v(uxv = y)$, and let $x \sqsupset y$ mean $\exists u(ux = y)$. The formulas $x \sqsubseteq y$ and $x \sqsupset y$ can be read the string x is a substring of y and the string y ends by x respectively. The model $\mathbb{B}$ above shows that the sentence $\forall x(x \neq \varepsilon \rightarrow \mathsf{a} \sqsubseteq x \vee \mathsf{b} \sqsubseteq x)$ is not provable in TC.

The following theorem gives some more examples of provable and unprovable sentences. Its purpose is to give the reader some feeling about provability in TC.

Theorem 1. *The following sentences* (a)–(d) *are provable in* TC,

(a) $\forall x(x\mathsf{a} \neq \varepsilon)$,

(b) $\forall x \forall y(xy = \varepsilon \rightarrow x = \varepsilon \;\&\; y = \varepsilon)$,

(c) $\forall x \forall y(x\mathsf{a} = y\mathsf{a} \rightarrow x = y)$,

(d) $\forall x \forall y(\mathsf{a} \sqsupset xy \rightarrow y = \varepsilon \vee \mathsf{a} \sqsupset y)$,

while the following sentence (e) *is* not *provable in* TC*:*

(e) $\forall x \forall y \forall z(xz = yz \rightarrow x = y)$.

Proof. (a) Assume $x\mathsf{a} = \varepsilon$. Then, by TC1 and TC2, we have $(\mathsf{b}x)\mathsf{a} = \mathsf{b}$. Irreducibility of b, i.e. TC5, yields $\mathsf{b}x = \varepsilon$ or $\mathsf{a} = \varepsilon$. The latter is excluded by TC4. Then from $\mathsf{b}x = \varepsilon$, $(\mathsf{b}x)\mathsf{a} = \mathsf{b}$, and TC1 we have $\mathsf{a} = \mathsf{b}$, a contradiction with TC6.

(b) If $xy = \varepsilon$ then $x(y\mathsf{a}) = \mathsf{a}$ using TC1 and TC2. So $x = \varepsilon$ or $y\mathsf{a} = \varepsilon$ by TC4. From (a) we have $x = \varepsilon$. Then $xy = \varepsilon$ yields $y = \varepsilon$.

(c) Let $x\mathsf{a} = y\mathsf{a}$. By the editors axiom TC3, there exists a w such that $xw = y \;\&\; w\mathsf{a} = \mathsf{a}$ or $yw = x \;\&\; w\mathsf{a} = \mathsf{a}$. Consider the first case, the second one is symmetric. From $w\mathsf{a} = \mathsf{a}$ and irreducibility of a we have $w = \varepsilon$. From that and $xw = y$ we indeed have $x = y$.

(d) Let $\mathsf{a} \sqsupset xy$, and let u be such that $u\mathsf{a} = xy$. The axiom TC3 yields a w satisfying $uw = x \;\&\; wy = \mathsf{a}$, or $xw = u \;\&\; w\mathsf{a} = y$. In the second case we obviously have $\mathsf{a} \sqsupset y$. So consider the first case. From $wy = \mathsf{a}$ we have $w = \varepsilon$ or $y = \varepsilon$. If $y = \varepsilon$ then we are done. If $w = \varepsilon$ then $y = \mathsf{a}$, and thus $\mathsf{a} \sqsupset y$.

(e) Let D be the set of all strings in $\{\mathsf{a}, \mathsf{b}, \mathsf{e}\}^*$ that have no occurrences of ae. Realize a and b by a and b respectively, and define $x + y$ accordingly: $x + y$ results from xy by repeating the substitution ae→e while possible. For example, bab + eb = babeb, but baa + eb = beb. One can check, in case of TC3 with a little effort, in case of the remaining axioms rather easily, that the structure $\mathbb{D} = \langle D, +, \varepsilon, \mathsf{a}, \mathsf{b}\rangle$ is a model of the theory TC. In $\mathbb{D}$ we have $\mathsf{a} + \mathsf{e} = \varepsilon + \mathsf{e}$. So the formula $x^\frown z = y^\frown z$ is not true in $\mathbb{D}$ if x, y, z are evaluated by a, the empty string, and e respectively, and thus the sentence (e) is not valid in $\mathbb{D}$. □

Another useful sentence is $\forall x \forall y(\mathsf{a} \sqsubseteq xy \rightarrow \mathsf{a} \sqsubseteq x \vee \mathsf{a} \sqsubseteq y)$. We leave its proof in TC as an exercise. More about the theory TC and about its models is in (Visser, 2009).

5 The theory F, interpretability

Theorem 2. *Robinson arithmetic* Q *is interpretable in* TC.

Proof. We only give the basic idea of the proof given in (Švejdar, 2009). The full proof is rather technical.

When constructing an interpretation, one first has to specify its *domain*, which in our case means to work in TC and select strings that will play the role of natural numbers. It appears that the following definition works:

$$\mathrm{Num}(x) \equiv \forall u(u \sqsubseteq x \;\&\; u \neq \varepsilon \rightarrow \mathsf{a} \sqsupset u),$$

a string x is a number if each non-empty substring of x ends by a. Note that, in the model $\mathbb{D}$ in the proof of Theorem 1(e), the string e starts by a (since $\mathsf{e} = \mathsf{a} + \mathsf{e}$). However, e is not a number because it is a non-empty substring of itself and cannot be written as $\mathsf{e} = z + \mathsf{a}$, i.e. does not end by a.

Having numbers, addition is interpreted as concatenation, zero is interpreted as the empty string ε, and the successor function S is interpreted as the function $x \mapsto x\mathsf{a}$. These definitions work because in TC one can prove that ε and a are numbers and that numbers are closed under concatenation. All axioms of Q about 0, S, and + translate to sentences provable in TC under this interpretation.

To interpret *multiplication*, a straightforward idea is to first define the notion of a witnessing sequence. A sequence of pairs $[u_0, v_0], .., [u_q, v_q]$ is a witnessing sequence for $x \cdot y$ if: $u_0 = v_0 = \varepsilon$, for each $i < q$ the pair $[u_{i+1}, v_{i+1}]$ equals $[u_i\mathsf{a}, v_i y]$, and $u_q = x$. Then one can define that $x \cdot y = z$ if there exists a witnessing sequence for $x \cdot y$ with $[x, z]$ as the last member. The problem here is that in TC it is not possible to prove that a witnessing sequence exists for each choice of x, y, and it is also not possible to prove that if it exists, it is uniquely determined. A way how to overcome this problem is interpreting not the full Robinson arithmetic Q, but rather its variant Q^- in which addition and multiplication are non-total functions. Then the result is obtained by combining the constructed interpretation of Q^- in TC with a fact known from (Švejdar, 2007) that Q is interpretable in Q^-. □

The theory Q^- used in the proof of Theorem 2 was also introduced by Grzegorczyk. The interpretation of Q in Q^- in (Švejdar, 2007) is constructed using the *Solovay method of shortening of cuts*. This method is now widely known, but was never published: it is only explained in a letter to Petr Hájek (Solovay, 1976). M. Ganea in (Ganea, 2007) gives a different proof of interpretability of Q in TC, but he also uses the detour via Q^-. Sterken and Visser give a proof not using Q^-, see (Visser, 2009).

A consequence of the fact that Q is interpretable in TC is essential undecidability of TC. All proofs of interpretability of Q in TC are somewhat involved, but still simpler than the direct proof of essential undecidability of TC given in (Grzegorczyk & Zdanowski, 2008). These interpretability proofs might use some ideas developed by Grzegorczyk and Zdanowski: that is certainly true about the author's proof in (Švejdar, 2009).

Since TC is interpretable in $I\Delta_0$, the theories TC and Q are mutually interpretable; thus they represent the same expressive and deductive power. This is a piece of information missing in (Grzegorczyk & Zdanowski, 2008).

An interesting alternative theory of concatenation is the *theory* F. It has the same language as TC, and its axioms are:

F1: $\forall x(x\varepsilon = \varepsilon x = x)$,

F2: $\forall x \forall y \forall z(x(yz) = (xy)z)$,

F3: $\forall x \forall y \forall z(yx = zx \lor xy = xz \rightarrow y = z)$,

F4: $\forall x \forall y (x\mathtt{a} \neq y\mathtt{b})$,

F5: $\forall x (x \neq \varepsilon \rightarrow \exists u (x = u\mathtt{a} \ \vee \ x = u\mathtt{b}))$.

Axioms F1 and F2 are the same as axioms TC1 and TC2 of TC. It is easy to verify that axiom F4 is provable in TC; axioms F3 and F5, as is evident from models $\mathbb{D}$ and $\mathbb{B}$ in the previous section, are not provable in TC. From the opposite point of view, axioms TC4–TC6 and sentences (a) and (b) in Theorem 1 are examples of sentences provable in F; we leave their proofs to the reader as an interesting exercise. Albert Visser, see (Visser, 2009), has constructed a model $\mathbb{M}$ of F such that $\mathbb{M} \not\models \forall x \forall y (\mathtt{a} \sqsubseteq xy \rightarrow \mathtt{a} \sqsubseteq x \vee \mathtt{a} \sqsubseteq y)$. Thus in F, one can have strings w_1 and w_2 such that $\mathtt{a} \sqsubseteq w_1 w_2$, $\mathtt{a} \not\sqsubseteq w_1$, $\mathtt{a} \not\sqsubseteq w_2$; Albert Visser describes this situation as creating a letter *ex nihilo.* A consequence of these remarks is that Thm(TC) and Thm(F) are incomparable sets of sentences.

It is claimed in (Tarski et al., 1953) that W. Smielew and A. Tarski proved essential undecidability of F by interpreting Q in F; however, no proof is given. Ganea (2007) constructed an interpretation of TC in F. In conjunction with Theorem 2, this gives a proof of the theorem of Smielew and Tarski. We give (a slight simplification of) Ganea's proof below in Theorem 3. Note however, that it is still an interesting historical problem what proof could Smielew and Tarski have had in mind. Ours (Ganea's) proof implicitly uses the Solovay's shortening technique, formulated long after the book (Tarski et al., 1953) was published. A. Visser has some possible explanation of this historical problem.

Theorem 3 (Ganea). TC *is interpretable in* F.

Proof. Work in F and define *tame* strings as follows:

$$\mathrm{Tame}(x) \equiv \forall v \forall z (z \sqsupset vx \rightarrow z \sqsupset x \vee x \sqsupset z),$$

where $\sqsupset$ has the same meaning as in TC.

(i) We first show (prove within F) that tame strings are closed under concatenation. So assume that x and y are tame, and let v and z be such that $z \sqsupset vxy$. We need to show that $z \sqsupset xy$ or $xy \sqsupset z$. Since y is tame, we have $z \sqsupset y$ or $y \sqsupset z$. If $z \sqsupset y$ then $z \sqsupset xy$ and we are done. So assume that $y \sqsupset z$ and take t such that $ty = z$. From $z \sqsupset vxy$ we have a u such that $uz = vxy$; thus $uty = vxy$. From axiom F3 we have $ut = vx$. Since x is tame, we have $t \sqsupset x$ or $x \sqsupset t$. Then $ty \sqsupset xy$ or $xy \sqsupset ty$. Since $ty = z$, we indeed have $z \sqsupset xy$ or $xy \sqsupset z$.

(ii) Next we show that if wy is tame, then also w is tame. So let v and z be such that $z \sqsupset vw$. We want to show that $z \sqsupset w$ or $w \sqsupset z$. From $z \sqsupset vw$ we have $zy \sqsupset vwy$. Since yw is tame, we have $zy \sqsupset wy$ or $wy \sqsupset zy$. Then a straightforward use of axiom F3 yields $z \sqsupset w$ or $w \sqsupset z$.

Now we are ready to verify that the domain of tame strings, together with the identical mapping of symbols ($\mathtt{a}$, $\mathtt{b}$, and ε to $\mathtt{a}$, $\mathtt{b}$, and ε respectively, concatenation to concatenation), defines an interpretation of TC in F. It is not difficult to verify that $\mathtt{a}$, $\mathtt{b}$, and ε are tame; this together with (i) means that the domain of tame strings is closed under all operations. The axiom TC1 translates to the sentence $\forall x(\mathrm{Tame}(x) \rightarrow x\varepsilon = \varepsilon x = x)$. This sentence is evidently provable in F. A similar argument shows that axioms TC2 and TC4–TC6 translate to sentences provable in F as well. This is so easy because TC2 and TC4–TC6 are universal sentences.

Thus it remains to prove the the translation of the editors axiom TC3 is provable in F. Note that TC3 is the only axiom of TC that is not a universal sentence; it contains an existential quantifier. Let x, y, u, v, be tame strings such that $xy = uv$. We have to show that there exists a *tame* w satisfying $xw = u \ \&\ wv = y$ or $uw = x \ \&\ wy = v$. Since y is tame, from $uv = xy$ we have $v \sqsupset y$ or $y \sqsupset v$. It is sufficient to consider the latter, the former is symmetric. We have a w such that $wy = v$. Then $uwy = uv$ and $uwy = xy$. From axiom F3 we have $uw = x$. So w is an interpolant. Since v is tame, from $wy = v$ and (ii) above we know that w is tame. □

Since F is easily interpretable in $\mathrm{I}\Delta_0$, from the other results mentioned in this paper we know that F and TC are deductively incomparable, but from interpretability point of view they represent the same degree of deductive strength. It may be of some interest to directly interpret F in TC.

Theorem 4. F *is interpretable in* TC.

Proof. Now in TC, work with *radical* strings, where

$$\mathrm{Rad}(x) \equiv \forall y \forall z(yx = zx \rightarrow y = z).$$

It is not difficult to show that radical strings include ε, $\mathtt{a}$, and $\mathtt{b}$, and that the domain of all radical strings that are empty or end in either $\mathtt{a}$ or $\mathtt{b}$ is closed under concatenation and defines an interpretation of F. □

6 On the Grzegorczyk's project

Let us repeat from the Introduction that Grzegorczyk's suggestion is to consider strings and concatenation on both formal and metamathematical level. On formal level, the theory of concatenation can serve as an alternative to Robinson arithmetic; on metamathematical level, dealing with texts is philosophically better justified because intellectual activities like reasoning and computing involve working with texts. Briefly, the motivations of that project can be summarized as follows:

- in Gödel's argument, the only use of numbers is coding of syntactical objects,
- then Gödel theorems are presented as a part of mathematics, but their significance is broader,
- when reasoning, communicating, or even computing, we deal with texts, not with numbers,
- on metamathematical level, the notion of computability can be defined without reference to numbers.

One could remark that mathematics in not necessarily identified by working with numbers; Gödel theorems could be presented as part of mathematics even if reformulated without numbers, and they transcede mathematics regardless whether their formulation involves strings or numbers. With this little remark in mind, one can say that the arguments *pro* Grzegorczyk's project are clear and easily acceptable. The definition of recursiveness without using numbers, as done in (Grzegorczyk, 2005), is very interesting.

However, it is also possible to find some arguments that speak *contra* that project, or at least for modifying or extending it. First, when reasoning or computing, we not only concatenate: we also *substitute*. Creating a grammatically correct sentence in a natural language can be described as substituting into patterns. In logic, we have substitution in formulation of predicate axioms. So one can think that the theory of concatenation, if enhanced by some notion of occurrence or substitution, could better serve its purpose. Second, when proving essential undecidability, one also needs an *order*. Known proofs usually (is it a mistake to say *always*?) contain some sort of *Rosser trick*, i.e., speak about an event that occurs *before* some other event. One can think that considering order is more natural in the environment of numbers than in the environment of strings. In fact, defining an order of strings is one of crucial and rather difficult steps in the essential undecidability proof of TC contained in (Grzegorczyk & Zdanowski, 2008).

Vítězslav Švejdar
Department of Logic, Charles University
nám. Jana Palacha 2, 116 38 Praha 1, Czech Republic
`vitezslav.svejdar@cuni.cz`
`http://www.cuni.cz/~svejdar/`

References

Ganea, M. (2007). Arithmetic on Semigroups. *J. Symbolic Logic*, *74*(1), 265–278.

Grzegorczyk, A. (2005). Undecidability without Arithmetization. *Studia Logica*, *79*(2), 163–230.

Grzegorczyk, A., & Zdanowski, K. (2008). Undecidability and concatenation. In A. Ehrenfeucht, V. W. Marek, & M. Srebrny (Eds.), *Andrzej Mostowski and foundational studies* (pp. 72–91). Amsterdam: IOS Press.

Quine, W. V. O. (1946). Concatenation as a basis for arithmetic. *J. Symbolic Logic*, *11*(4), 105–114.

Solovay, R. M. (1976). *Interpretability in set theories.* (Unpublished letter to P. Hájek, Aug. 17, 1976, `http://www.cs.cas.cz/ hajek/RSolovayZFGB.pdf`.)

Švejdar, V. (2007). An Interpretation of Robinson Arithmetic in its Grzegorczyk's Weaker Variant. *Fundamenta Informaticae*, *81*(1–3), 347–354.

Švejdar, V. (2009). On interpretability in the theory of concatenation. *Notre Dame J. of Formal Logic*, *50*(1), 87–95.

Tarski, A., Mostowski, A., & Robinson, R. M. (1953). *Undecidable theories.* Amsterdam: North-Holland.

Visser, A. (2009). Growing Commas: A Study of Sequentiality and Concatenation. *Notre Dame J. of Formal Logic*, *50*(1), 61–85.

The Role of Negation in Proof-theoretic Semantics: a Proposal

Luca Tranchini*

Proof-theoretic semantics, as developed by authors such as Dummett and Prawitz, tries to account for the meaning of logical constants through the use made of them in practice. The typical context in which they figure is deduction, so the program becomes the one of showing how the rules governing deductive practices fix the meaning of logical constants. The theoretical requirement rules have to satisfy is *harmony*, which is endorsed in Gentzen's inversion principle.

One of the distinctive features of human language is compositionality, that is the possibility of producing sentences of arbitrary complexity by means of logical operators. Hence proof-theoretic semantics can aim at being the core of a fully-fledged theory of meaning, that is of an explication of speakers language competence.

1 Verificationism: proof and assertion

The verificationist theory of meaning, is grounded on the choice of assertion as the basic linguistic act. Assertion is taken to be governed by the following principle

> The assertion of a sentence is warranted only if its truth is recognized

where the intuitive notion of truth recognition is to be explained by means of the notion of proof.

Clearly, the proofs that count as evidence for the truth of sentences are only closed proofs, i.e. proofs in which the conclusion does not depend on any assumption.

Open proofs are only mediately connected with linguistic acts. Take an open proof of B from A

*This work has been supported by the ESF EUROCORES programme "LogiCCC — Modelling Intelligent Interaction" (DFG grant Schr275/15–1).

the sentence B can be asserted only if evidence for A is available. If A is atomic, evidence will consist in the opportune computation, if A is a mathematical sentence. Whenever A is an empirical sentence we can think of evidence for it as an opportune empirical observation.

Technically, to account for these "atomic proofs" in a standard natural deduction system we extend the vocabulary with a set $\mathfrak{P}$ of propositional constants, standing for the atomic sentences of language. A subset of such sentences, $\mathfrak{T}$ will consists of the sentences for which extra-deductive evidence is available. The set of open assumptions in a deduction will be restricted to those not belonging to $\mathfrak{T}$.

If A is a complex sentence, one can try to obtain a proof of it from atomic assumptions in $\mathfrak{T}$. Still, it is not always possible to do so. Nonetheless, even in cases such as these we can obtain a closed proof from the open one, even though the conclusion of the closed proof is not the conclusion of the open one, but a more complex sentence. Typically, implication is the device by means of which an open proof is taken into a closed one having as conclusion the implication of the assumption and the conclusion of the open proof:

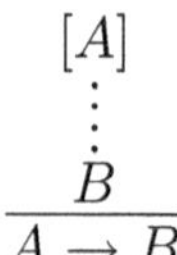

In general, we can say that verificationism focuses on the role of sentences as conclusions of deductive processes. As a consequence, the rules that are taken to fix the meaning of logical constants are introduction rules. For, they specify the conditions under which a sentence having the relevant constant as principal operator can be introduced as conclusion of a derivation. Accordingly, the notion of canonical proof (that is of a proof in which introduction rules play a prominent role) has been taken as the *explicans* of the notion of meaning. That is, to know the meaning of a sentence is to know what counts as a canonical proof of it.

2 Negation

To account for negation in verificationism, the symbol $\bot$, standing for absurdity, is introduced. The negation of a sentence A, $\neg A$, is defined, in full analogy with the **BHK** clause, as $A \to \bot$. So we have two rules for negation (an introduction and an elimination), which are nothing but special cases of the implication rules.

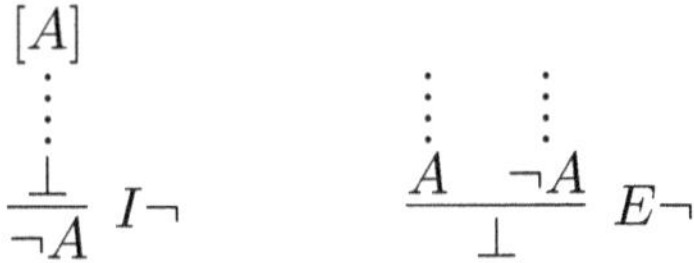

Obviously, the inversion principle holds for these rules as well (as a consequence of its validity for implication).

The problem

Such a characterization grasps all properties of intuitionistic negation except the fact that no construction offers evidence for the absurdity. So, further rules have to be added, to fix the intended meaning of $\bot$.

It is no simple task to explicitly express with a rule the fact that no construction satisfies the absurdity: rules specify ways of *obtaining* proofs, while our aim is to specify the *absence* of proofs.

Intuitionistic Natural Deduction **NJ** is obtained by extending minimal logic with the following rule, so called *ex falso quodlibet*:

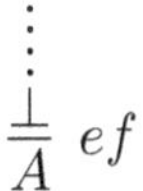

Once the absurdity has been derived, it is possible to derive everything.

The rule is intended to hold for any sentence A. Still, without loss of generality, it is useful to restrict it to atomic sentences. As a consequence, $\bot$ can be taken as an abbreviation of a self-contradictory atomic sentence, for instance '$0 = 1$', deductively characterized by the fact that it entails all other atomic sentences.

As with the *ex falso* we can formally seize intuitionistic logic, it is natural to think of it as grasping the intended meaning of the absurdity. Yet, as some authors have noticed, it is dubious that the *ex falso* conveys to $\bot$ the desired meaning. The fact that it is to be read as absurdity seems to depend on which atomic sentences are provable. Indeed if all atomic sentences were provable, there would be nothing wrong in asserting $\bot$ as only true sentences could be inferred from it. But if $\bot$ has to be considered as standing for the absurdity, then it should not be assertible in any situation.

A possible way out (Dummett, 1991) is to ban the possibility that all atomic sentences can be simultaneously asserted, that is to assume that at least two atomic sentences are mutually incompatible. Nonetheless, this restriction sounds definitely *ad hoc*: we have no reason to propose it, apart from the need of warranting that the *ex falso* conveys the expected meaning to $\bot$ and, consequently, to negation.

Brouwer on absurdity and negation

In Brouwer's early writings we find the idea that the negation of a sentence is warranted when

> we arrive by a construction at the arrestment of the process which would lead to [a construction for the sentence].[1]

To clarify this point, one can imagine a mathematician (or, rather, the idealized mathematician) attempting to produce a construction for a sentence A. Unfortunately, it is impossible to obtain a construction for A. Hence, each attempt reaches a certain point after which it is not possible to carry out the construction, a point beyond which, in Brouwer's words, "the construction no longer goes."

According to Brouwer, when the mathematician finds herself (i.e., she produces a construction showing that she is) in such a situation she can declare the sentence false or, equivalently, accept its negation as warranted.

It is only later that the idea of arrest in the process of construction is substituted by the one of contradiction, glorified in the **BHK** specification of the semantics for the intuitionistic logical constants. The treatment of contradiction as a sentence, implicit in the intuitionistic informal semantics has been fully fledged, as we briefly showed, with the marriage between intuitionism and natural deduction, through the opportune reading of the *ex falso* rule.

Tennant on $\bot$

Recently, Tennant (1999) tried to challenge the verificationist account of what is a proof for the negation of a sentence. Even if he doesn't make explicit reference to Brouwer, it is quite natural to put the conception of $\bot$ he proposes side by side with the idea that a contradiction is nothing but a dead end in the process of construction.

Tennant starts from the refusal of considering proofs of the negation of a sentence as methods to obtain a proof of a false sentence, $\bot$. Rather, he proposes considering $\bot$ as a marker of a dead end in the process of construction. Clearly, $\bot$ is devoid of sentential content, i.e., it is no more an abbreviation for '$0 = 1$'. Hence, we are forced to withdraw the interpretation of $\neg A$ as $A \to \bot$: as $\bot$ has no sentential content we can't apply to it sentential operators, in particular implication. We can conclude with Tennant's that,

> accordingly, an occurrence of $\bot$ is appropriate only within a proof, as a kind of knot —the knot of patent absurdity, or of self-contradiction.[2]

[1] (Brouwer, 1908, p. 109).
[2] (Tennant, 1999, pp. 203–204).

However, Tennant does not completely embody Brouwer's solution. For, according to Brouwer, proving the negation of A means finding a dead end in the route toward the proof of A. On the other hand, Tennant thinks that the role of A is not that of an unreachable goal, but rather a starting point. We prove the negation of A when we reach a dead end *starting* from A. It is important to observe that this is not to say that we start from a hypothetical construction for A, as in Heyting interpretation of the meaning of negation. Rather we are performing an activity which is different from the production of proof. Such an activity is not oriented by the conclusion that we want to reach, but rather by the point from which we start. Tennant introduces a new primitive notion to refer to this alternative activity: disproof. The activity of construction is then split in two different subspecies: the production of proofs and the production of disproof. When we attempt to disproof a sentence we do not start from a proof of it that then turns out to be impossible. We simply look for a disproof of it.

As a consequence, the **BHK** negation clause is reformulated as:

- A proof of $\neg A$ is a disproof of A

dropping the $\bot$ clause. So, instead of analyzing negation in terms of implication and absurdity, we try to do this in terms of disproofs.

Tennant presents his notion of disproof without any reference to related work. Though it seems quite natural to compare these ideas with what Dummett and Prawitz said about the possibility of developing a theory of meaning centered around the notion of refutation.[3]

3 Falsificationism: refutation and denial

According to both Dummett and Prawitz,[4] it is possible to think of theories of meaning alternative to verificationism, in particular to one in which the meaning of sentences is specified by what counts as their refutation. The primitive character of refutations can be endorsed by considering the linguistic act parallel to the one of assertion, the linguistic manifestation of the possession of a refutation of a sentence: denial. The relationship of denial to refutation is governed by the principle:

The denial of a sentence is warranted only if its falsity is recognized

where the recognition of the falsity of a sentence amounts to the possession of a refutation of it.

[3] While Tennant speaks of disproofs, we prefer refutation. As will be cleared, the intuitions behind the two notion are common, even though the detailed treatment is sensibly different.

[4] The ideas presented in this section come from (Dummett, 1991, Ch. 13) and (Prawitz, 1987, § 6).

This theoretical perspective of Popperian flavor is grounded on the intuition according to which, in accepting a sentence, a speaker must also be ready to accept all its consequences. Whenever one of its consequences turns out to be unacceptable, so too must the sentence upon which it depends be rejected. Hence the central notion of a theory of meaning in which denial plays a basic role will be the one of consequence of a sentence. Thus, the meaning of the logical constants is fixed by elimination rules, as they typically specify how a sentence can be used as assumption in a derivation.

According to Dummett and Prawitz, one needs not to introduce new technical tools to account for refutations. Rather the notion of refutation can be defined in a standard natural deduction framework providing an alternative interpretation of the deductive system.

According to the verificationist reading, one can easily construct proofs of more complex sentences starting from proofs of simpler ones with introduction rules. So, according to falsificationism, one can construct refutations of more complex sentences from refutations of simpler ones with elimination rules. For example taken a refutation of A:

$$\begin{array}{c} A \\ \vdots \end{array}$$

one can obtain a refutation of $A \wedge B$ with the help of the $\wedge$ elimination rule:

$$\begin{array}{c} \dfrac{A \wedge B}{A}\, E\wedge \\ \vdots \end{array}$$

The core of the deductive processes will then be the assumption, that acts like a starting point of the derivation and that one tries to refute. As a consequence, the rules that are taken to fix the meaning of logical constants are elimination rules. For, they specify the conditions under which a sentence having the relevant constant as principal operator can be used as assumption in a derivation. Accordingly, the notion of canonical refutation (that is of a deduction in which elimination rules play a prominent role) has been taken as the *explicans* of the notion of meaning. That is, to know the meaning of a sentence is to know what counts as a canonical refutation of it.

As in the verificationist framework, so here not all derivations are directly linked to the basic linguistic act. Again, an open derivation of B from A

receives a hypothetical reading: if one comes into possession of a refutation of B (i.e., if she is in the position of denying B), then she will also be in the position of denying A.

Just as in the verificationist case, we introduce a notion of evidence for atomic sentences to which we refer as extra-deductive refuting evidence. Suppose B is the sentence 'The cup on the table is blue', the empirical observation that the cup on the table is red can be taken as refuting evidence for B.

Technically, we define a subset of the propositional constants, $\mathfrak{F}$, containing the atoms for which an extra-deductive refutation is available. Refutations ending with atoms belonging to $\mathfrak{F}$, having as open assumptions instances of only one sentence, will allow the denial of that sentence.

The disanalogy between the two perspectives consists in the lack of a connective acting in falsificationism as implication does in verificationism. Such a connective should allow the denial of a sentence also in situations in which only an open deduction is at hand. As implication is said to discharge the open assumption, so the connective in question could be said to "discharge the conclusion" of the deduction.

But does the standard language possess a tool which can be taken in some sense to discharge conclusions? Negation can be (partially) thought of in these terms. Consider a derivation of A

The negation elimination rule can be seen as a way of closing the conclusion of the derivation by introducing a more complex assumption:

$$\dfrac{\begin{matrix}\vdots\\ A\end{matrix} \quad \neg A}{\bot}\ .$$

This is actually in full analogy with the way in which implication works: it closes an assumption and introduces a more complex conclusion.

4 Toward a unified framework

The two theories of meaning, verificationist and falsificationism, have been treated by both Dummett and Prawitz as two different (concurrent) theoretical enterprises. That is, a semantics for a given language can be developed either according to the verificationist or the falsificationist standpoint.[5]

On the contrary, Tennant's suggestions on the role of $\bot$ in a deduction are very near to the falsificationist perspective. By looking at his proposal in more detail, one realizes that it is nothing but a mixture of the two views on meaning.

[5] Dummett actually gives reasons for developing simultaneously both perspectives. But even in such a case the two theories are distinct.

The limits of Tennant's approach

As we saw, Tennant's suggests to read $\bot$ as a marker of dead ends in deductions. In other words, a deduction ending with $\bot$ is taken as a deduction with no conclusion. Hence, he proposes to interpret natural deduction systems as providing the means for producing open proofs, closed proofs and disproofs.

> Suppose one has a logical system for which the existence of proofs is indicated by the usual turnstile $\vdash$, a relation of exact deducibility holding between premises on the left and a conclusion on the right. The intuitive meaning of '$X \vdash A$' is that there is a proof whose conclusion is A and whose premises (undischarged assumptions) form the set X. [...]
>
> There are two extreme cases.
>
> 1. X is empty. Then '$\vdash A$' means A is a theorem. That is, there is a proof of A 'from no assumptions'. [...]
> 2. A is 'empty'. Then '$X \vdash$' means that there is a disproof of X, that is, a deduction showing that X is inconsistent.

Accordingly, instead of the usual inductive definition of proof, Tennant's gives a simultaneous definition of the notions of proof and disproof.

Still, in the light of the considerations on falsificationism, Tennant's approach can be criticized for the asymmetry in the treatment of the two notions. In particular, to treat open deductions as open proofs means to treat them as "incomplete proofs": they are means of obtaining closed proofs of the conclusions, provided closed proofs of the assumptions. But why should they not be considered as "incomplete refutations", that is as means of rejecting the premises once refutations of the conclusions are provided?

This asymmetry can also be seen in Tennant's way of dealing with rules, in giving the definition of proof and disproof. Introduction rules can be used only to produce proofs. Elimination rules, on the other hand can be used to produce either proofs or disproofs, depending on whether the deductions of the minor premises are proofs or disproofs. Here are the two cases for disjunction:

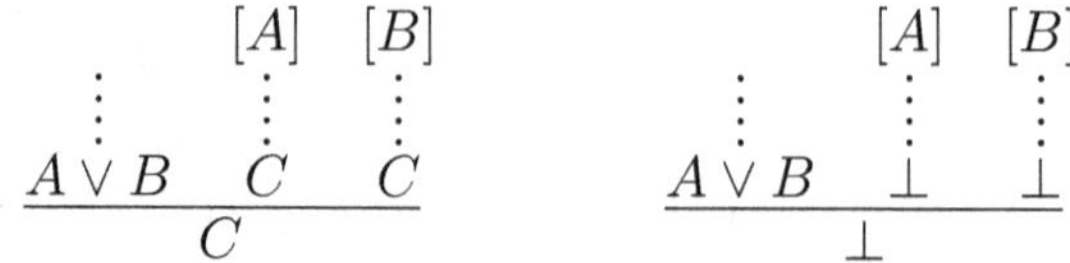

The point is that also introduction rules can be used in producing refutations. Just like in the verificationist perspective one produces non-canonical proofs with elimination rules, so in falsificationism one produces non-canonical refutations with introduction rules.

Finally, it is implicit in Tennant's line of argument that the role of $\perp$ in disproofs is analogous to the role of discharged assumptions in proofs. As a closed proof is a deduction with no open assumptions, so a refutation is a deduction with no conclusion. Furthermore, even if Tennant does not consider it, we saw that in order to use a natural deduction system for meaning-theoretical purposes one also has to account for extra-deductive evidence for atomic sentences. There is a deep analogy of the role of extra-deductive probative and refuting evidence for atomic sentences and (respectively) the role of discharging the assumptions and reaching a dead end in a deduction. For all these are the means through which an open deduction is taken into a closed one (either a proof or a refutation).

All these considerations suggest the possibility of re-framing the natural deduction system in order to explicitly show these symmetries.

Top-closed and bottom-closed derivations

Both perspectives on meaning distinguish between derivations that immediately allow a linguistic performance and those that do not. In verificationism we have a distinction between closed and open proofs. It seems natural to adapt this terminology to refutations, so that we have open and closed refutations.

As we saw, Tennant proposes to treat (what in the standard framework are considered) derivations of conclusion $\perp$ as disproofs. For, $\perp$ has no sentential content and hence can't be taken to be the conclusion of a deductive process. Rather, it registers the fact that the deductive path leading to the conclusion of the derivation is a dead end, or in other words, it is *closed.* Can these two notion of "closure", the one registered by $\perp$ and the one of deductive processes linked to linguistic acts, be taken into one?

To explicitly state the analogy we introduce the sign $\top$ to mark assumption closure. So whenever an assumption is closed, we will mark it $\top$. In the case of assumptions discharge through implication this simply amounts to a notational change. Instead of putting the sentence in brackets (or overlining it), we put the sign $\top$ over it. So, the introduction rule for implication will appear as:

$$\dfrac{\begin{matrix}\top \\ A \\ \vdots \\ B\end{matrix}}{A \to B}$$

As we observed there are two different ways in which an open deduction can be taken into a closed one. One can close one of the edges (assumptions or conclusion) by logical means, in verificationism with implication, in falsificationism with negation; alternatively one can try to reach the atomic components of the sentence to be proved or refuted, to see if there is extra-deductive evidence purporting or refuting such components.

If we consider verificationism, the notion of closure (by means of which we usually refer to discharged assumptions) applies quite well also to atomic sentences for which we have extra-deductive probative evidence: an assumption is closed when the conclusion of the deductive process does not depend on it. And clearly, not only the assumptions discharged through implication are closed, but also the atomic ones for which extra-deductive evidence is available. This suggests the idea of extending the use of $\top$ to mark the closure of the atomic assumptions as well. According to the way in which we introduced atomic sentences in natural deduction in section 1, we can use $\top$ to explicitly mark the atomic sentences belonging to the set $\mathfrak{T}$ of verified atoms. To do this, we add a new rule to the natural deduction system:

if $A \in \mathfrak{T}$ then

$$\begin{matrix}\top \\ A\end{matrix}$$

is a derivation of conclusion A from no assumptions.

For example, suppose the weather is windy: in such a case, the conclusion of the derivation

$$\dfrac{\dfrac{\overset{\top}{\text{It rains}} \quad \overset{\top}{\text{It is windy}}}{\text{It rains and it is windy}}\, I\wedge}{\text{If it rains then it rains and it is windy}}\, I\rightarrow$$

can be asserted, because it does not depend on any assumption, even if the two assumptions are closed in different ways: the first one is discharged by the application of the $I \rightarrow$ rule; the second one is closed by the availability of the empirical evidence for it.

In analogy with this, in falsificationism we have two ways of taking an open deduction into a refutation: either refuting evidence is provided for the conclusion; or alternatively, we can use some language devices to "discharge" the conclusion in the course of the derivation.

If we take a close look at the first possibility, Tennant's idea of $\bot$ as registering a knot of inconsistency fits this situation quite well. For, $\bot$ can be taken to register an incompatibility between the output of the deductive process and the available evidence. This suggests the possibility of extending the use of $\bot$, by marking with it the atomic conclusions of derivations, for which we are in possession of extra-deductive refuting evidence. We can formally achieve this with a rule analogous to the one for atomic assumptions:

if $A \in \mathfrak{F}$ and

is a derivation of conclusion A from assumptions Γ then

is a derivation of no conclusion from assumptions Γ.

For example, suppose the cup on the table is red: in such a case we mark the conclusion of the following derivation with $\bot$:

$$\dfrac{\text{The cup on the table is blue and it is full of tea}}{\text{The cup on the table is blue}}$$
$$\bot$$

This extension of the use of $\bot$ makes it possible to schematically represent the core processes of falsificationism as

This pattern stands for a refutation (and hence allows the denial) of a given sentence A. We will also refer to such deductive patterns as bottom-closed derivations. Once introduced the sign $\top$ in order to mark closed assumptions in deductions, it is possible to represent the core processes of verificationism with the scheme:

standing for a proof (and hence allowing the assertion) of the sentence A. We will refer to such deductive patterns as top-closed derivations.

New horizons for proof-theoretic semantics

As we previously underscored what we are proposing is a unified framework in which both proofs and refutations can be accounted for. To do this we have to add a set of propositional constants $\mathfrak{P}$ to a standard natural deduction system and both a subset $\mathfrak{T}$ of verified atoms and a subset $\mathfrak{F}$ of refuted atoms have to be specified. At this point we have that top-closed derivation and bottom-closed derivations count as closed proof and closed refutations for sentences, i.e., they allow the assertion and denial of sentences. It remains only a disanalogy between the two kinds of derivations, namely that while to a top-closed derivation always corresponds the assertion of the conclusion, to a bottom-closed derivation corresponds a denial only if the assumptions of the deduction are occurrences of the same sentence. This is due to the fact that the natural deduction framework allows at most one conclusion but there is no limit on the number of possible assumptions.

Beside this, what looks really problematic for the full development of this perspective is the definition of validity. For, in verificationism an open deduction is valid if, provided closed derivations of the assumptions, it reduces to a closed proof; in falsificationism an open deduction is valid if, provided a closed derivation of the conclusion, it reduces to a closed refutation of the assumption(s). In other words, in both perspectives the categorical notion of closed derivation has primacy over the hypothetical notion of open derivation. The point is that it is not clear how the notion of validity is to be shaped in a system in which we have *two* distinct categorical notions.

The direction in which the solution can be found is the rejection of the proof-theoretic dogma according to which the categorical notion has primacy over the hypothetical one. By doing this we could really embody Tennant's intuition according to which closed proofs are just limit cases of open ones. Intuitively, this means that the ground concept of proof-theoretic semantics is the recognition of deductive links among sentences, that only in very special occasions can be taken to be oriented by the conclusion or by the assumption. This idea can be seen at work in reading the rule for making assumptions in natural deduction:

for any A

$$A$$

is a deduction having A as conclusion and A as assumption

Both verificationism and falsificationism are forced to read the rule as producing "incomplete" derivations, in the sense of either an incomplete proof of A or an incomplete refutation of A. From the unified perspective, the rule for assumption is interpreted simply as 'Consider A': in considering A we are neither committed to the expectation of a proof nor to the expectation of a refutation of it, we are open to see what will happen at later stages of the development of the deductive process.

Formal models which explicitly endorse this intuition are sequent calculi. In such systems the full symmetry between assumptions and conclusion, i.e. assertion and denial, is embodied in the symmetry between left and right side of the turnstile. Coming back to validity, it is interesting to note that no question of validity has ever been addressed for sequent calculi and it is not completely clear how to formulate it. It would not be surprising that rather than a global definition of validity what is needed are simply local criteria to be imposed on rules.[6] But we do not push the issue further.

Absurdity and consistency

As we saw, in order to fix the meaning of $\bot$ *via* deductive rules, the verificationist has to require that all atomic sentences of language can't be simultaneously asserted. Otherwise nothing bans the possibility of asserting $\bot$, violating the **BHK** clause that states that $\bot$ can't be asserted in any situation.

The interpretation we are proposing clearly makes the problem of the assertion of $\bot$ disappear, as $\bot$ is no longer to be considered a sentential content capable of being asserted (or denied). Nonetheless the intuition that $\bot$ can't be asserted can be reformulated as follows. We noted that a sentence can be asserted when we are in possession of a derivation, having the sentence as conclusion, with no open assumptions. So, the expression "$\bot$ can be asserted" appears as a rough way to refer to a situation that we can schematically represent in this way:

$$\begin{array}{c}\top \\ \vdots \\ \bot\end{array}$$

How is this pattern to be read? It looks like a derivation in which both assumptions and conclusion have been closed. To better understand it, consider a sentence A figuring in the derivation:

$$\begin{array}{c}\top \\ \vdots \\ A \\ \vdots \\ \bot\end{array}$$

If we split up this deductive pattern we find ourselves with the following:

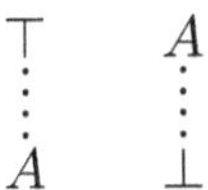

According to the reading of $\bot$ and $\top$, these derivations amount to a proof of A and to a refutation of A. That is, the possession of both the top- and bottom- closed derivations allows both the assertion and the denial of A.

If we take the sentence A to be an atomic sentence, e.g., 'The cup on the table is red', the top- and bottom-closed deductive pattern is available

[6] This direction is strongly called for by Schroeder-Heister (see for instance (Schröder-Heister, 2009)).

only if we are in possession both of supporting and refuting evidence for it. Obviously, the fact that the sentence 'The cup on the table is red' can be neither proved nor refuted does not depend on deduction, but rather on the fact that it is not possible to see a red cup and a green cup on the table at the same time.

Usually consistency is taken to be the impossibility of asserting the absurdity. In the framework we are developing an alternative notion of consistency can be put forward: namely, the impossibility of being in the position of asserting and denying a sentence at the same time. This notion of consistency amounts to the impossibility of obtaining deductive patterns having both assumptions and conclusions that are closed.

Just like Dummett, to preserve consistency we have to impose a restriction on atomic sentences: namely, we have to require that every atomic sentence can't be both asserted and denied (we will refer to this as *atomic consistency*). Our restriction can't be justified on logical basis, just as Dummett's one. Nonetheless, it is much more plausible to require that each atomic sentence can't be asserted and denied at the same time rather than to require that there must be mutually incompatible atomic sentences. In particular, such a restriction could be fully argued for, on the background of considerations on human cognition.

5 Conclusions

Traditionally, the possibilities of developing an account of assertion and an account of denial have been considered two different enterprises. To give an account of the meaning of negation we suggested to develop a unique framework in which the central role is played by the notion of open deduction. By means of the $\top$ and $\bot$ signs we can give an account in which deductive patterns count as proofs and refutations of sentences, i.e., allow their assertion and denial. As we saw, introduction rules are at the core of the process of proof, while eliminations are at the core of the process of refutation.

At this point we can reconsider the alternative to the **BHK** clause for negation Tennant proposed:

A proof for $\neg A$ is a refutation of A

In the light of the conventions introduced we can schematize the equivalence as:

$$\begin{array}{c} A \\ \vdots \\ \bot \end{array} \quad = \quad \begin{array}{c} \top \\ \vdots \\ \neg A \end{array}$$

It seems that we are faced with a sort of geometrical operation on derivation: by rotating 180° a refutation of A can be turned into a proof of $\neg A$.

Hence, negation appears as a linguistic device that states in an explicit way the implicit harmony, embodied in the inversion principle, holding between grounds for a sentence and consequences of a sentence. Indeed, inversion governs the relationship between introduction and elimination rules and negation the one between proofs and refutations which are directly connected to the two sets of rules.

A further question naturally arises, namely whether the standard rules for negation do properly seize this crucial feature of the connective. If the answer to be given were negative, than moving apart from standard intuitionistic logic would be necessary.

In conclusion, we believe to have isolated the role of negation in the architecture of deductive activity as being radically different from that of other connectives. Even though the development of a unified framework, in which to account for both the activities of proof and refutation, seems to require further investigation, we believe it to be an important step to fully develop a proof-theoretic account of the meaning of logical constants, as the analysis of negation emerging from it shows.

Luca Tranchini
Wilhelm-Schickard Institut für Informatik, Tübingen University
Sand 13, 72076 Tübingen, Germany
luca.tranchini@gmail.com

References

Brouwer, L. E. J. (1908). The unreliability of the logical principles. In A. Heyting (Ed.), *Collected works* (Vol. I, pp. 443–446).

Dummett, M. (1991). *The logical basis of metaphysics.* London: Duckworth.

Prawitz, D. (1987). Dummett on a theory of meaning and its impact on logic. In B. M. Taylor (Ed.), *Micheal Dummett.*

Schröder-Heister, P. (2009). Hypothetical reasoning: A critique of Dummett-Prawitz-style proof-theoretic semantics. In *The Logica Yearbook 2008.* (Sequent Calculi and Bidirectional Natural Deduction: On the Proper Basis of Proof-Theoretic Semantics.)

Tennant, N. (1999). Negation, absurdity and contrariety. In D. M. Gabbay & H. Wansing (Eds.), *What is negation?* Kluwer Academic Publishers.

Oiva Ketonen's Logical Discovery

Michael von Boguslawski

1 Short biography of Oiva Ketonen

Oiva Toivo Ketonen was born January 21, 1913, in the municipality of Teuva in the Southern Ostrobothnia region of Finland.[1] He was child number eight in a family that raised altogether 13 children. Already at a young age the law-governedness of nature made a deep impression on him and apparently planted the seed for an interest in the natural sciences. Ketonen was the only one of the family's children to get any form of higher education.

Ketonen graduated from Kristiinakaupungin Lukio (roughly equivalent to high school) in 1932, and enrolled into the department of history and linguistics (where philosophy in Helsinki was taught at that time) at the university of Helsinki. The professor of philosophy at that time was Eino Kaila, who had close connections with the *Wiener Kreis* and it was due to his personal efforts that logic arrived in Finland. Ketonen switched to the department of mathematics a year later, despite having doubts that mathematics alone would satisfy his academic interests. Ketonen's teacher in mathematics became Rolf Nevanlinna, the famous complex function theoretician, and we can tell from preserved correspondence that Nevanlinna was extremely impressed by Ketonen's mathematical abilities.

There was only one text-book on logic available in Finnish at that time — Thiodolf Rein's *Muodollinen logiikka* — *Formal logic* (free translation from Finnish) which treated only Aristotelian logic. There was a change in the curriculum, however, and Bertrand Russell's *The problems of philosophy* and Kaila's *Nykyinen maailmankäsitys* — *The present world-view* (free translation from Finnish), among others, were introduced. The teaching of logic was, according to Ketonen, confined to the basics and could not as

[1] An extended version of this article will appear in the Year book of the Vienna Circle institute. I would also like to thank Oiva's son Timo for generously providing me with a copy of Oiva's unfinished autobiography.

such, Ketonen speculates, cause any interest. We read in Ketonen's study book that he did not take a single course in logic.

According to Timo Ketonen, Oiva's early interests were algebra and number theory, which paved the way for the huge interest in Gödel's first incompleteness theorem, of which he was made aware by his fellow student, Max Söderman. Nevanlinna also later mentioned the theorem.[2] Gödel's fantastic result was probably what ignited Ketonen's interest in formal logic. Ketonen writes in the autobiography that he frequently went to evening meetings of what he called "The philosophical club." These meetings seem to have been quite unofficial, usually the group gathered at the home of one of the professors, e.g., Kaila or Yrjö Reenpää and logic was among the topics discussed. They also gathered at least once at Söderman's home. In the study diary we can read that he later also spent some evenings attending what he calls "mathematical-logical conferences". It is unclear at this moment whether these conferences and the meetings of the "philosophical club" were the same.

Nevanlinna tried to convince Ketonen to take up function theory — another witness of Nevanlinna's faith in Ketonen's abilities — but Ketonen, after some contemplation, decided to work on logic. He wrote his master's thesis on axiomatic logic, arithmetic, and Gödel's theorem. The first part was published (Ketonen, 1938) and used by Kaila as a text book for logic courses. Ketonen had received the impression from Nevanlinna that some mathematicians suspected that there was some fault in Gödel's proof, and that this fault might be worth uncovering. Ketonen believed that as a result of his work with the thesis, he succeeded in streamlining Gödel's proof somewhat.[3]

Ketonen kept working on Gödel's results and made a small improvement to Gödel's completeness theorem for the predicate calculus in 1941 (Ketonen, 1941). Gödel showed that that either a proposition A is provable, or it is impossible that there does not exist a counterexample. Ketonen improved this result so that this counter example can be found directly. Söderman, who resided in Vienna at the time, reported Ketonen's result to Gödel, who admitted that it was indeed an improvement (von Plato, 2004).

2 The dissertation — Untersuchungen zum Prädikatenkalkül

According to his autobiography, Ketonen had decided already in the spring of 1938 to go for a dissertation immediately. He went to the university in Göttingen, most probably with the aid of Nevanlinna's contacts, who had

[2] How well Nevanlinna was acquainted with logic, and what he thought of the at the time completely new discipline, remains debated.

[3] We hope to investigate this streamlining in a later work.

worked at the university as a visiting professor in 1936–1937. Kaila had met Gentzen in Münster in 1936 as well. Ketonen also went to Münster where he met — among others — Heinrich Scholz with whom there was some correspondence. Shockingly, the very same night that Ketonen arrived in Göttingen, 9–10 November 1938, later became infamous as the "Kristallnacht" — "crystal night". In Göttingen, in the autumn of 1938, Ketonen became Gerhard Gentzen's presumably first — and also last — student, although Ketonen had to wait until Christmas to receive a problem from Gentzen to work on. He recalls Gentzen as a sympathetic young man who "did not talk much" but mentioned that his chief assignment as Hilbert's assistant was the reading (apparently aloud) of "popular" scientific publications to his professor.

The dissertation (Ketonen, 1944), *Untersuchungen zum Prädikatenkalkül*, is divided into three parts. The first part presents and improves Gerhard Gentzens *sequent calculus* by introducing invertible rules for the calculus' propositional parts,[4] part two discusses a certain Skolem normalization of derivations, and the third part applies the results from parts one and two to produce a proof of the underivability of Euclid's parallell postulate from the rest of the Skolem-axioms for Euclidean geometry. Ketonen was the first to continue Skolem's work on geometry (von Plato, 2007b). The invertibility result will now be presented in detail.

Invertibility of rules in Gentzen's LK

A *sequent* is of the form $A_1, A_2, \ldots, A_m \rightarrow B_1, B_2, \ldots, B_n$. Capital latin letters $A, B, C, \ldots$ will be used to denote formulas, capital greek letters $\Gamma, \Delta, \Theta, \ldots$ will be used to denote the (possible) *context* of a derivation. Contexts are treated as lists of formulas. The formulas to the left of the sequent arrow $\rightarrow$ make up the antecedent, the formulas to the right the succedent. The sequent arrow can conveniently be read as "gives". Thus the sequent $A\&B \rightarrow C$ means that from the assumptions A and B together, the conclusion C follows. The sequent is read as "A and B gives C". A sequent should be viewed as a generalization of the concept of derivability, with one or more assumptions in the antecedent giving one or more possible cases in the succedent. We use the parentheses in the usual way, and all the connectives $\neg, \vee, \&$, and $\supset$. For the false sentence (and to denote a contradiction), Gentzen uses a special symbol but we will not need it here. The only axiom is the *initial sequent* $A \rightarrow A$. To be able to carry out derivations and proofs within the system, we need logical and structural rules. The logical rules manipulate connectives whereas the structural rules manipulate formulas. Derivations are in "tree-form" and begin from initial

[4] Obviously, invertibility in Ketonen's sense cannot hold for the predicate part.

sequents (and possibly contexts) at the end of branches, and end with the proven sequent at the bottom of the tree, the "root".[5] Below[6] are given the structural and logical rules of Gentzen's first system of sequent calculus, which we today call Gentzen LK:

Structural rules for Gentzen LK

$$\frac{\Gamma \to \Theta}{A, \Gamma \to \Theta}\,\mathrm{LW} \qquad \frac{\Gamma \to \Theta}{\Gamma \to \Theta, A}\,\mathrm{RW}$$

Left weakening — *Right weakening*

$$\frac{A, A, \Gamma \to \Theta}{A, \Gamma \to \Theta}\,\mathrm{LC} \qquad \frac{\Gamma \to \Theta, A, A}{\Gamma \to \Theta, A}\,\mathrm{RC}$$

Left contraction — *Right contraction*

$$\frac{\Delta, B, A, \Gamma \to \Theta}{\Delta, A, B, \Gamma \to \Theta}\,\mathrm{LE} \qquad \frac{\Gamma \to \Theta, B, A, \Lambda}{\Gamma \to \Theta, A, B, \Lambda}\,\mathrm{RE}$$

Left exchange — *Right exchange*

$$\frac{\Gamma \to \Theta, B \quad B, \Delta \to \Lambda}{\Gamma, \Delta \to \Theta, \Lambda}\,\mathrm{Cut}$$

Cut

Logical rules for Gentzen LK

$$\frac{\Gamma \to \Theta, A \quad \Gamma \to \Theta, B}{\Gamma \to \Theta, A\&B}\,\mathrm{R\&} \qquad \frac{A, \Gamma \to \Theta \quad B, \Gamma \to \Theta}{A \vee B, \Gamma \to \Theta}\,\mathrm{L}\vee$$

Right conjunction — *Left disjunction*

$$\frac{A, \Gamma \to \Theta}{A\&B, \Gamma \to \Theta}\,\mathrm{L\&_1} \qquad \frac{B, \Gamma \to \Theta}{A\&B, \Gamma \to \Theta}\,\mathrm{L\&_2}$$

Left conjunction 1 — *Left conjunction 2*

$$\frac{\Gamma \to \Theta, A}{\Gamma \to \Theta, A \vee B}\,\mathrm{R}\vee_1 \qquad \frac{\Gamma \to \Theta, B}{\Gamma \to \Theta, A \vee B}\,\mathrm{R}\vee_2$$

Right disjunction 1 — *Right disjunction 2*

$$\frac{A, \Gamma \to \Theta}{\Gamma \to \Theta, \neg A}\,\mathrm{R}\neg \qquad \frac{\Gamma \to \Theta, A}{\neg A, \Gamma \to \Theta}\,\mathrm{L}\neg$$

Right negation — *Left negation*

$$\frac{A, \Gamma \to \Theta, B}{\Gamma \to \Theta, A \supset B}\,\mathrm{R}\supset \qquad \frac{\Gamma \to \Theta, A \quad B, \Delta \to \Lambda}{A \supset B, \Gamma, \Delta \to \Theta, \Lambda}\,\mathrm{L}\supset$$

Right implication — *Left implication*

---

[5] A derivation may of course have only one branch, i.e., have only one initial sequent from which some other sequent is proven.

[6] See (Gentzen, n.d.) for details.

With *invertibility* is meant that if a sequent matches the conclusion of a rule, and if it is derivable, then the corresponding premisses are derivable. Gentzen's LK is not invertible. Consider rule $R\lor_2$, for example. If it were invertible, then the sequent $A \to B$ would be derivable because $A \to A \lor B$ is derivable from the initial sequent $A \to A$. $A \to B$ is not at all a tautology so it clearly should not be derivable without assumptions in a complete and consistent system. Thus, the logical rules for left conjunction and right disjunction need to be replaced with invertible ones, and Ketonen notes that the rule for left implication will have to be replaced with a rule which has the same contexts in its two premisses:

Ketonen's invertible rules for Gentzen LK

$$\frac{A, B, \Gamma \to \Delta}{A\&B, \Gamma \to \Delta}\,{\scriptstyle L\&} \qquad \frac{\Gamma \to \Delta, A, B}{\Gamma \to \Delta, A \lor B}\,{\scriptstyle R\lor} \qquad \frac{\Gamma \to \Delta, A \quad B, \Gamma \to \Delta}{A \supset B, \Gamma \to \Delta}\,{\scriptstyle L\supset}$$

The proofs of the invertibility of the rules are easy and short. The ones given here differ somewhat from those given by Ketonen, specifically so that when Ketonen introduces the conclusion of a rule the invertibility of which is to be proved through an instance of its non-invertible counterpart, we simply introduce the conclusion after the vertical dots that indicate some possible derivation.

Proof of invertibility of rule L&

$$\frac{\dfrac{\dfrac{\dfrac{A \to A}{B, A \to A}\,{\scriptstyle LW}}{A, B \to A}\,{\scriptstyle LE} \qquad \dfrac{B \to B}{A, B \to B}\,{\scriptstyle LW}}{A, B \to A\&B}\,{\scriptstyle R\&} \qquad \begin{array}{c}\vdots \\ A\&B, \Gamma \to \Omega\end{array}}{A, B, \Gamma \to \Omega}\,{\scriptstyle Cut}$$

The proof of the invertibility of rule $R\lor$ is simply a horizontal "mirror image" of the proof above. In order to prove the invertibility of rule $L\supset$, we show that both premisses are derivable from the conclusion by cut:

$$\frac{\dfrac{\dfrac{A \to A}{A \to A, B}\,{\scriptstyle RW}}{\to A, A \supset B}\,{\scriptstyle R\supset} \qquad \begin{array}{c}\vdots \\ A \supset B, \Gamma \to \Omega\end{array}}{\Gamma \to \Omega, A}\,{\scriptstyle Cut,\ RE}$$

$$\frac{\dfrac{\dfrac{B \to B}{A \to A, B}\,{\scriptstyle LW}}{B \to A \supset B}\,{\scriptstyle R\supset} \qquad \begin{array}{c}\vdots \\ A \supset B, \Gamma \to \Omega\end{array}}{B, \Gamma \to \Omega}\,{\scriptstyle Cut}$$

With the invertible rules, we can carry out a "root-first" proof search in an algorithmic fashion, beginning with the sequent we want to prove, and then applying the rules in reverse until we reach a situation with only initial sequents (and possibly contexts). This proof search will terminate, so it can in theory be done by a computer. Indeed it is possible that Ketonen's sequent calculus is the first system that would permit a computer to produce proofs. It does not matter in which order the rules are applied in reverse, as the two proofs of $\to (A \supset B) \supset (\neg B \supset \neg A)$ below illustrate:

$$\dfrac{\dfrac{\dfrac{\dfrac{\dfrac{A \to A}{\neg B, A \to A}\,\text{LW}}{\neg B \to \neg A, A}\,\text{R}\neg}{\to \neg B \supset \neg A, A}\,\text{R}\supset \qquad \dfrac{\dfrac{\dfrac{B \to B}{B \to B, \neg A}\,\text{RW}}{B, \neg B \to \neg A}\,\text{L}\neg}{B \to \neg B \supset \neg A}\,\text{R}\supset}{A \supset B \to \neg B \supset \neg A}\,\text{L}\supset}{\to (A \supset B) \supset (\neg B \supset \neg A)}\,\text{R}\supset$$

$$\dfrac{\dfrac{\dfrac{\dfrac{\dfrac{A \to A}{\neg B, A \to A}\,\text{LW}}{\neg B \to \neg A, A}\,\text{R}\neg \qquad \dfrac{\dfrac{B \to B}{B \to B, \neg A}\,\text{RW}}{B, \neg B \to \neg A}\,\text{L}\neg}{A \supset B, \neg B \supset \to A}\,\text{L}\supset}{A \supset B \to \neg B \supset \neg A}\,\text{R}\supset}{\to (A \supset B) \supset (\neg B \supset \neg A)}\,\text{R}\supset$$

The modification of the rules does not hamper the properties of the system, the *Hauptsatz*, for example, still holds. Kurt Schütte and Haskell Curry gave cut-free proofs of invertibility in 1950 and 1963 respectively, Curry with the added result that inversions are height preserving.[7]

Reactions to the thesis and follow-up

Paul Bernays (Bernays, 1945) wrote a favorable review of Ketonen's thesis in *The Journal of Symbolic Logic* in 1945, and Kleene notes (Kleene, 1952) that he knows of Ketonen's calculus only through this review. We know through several sources, for example (von Wright, 1951), that several researchers including Richard Feys, and the already mentioned Curry, Kleene, and Bernays held Ketonen's work in high regard. Curry reportedly (von Plato, 2004) held Ketonen's work to be the best thing in proof theory since Gentzen, and the present writer has seen a letter from Curry to Ketonen where the former asks for everything Ketonen has written on logic, even in Finnish. Arend Heyting wrote a review of the thesis in 1947, but apparently failed to see its main point and appears instead to view it as a work on geometry rather than on proof theory. The first international reference to Ketonen's work seems to be by Karl Popper in 1947 (Popper, 1947), and Beth uses parts of Ketonen's calculus in his tableau method[8] but cites Kleene and Gentzen, but not Ketonen.

---

[7] See (von Plato, 2007a).

[8] See for example (Beth, 1962).

No more original work on logic by Ketonen appeared after the thesis, and exactly why this is so is not completely clear. He is known to have been working on forcing in set theory and even relativity theory (inspired by previous work on the subject by Kaila) but did not publish any own results even if survived correspondence suggests that he indeed had worked out some results of his own. He has also worked on the interpretation of consistency proofs, many-valued logics, and the application of some of the results of the thesis on epistemology (von Wright, 1951). A possible reason as to why he did not continue with logic could be the severe disappointment he experienced with philosophy of science in general during his visit to the United States in the 1950's (Ketonen & von Wright, 1950) and, as is hinted at in the autobiography, the effects that the second World War brought with it which possibly steered also his philosophical interests away from the world of mathematics towards broader philosophical enquiries. Only one work on logic after the dissertation has been found as a very rough manuscript of about ten pages, written on a typewriter but with several hand-written corrections, and containing some notes on epistemology and geometry, but is nothing like such a polished version mentioned by von Wright (Ketonen, 1944–1950). We know for certain however, from survived correspondence, that at least still in the late 1960's Ketonen tried to stay up-to-date with recent logical research. He also gave lectures in basic logic for students at the university. When he was asked in his later years why he had abandoned logic, Ketonen always remarked abruptly "logic gives me such headache". One could perhaps speculate that logic became something of a spare-time activity, while the main attention was on university politics, his professorship that he held for over 25 years, between 1951–1977, and on a philosophy incorporating elements which fall outside those of the natural sciences.

Michael von Boguslawski
Department of philosophy, University of Helsinki
Siltavuorenpenger 20 A, P. O. Box 9, 00014 Helsinki, Finland
`michael.vonboguslawski@helsinki.fi`

References

Bernays, P. (1945). Review: Oiva Ketonen, Untersuchungen zum Prädikatenkalkül. *The Journal of Symbolic Logic*, *10*(4), 127–130.

Beth, E. (1962). *Formal methods.* Dordrecht: D. Reidel.

Gentzen, G. (n.d.). Untersuchungen über das logische Schliessen. In M. E. Szabo (Ed.), *The collected papers of Gerhard Gentzen.*

Ketonen, O. (1935–1936). *Lahjomaton tilintekijä (The unbribable accountant).* (Study Diary.)

Ketonen, O. (1937). *Tutkimuksia formaalisen todistamisen ristiriidattomuudesta (Investigations into the consistency of formal proving).* (Manuscript of M. A. thesis.)

Ketonen, O. (1938). Todistusteorian perusaatteet. *Ajatus*, *IX*, 28–108.

Ketonen, O. (1941). Predikaattilogiikan täydellisyydestä. *Ajatus*, *X*, 77–92.

Ketonen, O. (1944). Untersuchungen zum Prädikatenkalkül. *Annales Acad. Sci. Fenn.*, *23*.

Ketonen, O. (1944–1950). *Tietomme apriorisista aineksista.* (Manuscript in *National Archive of Finland.*)

Ketonen, O. (2000). *Unfinished autobiography.*

Ketonen, O., & von Wright, G. (1950). *Correspondence between Oiva Ketonen and Georg Henrik von Wright.* (Stored both in the national library in Helsinki, and in the national archive of Finland.)

Kleene, S. (1952). Permutability of inferences in Gentzen's calculi LK and LJ. *Memoires of the American Mathematical Society*(10), 1–26.

Menzler-Trott, E. (2007). *Logic's lost genius: the life of Gerhard Gentzen* (Vol. 33). Providence, RI: American Mathematical Society.

Nevanlinna, R. (1938). *Letters to Oiva Ketonen.* (Kept by Timo Ketonen.)

Popper, K. (1947). New foundations for logic. *Mind*, *56*, 193–235.

von Plato, J. (2004). Ein leben, ein Werk. Gedanken über das wissenschaftliche Schaffen des finnischen Logikers Oiva Ketonen. In *Form, Zahl, Ordnung — Studien zu Wissenschafts- und Technikgeschichte* (pp. 427–435). Stuttgart: Franz Steiner Verlag.

von Plato, J. (2007a). Gentzen's logic. In (Vol. 33). Providence, RI: American Mathematical Society.

von Plato, J. (2007b). In the shadows of the Löwenheim—Skolem theorem: Early combinatorial analyses ofmathematical proofs. *The Bulletin of Symbolic Logic*, *13*(2), 189–225.

von Wright, G. (1951). *Expert opinion on Oiva Ketonen's application for professorship in theoretical philosophy.* (Kept in the central archives of the university of Helsinki.)

www.ingramcontent.com/pod-product-compliance
Lightning Source LLC
LaVergne TN
LVHW012328060326
832902LV00011B/1767

* 9 7 8 1 9 0 4 9 8 7 4 6 8 *